Methods in Molecular Biology

Series Editor
John M. Walker
School of Life and Medical Sciences
University of Hertfordshire
Hatfield, Hertfordshire, UK

For further volumes:
http://www.springer.com/series/7651

For over 35 years, biological scientists have come to rely on the research protocols and methodologies in the critically acclaimed Methods in Molecular Biology series. The series was the first to introduce the step-by-step protocols approach that has become the standard in all biomedical protocol publishing. Each protocol is provided in readily reproducible step-by-step fashion, opening with an introductory overview a list of the materials and reagents needed to complete the experiment, and followed by a detailed procedure that is supported with a helpful notes section offering tips and tricks of the trade as well as troubleshooting advice. These hallmark features were introduced by series editor Dr. John Walker and constitute the key ingredient in each and every volume of the Methods in Molecular Biology series. Tested and trusted, comprehensive and reliable, all protocols from the series are indexed in PubMed.

Manuscript Instructions https://drive.google.com/file/d/1K4Az5vpAFVzYadJ3jKx2-QSo7Dqa_SNA/view?usp=sharing

Functional Disulphide Bonds

Methods and Protocols

Second Edition

Edited by

Philip J. Hogg

School of Life Sciences, University of Technology Sydney and Centenary Institute, University of Sydney, Sydney, NSW, Australia

Humana Press

Editor
Philip J. Hogg
School of Life Sciences
University of Technology Sydney and Centenary Institute
University of Sydney
Sydney, NSW, Australia

ISSN 1064-3745 ISSN 1940-6029 (electronic)
Methods in Molecular Biology
ISBN 978-1-0716-5157-5 ISBN 978-1-0716-5158-2 (eBook)
https://doi.org/10.1007/978-1-0716-5158-2

1st edition: Springer Science+Business Media, LLC, part of Springer Nature 2019

This Humana imprint is published by the registered company Springer Science+Business Media, LLC, part of Springer Nature.
The registered company address is: 1 New York Plaza, New York, NY 10004, U.S.A.

Preface

Proteins are responsible for all of life's processes, and two types of covalent bonds define their structure: the peptide bonds that link the amino acid residues and the disulfide bonds that link pairs of cysteine amino acids. There has been a paradigm shift in the role of disulfide bonds in protein function over the last decade. Since the identification of these bonds in the 1960s, they were long thought to be fully formed and inert in the mature protein. We now know that this is often not the case.

Recent investigations have demonstrated that disulfide bonds are not fully formed in native proteins. Many proteins are constitutively produced and function as an array of partially disulfide-bonded states. Also, a subset of the formed disulfide bonds—the allosteric disulfides—controls the function of the mature protein when cleaved by different factors. More than 50 allosteric disulfide bonds have been reported. These discoveries are changing our understanding of how proteins function and have profound implications for drug development and drug resistance.

This book describes the current techniques used to study disulfide bonds. To set the scene, how disulfide bonds are classified, how to study their evolution, and in silico methods of analysis are described. This is followed by the chemical, biochemical, biophysical, and biological techniques used to study functional disulfides and probe their behavior. The book then highlights techniques employed in the study of the factors that cleave allosteric disulfide bonds. It finishes with examples of how labile disulfide bonds are employed for new diagnostics and therapeutics.

Sydney, NSW, Australia *Philip J. Hogg*

Preface

Contents

Contributors

KENTA ARAI • *Institute of Advanced Biosciences, Tokai University, Hiratsuka-shi, Kanagawa, Japan*

ROELOF H. BEKENDAM • *Division of Hemostasis and Thrombosis, Department of Medicine, BIDMC, Harvard Medical School, Boston, MA, USA; Division of Hematology & Hematologic Malignancies, Department of Medicine, BIDMC, Harvard Medical School, Boston, MA, USA*

MORGANE BOURCY • *Platelet and Thrombosis Research Laboratory, ANZAC Research Institute, Sydney, Australia; University of Liege, Liège, Belgium*

DIEGO BUTERA • *School of Life Sciences, University of Technology Sydney and Centenary Institute, University of Sydney, Sydney, NSW, Australia*

FENGWU CHEN • *Cyrus Tang Hematology Center, Cyrus Tang Medical Institute, Soochow University, Suzhou, China; The State Key Laboratory of Membrane Biology, School of Life Sciences, Tsinghua University, Beijing, China*

VIVIEN M. CHEN • *Platelet and Thrombosis Research Laboratory, ANZAC Research Institute, Sydney, Australia; Haematology, Concord Hospital, Sydney, Australia; University of Sydney, Sydney, Australia*

JOYCE CHIU • *School of Clinical Medicine, Faculty of Medicine and Health, University of New South Wales, Sydney, NSW, Australia*

RACHEL CODD • *School of Medical Sciences (Pharmacology), Sydney Medical School, The University of Sydney, Sydney, NSW, Australia*

KRISTINA M. COOK • *University of Sydney School of Medicine, Charles Perkins Centre, Sydney, NSW, Australia*

ALEXANDER DUPUY • *Central Clinical School, Faculty of Medicine and Health, University of Sydney, Sydney, NSW, Australia; School of Biomedical Engineering, Faculty of Engineering, University of Sydney, Sydney, NSW, Australia; Sydney Nano Institute, University of Sydney, Sydney, NSW, Australia*

CHAO FANG • *Department of Pharmacology, School of Basic Medicine, Tongji Medical College and State Key Laboratory for Diagnosis and Treatment of Severe Zoonotic Infectious Diseases, Huazhong University of Science and Technology, Wuhan, Hubei, China; The Key Laboratory for Drug Target Researches and Pharmacodynamic Evaluation of Hubei Province, Wuhan, Hubei, China*

ROBERT FLAUMENHAFT • *Division of Hemostasis and Thrombosis, Department of Medicine, BIDMC, Harvard Medical School, Boston, MA, USA*

MICHAEL P. GOTSBACHER • *School of Medical Sciences (Pharmacology), Sydney Medical School, The University of Sydney, Sydney, NSW, Australia*

JENNIFER GUILLE • *Department of Nuclear Medicine and PET, Prince of Wales Hospital, Randwick, NSW, Australia*

PHILIP J. HOGG • *School of Life Sciences, University of Technology Sydney and Centenary Institute, University of Sydney, Sydney, NSW, Australia*

MICHIO IWAOKA • *Institute of Advanced Biosciences, Tokai University, Hiratsuka-shi, Kanagawa, Japan; Department of Chemistry, School of Science, Tokai University, Hiratsuka-shi, Kanagawa, Japan*
QIUYU JING • *School of Biomedical Sciences, Li Ka Shing Faculty of Medicine, The University of Hong Kong, Hong Kong SAR, China; Centre for Oncology and Immunology, Hong Kong Science Park, Hong Kong SAR, China*
LINING ARNOLD JU • *School of Biomedical Engineering, Faculty of Engineering, University of Sydney, Sydney, NSW, Australia; Sydney Nano Institute, University of Sydney, Sydney, NSW, Australia; Charles Perkins Centre, The University of Sydney, Camperdown, NSW, Australia*
DIVESH KUMAR • *Department of Nuclear Medicine and PET, Fiona Stanley Hospital, Murdoch, WA, Australia*
WICK LAKSHANTHA • *Centenary Institute, University of Sydney, Sydney, NSW, Australia*
MARK LARANCE • *Charles Perkins Centre and School of Medical Sciences, The University of Sydney, Sydney, Australia*
YAOFENG LI • *Department of Pharmacy, Renmin Hospital of Wuhan University, Wuhan, Hubei, China*
KENNY LU • *School of Life Sciences, University of Technology Sydney and Centenary Institute, University of Sydney, Sydney, NSW, Australia*
LAURA MOLDOVAN • *School of Biomedical Engineering, The University of Sydney, Darlington, NSW, Australia; Charles Perkins Centre, The University of Sydney, Camperdown, NSW, Australia; Heart Research Institute, Newtown, NSW, Australia*
OSAMEDE C. OWEGIE • *Bloodworks Northwest Research Institute, Seattle, WA, USA*
LEONARDO PASALIC • *ICPMR Westmead Hospital, Sydney, Australia; University of Sydney, Sydney, Australia*
FREDA H. PASSAM • *Central Clinical School, Faculty of Medicine and Health, University of Sydney, Sydney, NSW, Australia; Royal Prince Alfred Hospital, Sydney, NSW, Australia*
ASTER PIJNING • *School of Life Sciences, University of Technology Sydney and Centenary Institute, University of Sydney, Sydney, NSW, Australia*
ASTER E. PIJNING • *School of Life Sciences, University of Technology Sydney, Sydney, NSW, Australia; Centenary Institute, University of Sydney, Sydney, NSW, Australia*
MARK SCHREUDER • *School of Life Sciences, University of Technology Sydney and Centenary Institute, University of Sydney, Sydney, NSW, Australia*
IVAN HO SHON • *Department of Nuclear Medicine and PET, Prince of Wales Hospital, Randwick, NSW, Australia; Discipline of Medicine, Randwick Clinical Campus, Medicine and Health, University of New South Wales, Sydney, Australia*
ERIC G. SUCHANEK • *Flux-Frontiers, Cincinnati, OH, USA*
CHUEN WEN TAN • *Haematology, Singapore General Hospital, Sydney, Australia; Platelet and Thrombosis Research Laboratory, ANZAC Research Institute, Sydney, Australia; Haematology, Concord Hospital, Sydney, Australia*
JINYU WANG • *Department of Pharmacology, School of Basic Medicine, Tongji Medical College and State Key Laboratory for Diagnosis and Treatment of Severe Zoonotic Infectious Diseases, Huazhong University of Science and Technology, Wuhan, Hubei, China; Department of Basic Science of Stomatology, School of Stomatology, Tongji Medical College, Huazhong University of Science and Technology, and Hubei Province Key Laboratory of Oral and Maxillofacial Development and Regeneration, Wuhan, Hubei, China*

JASON W. H. WONG • *School of Biomedical Sciences, Li Ka Shing Faculty of Medicine, The University of Hong Kong, Hong Kong SAR, China; Centre for Oncology and Immunology, Hong Kong Science Park, Hong Kong SAR, China*

YI WU • *Cyrus Tang Hematology Center, Cyrus Tang Medical Institute, Soochow University, Suzhou, China*

AIZHEN YANG • *Cyrus Tang Hematology Center, Cyrus Tang Medical Institute, Soochow University, Suzhou, China*

MOUA YANG • *Bloodworks Northwest Research Institute, Seattle, WA, USA; Division of Hematology and Oncology, Department of Medicine, University of Washington School of Medicine, Seattle, WA, USA*

KAITLYN ZHANG • *School of Biomedical Engineering, Faculty of Engineering, University of Sydney, Sydney, NSW, Australia; Sydney Nano Institute, University of Sydney, Sydney, NSW, Australia*

Chapter 1

Classification of Protein Disulfide Bonds

Philip J. Hogg

Abstract

Protein disulfide bonds are the links between pairs of cysteine residues in the polypeptide chain. These bonds are classified based on the sign of the five dihedral angles that define the cystine residue. Twenty disulfide configurations are possible using this convention and all 20 are represented in protein structures. Force distribution analysis of the pairwise forces between the cysteine residues of the different configurations identified two of the 20 as having significant strain: the –RHstaple and –/+RHhook disulfide bonds. These two disulfide configurations are associated with allosteric function in proteins. An online tool is available that provides a comprehensive analysis of protein disulfide bonds in Protein Data Bank structures, including configuration, strain energy, and solvent accessibility.

Keywords Allosteric disulfide bonds, Structural biology, Cysteine, Disulfide bond analysis tool

1 Introduction

Disulfide bonds are traditionally thought of as having a structural role in proteins. However, certain disulfide bonds are reversible in nature, meaning that they can be oxidized and reduced in their lifetime. Functional disulfide bonds can be involved in facilitating reduction, oxidation, or isomerization of other disulfide bonds in protein substrates. These disulfide bonds, for instance, those found at the active site of oxidoreductases, have been termed catalytic disulfides. Recently, a new class of functional disulfide bond was recognized and termed allosteric [1, 2] (Fig. 1). Allosteric disulfides are defined as affecting the activity of the protein in which they reside by being cleaved or formed. Since the term was introduced in 2006 [3], over 50 allosteric disulfide bonds have been described in a range of different proteins and species (Table 1). Allosteric disulfide bonds have been shown to be involved in a number of biological processes, including thrombosis and hemostasis, viral infections, and immune reactions [1].

The recognition of subsets of disulfide bonds with different biological functions created a need for a systematic way of analyzing

Philip J. Hogg (ed.), *Functional Disulphide Bonds*, Methods in Molecular Biology, vol. 3016,
https://doi.org/10.1007/978-1-0716-5158-2_1,

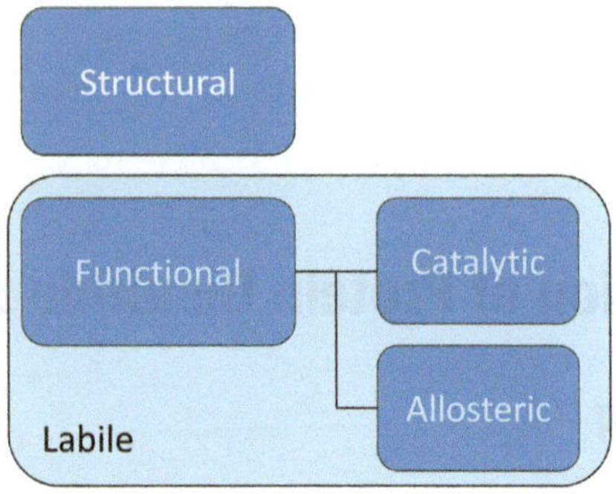

Fig. 1 Different classes of disulfide bonds

these structures. Different disulfide bond classification schemes have been introduced [4–6]. To date, the most widely accepted and used classification system of disulfide bonds is based on bond geometry, separating disulfide bonds into different configurations. This classification system has enabled systematic study of these protein motifs and identification of new functional disulfides [7–9].

2 The Geometry of Disulfide Bonds Is Used to Classify Them

Structurally, disulfide bonds are defined by six atoms of the two contributing cysteine residues, forming five dihedral or χ (chi) angles. The dihedral angle is defined as the angle between two planes through two sets of three atoms that have two atoms in common (Fig. 2). Dihedral angles are positive, if the angle is clockwise, and negative, if the angle is counterclockwise with respect to the position of the reference atom.

A classification system for disulfide bonds was devised based on the sign of the five χ angles: $\chi1$, $\chi2$, $\chi3$, $\chi2'$, and $\chi1'$. The sign of the $\chi2$, $\chi3$, and $\chi2'$ angles defines whether a bond is a spiral (–,–,–), hook (+,–,– or –,–,+), or staple (–,+,– or +,–,+) (Fig. 3). The middle χ angle, $\chi3$, which is formed between the two sulfur atoms, further determines whether the configuration is termed right-handed (RH), if the angle is positive, or left-handed (LH), if the angle is negative. Last, the signs of the first and last χ angles, $\chi1$ and $\chi1'$, determine whether the bond is termed – (if both are negative), + (if both are positive), –/+, or +/–. Together, these three levels of classification define 20 possible disulfide bond configurations, which are all found in proteins.

Though all 20 disulfide bond configurations have been found in vivo, certain configurations are overrepresented and others are rare. Accounting for about one quarter of all disulfide bonds, the –LHspiral is the most common disulfide configuration (Fig. 3). This configuration is associated with stable structural disulfide bonds [3].

Table 1

Examples of allosteric disulphides grouped according to the type of change in protein function associated with cleavage of the bonds. Cleavage of the indicated disulphide in the protein can result in increase or decrease in ligand binding to the protein, change in the efficiency of substrate hydrolysis if the protein is an enzyme, increase or decrease in subsequent proteolysis of the protein, or formation of inter-molecular disulphide bonds and homo- or hetero-oligomerisation of the protein. In some instances, the conformation of the disulphide is not known (n.d.). The PDB identifiers are an example structure containing the allosteric disulphide

Change	Protein	Disulphide	Conformation	PDB	Reference
Ligand binding	β2-glycoprotein I	C288-C326	–/+RHhook	1c1z	
	β3 integrin	C177-C184	–/+RHhook	3fcs	
	β2 integrin	C169-C176	–/+RHhook	3k6s	
		C224-C264	+LHhook	3k6s	
	αM integrin	C636-C693	–RHstaple	3k6s	
	HIV gp120	C126-C196	–RHstaple	1yym	
		C296-C331	–RHstaple	1yym	
		C385-C418	–RHstaple	1yym	
	CD4	C130-C159	–RHstaple	3cd4	
	CD44	C81-C101	–LHhook	2jcp	
	Glycoprotein 1b	C4-C17	–RHstaple	1m0z	
		C209-C248	–LHhook	1m0z	
	Interleukin-4	C46-C99	+/–RHhook	1cyl	
	Interleukin receptor subunit gamma	C160-C209	–LHhook	1erj	
	Tissue factor	C186-C209	–RHstaple	1boy	
	von Willebrand factor	C1669-C1670	+/–LHstaple	3gxb	
	Thromboxane receptor	C11-C102	–RHhook	6iiu	
		C105-C183	–LHhook	6iiu	
	OpcA	C410-C416	–RHstaple	9emm	
	HBV S glycoprotein	C301-C310	–/+RHhook	Predicted	
	TRPV1	C387-C391	–LHhook	8gf8	
	AT2A2 (SERCA2)	C875-C887	–LHspiral	7bt2	
	KIR2DLR	C10-C28	–RHstaple	3wyr	
	Thrombospondin-1	C974-n.d.	n.d.		
	Vitronectin	C137-C161	n.d.		
		C274-C453	n.d.		
	Histidine-rich glycoprotein	C306-C309	n.d.		
		C390-C434	n.d.		
		C409-C410	n.d.		
	STIM1	C49-C56	n.d.		

(Continued)

Table 1
(Continued)

Change	Protein	Disulphide	Conformation	PDB	Reference
Substrate hydrolysis	Arylsulfate sulfotransferase	C418-C424	–RHstaple	3elq	
	Botulinum neurotoxins	C436-C445	–RHstaple	1epw	
	QueF	C55-C99	–LHhook	5udg	
	Acetyl-CoA acetyltransferase	C88-C378	+/–LHspiral	4xl2	
	hTryptaseβ	C191-C220	–RHhook	2fpz	
	Methionine aminopeptidase 2	C228-C448	–RHstaple	1b59	
	Prolyl cis-trans isomerase, AtFKBP13	C106-C111	–RHstaple	1y0o	
	Transglutaminase 2	C370-C371	–RHstaple	2q3z	
		C230-C370	+RHstaple	3ly6	
	Lon protease	C617-C691	–/+RHhook	1wu4	
	CgDapF	C83-C221	–LHhook	5h2g	
	SAMHD1	C341-C350	–/+RHhook	4mz7	
	FdhAB	C845-C872	n.d.		
	Adenosine-5'-phosphosulfate kinase	C86-C119	n.d.		
	Aurora A kinase	C290-C290	n.d.		
Proteolysis	Angiotensinogen	C18-C138	–/+RHhook	2wxw	
	Factor XI	C362-C482	–/+RHhook	2f83	
	MICA	C202-C259	–/+LHhook	1hyr	
	Plasmin(ogen)	C512-C536	–LHhook	4dur	
		C462-C541	–/+RHhook	4dur	
	Activin-A	C35-C322	n.d.		
		C35-C314	n.d.		
	REDD1	C150-C157	n.d.		
Oligomerisation	CD4	C130-C159	–RHstaple	3cd4	
	VEGFC & VEGFD	C156-C165	–LHhook	2x1w	
	LYVE-1	C201-n.d.	n.d.		

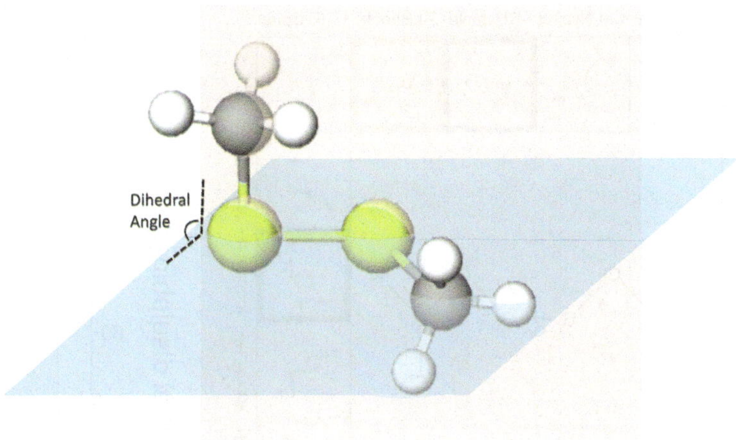

Fig. 2 Representation of the dihedral angle

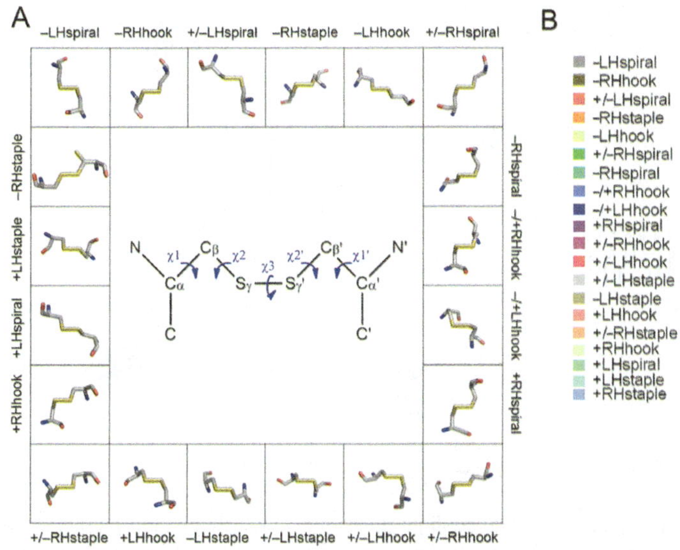

Fig. 3 Classification of disulfide bonds. (**A**) Classification of disulfide bonds using the five dihedral angles of the cystine residue (center). There are 20 possible disulfide conformations using these angles and example structures of each from the PDB are shown. (**B**) Prevalence of the different bond configurations in 13,030 unique disulfide bonds in PDB protein structures [7]

Not coincidentally, this configuration is also the one that is the most energetically favorable, having the lowest average dihedral strain energy (DSE). The DSE of a disulfide bond is predicted from the five χ angles and has been shown to be a semi-quantitative measure of the amount of strain on a disulfide bond [10–13]. Other measures of disulfide bond strain are the length of the sulfur–sulfur bond, which is 2.038 Å on average and strained when stretched, and the stretching of the α-angles (Fig. 4).

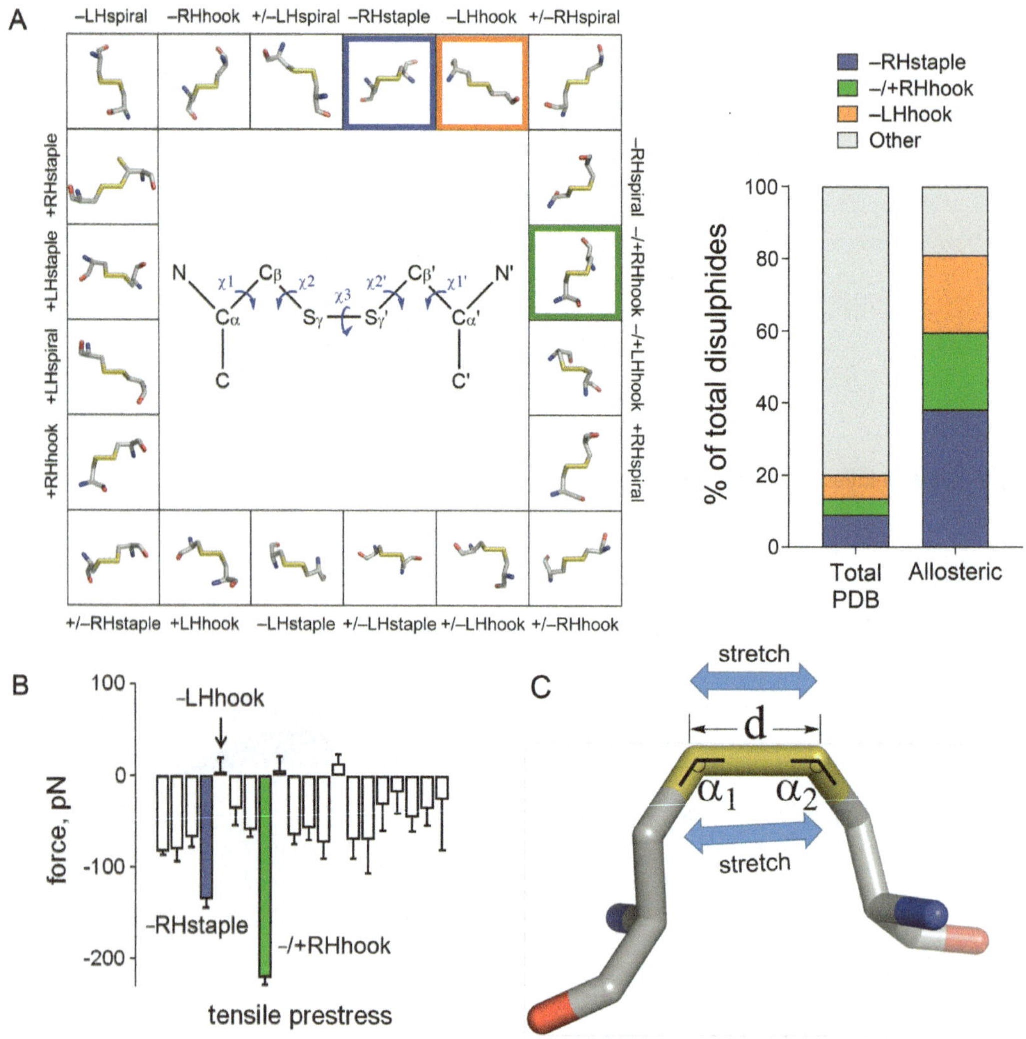

Fig. 4 Allosteric disulfide bonds are enriched for three bond configurations, two of which are highly strained. (**A**) Incidence of the three allosteric disulfide conformations. The –RHstaple, –/+RHhook, and –LHhook conformations constitute ~20% of all nonredundant disulfide bonds in the PDB, but ~80% of the structurally defined allosteric bonds ($n = 42$, Table 1). (**B**) Force distribution analysis was used to calculate the pairwise forces between the cysteine residues of the 20 different disulfide configurations [14]. The –RHstaple and –/+RHhook configurations carry significant tensile prestress and so are more easily cleaved. Mean tensile prestress of the 20 disulfide conformations is adapted from Fig. 2D of Zhou et al. [14]. (**C**) The tensile prestress of the –RHstaple and –/+RHhook bonds is a result of stretching of the sulfur–sulfur bond length (indicated by d) and neighboring α angles

3 Certain Bond Configurations Are More Strained Than Others

Allosteric disulfide bonds are defined as influencing the functioning of the protein in which they reside by being reduced or oxidized. Three disulfide bond configurations are associated with allosteric function. The –RHstaple is the most common allosteric configuration to date, accounting for 16 of the 42 structurally determined allosteric disulfide bonds to date (Table 1). These bonds most often link adjacent strands in the same antiparallel β sheet, or constrain β loops. There are nine examples each of the −/+RHhook and −LHhook configurations (Table 1, Fig. 4A). The reasons why the −RHstaple and −/+RHhook configurations are associated with allosteric function are beginning to emerge.

Force distribution analysis of the pairwise forces between the cysteine residues of the 20 different disulfide configurations has been analyzed [14]. The force terms involving the sulfur atoms, which are the relevant measure for disulfide reactivity, of the −RHstaple and −/+RHhook disulfide bonds are significantly larger than the other 18 configurations (Fig. 4B). In other words, the proteins structure prevents the −RHstaple and −/+RHhook disulfides from fully relaxing into the state of minimal potential energy, so the bonds carry a tensile prestress [3, 14]. The stress is largely a result of stretching of the S–S bond and neighboring α angles [14] (Fig. 4C). The tensile prestress fine-tunes disulfide cleavage. Stretching of S–S bonds has been demonstrated experimentally to increase their susceptibility to cleavage [15–18].

The prestress of the −RHstaple and −/+RHhook disulfides may have contributed to their evolution as allosteric bonds. Their propensity for cleavage and an associated positive functional change may have resulted in evolutionary pressure to retain the bonds. The −LHhook configuration, though, is no more stressed than the other 18 configurations. It is not known why this bond type is associated with allosteric function. It is possible the geometry of the −LHhook configuration favors attack of the disulfide by factors that cleave allosteric disulfides, such as the oxidoreductases.

It is important to note that a protein's structure and the geometry of the disulfide bonds that link the polypeptide backbone can change shape [19]. Isomerization of certain allosteric disulfide bonds is likely important for protein function. Cleavage of the bond may only occur when the disulfide adopts a particular configuration in response to events, such as ligand binding or mechanical forces. Some allosteric bonds, however, appear to exist in one configuration only. This is the case for the tissue factor [20, 21], CD4 [14, 22], von Willebrand factor [23], and integrin β3 [24] allosteric disulfides.

Besides configuration, other structural features of disulfide bonds could be indicative of their function. This includes solvent accessibility, hydrophobicity, and net charge index [25].

4 An Analysis Tool for Disulfide Bonds

The biophysical properties of disulfide bonds, such as configuration, strain energy and solvent accessibility, can be informative about function. However, extracting structural information about specific disulfide bonds from the Protein Data Bank is difficult and requires

Table 2
Disulfide bond features included in the analysis tool

Column	Description
PDB ID	PDB identifier from Protein Data Bank or file name for uploaded structures
Header	Header from the PDB file
Compound	Compound from the PDB file—typically contains the name of the protein
Organism	Species name from the "compound" described in the PDB structure
Experiment	Type of 3D structure (i.e., X-ray or NMR)
Cys1 chain	Chain of Cys1 from disulfide bond in structure
Cys1 residue	Residue number of Cys1 from disulfide bond in structure
Cys1 secondary structure	Secondary Cys1 from disulfide bond derived from DSSP^
Cys1 solvent accessibility, ~$Å^2$	Solvent accessibility of Cys1 from disulfide bond derived from DSSP, ~$Å^2$
Cys2 chain	Chain of Cys2 from disulfide bond in structure
Cys2 residue	Residue number of Cys2 from disulfide bond in structure
Cys2 secondary structure	Secondary Cys2 from disulfide bond derived from DSSP^
Cys2 solvent accessibility, ~$Å^2$	Solvent accessibility of Cys2 from disulfide bond derived from DSSP, ~$Å^2$
Resolution	Resolution of structure (for X-ray only)
R-factor	R factor of structure (for X-ray only)
$\chi1$ angle, degree	The angle between the α-carbon and β-carbon of Cys1, degree
$\chi2$ angle, degree	The angle between the β-carbon and sulfur of Cys1, degree
$\chi3$ angle, degree	The angle between the sulfur of Cys1 and the sulfur of Cys2, degree
$\chi2'$ angle, degree	The angle between the β-carbon and sulfur of Cys2, degree
$\chi1'$ angle, degree	The angle between the α-carbon and β-carbon of Cys2, degree
α1 angle, degree	The angle between the Cys1 β-carbon and the S–S bond, degree
α2 angle, degree	The angle between the Cys2 β-carbon and the S–S bond, degree
Dihedral strain energy, kJ/mol	Estimated dihedral strain energy based on disulfide bond angles, kJ/mol
Distance between α-carbon atoms of the Cys residues, Å	Distance between α-carbon atoms of the Cys residues, Å (shown as "c" in Fig. 5)
Distance between the sulfur atoms of the Cys residues, Å	Distance between the sulfur atoms of the Cys residues, Å (shown as "d" in Fig. 5)
Disulfide bond classification	Classification of disulfide bond based on χ angles

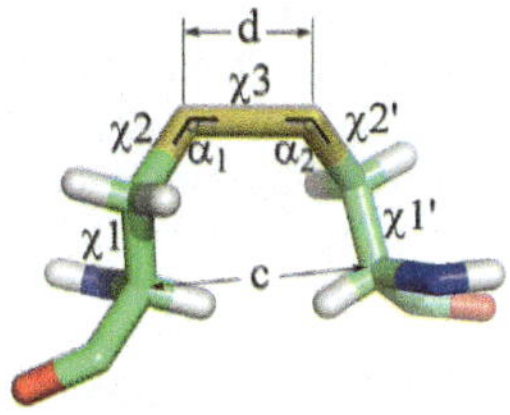

Fig. 5 The geometry of a disulfide bond. The distance between the α carbons is shown as "c." The distance between the two sulfur atoms is shown as "d." The five χ angles as well as the α angles are also shown

expert knowledge of 3D structure modeling and programs like PyMol and RasMol. To overcome this, a simple software tool was developed that allows users to extract information about all disulfide bonds in proteins by simply entering the PDB accession number [26]. The disulfide bond analysis tool is accessed at https://ssp-main.fly.dev/. In addition to accepting PDB IDs as input, any file complying with the PDB format (.pdb or .ent) can also be uploaded and lists 15 different features and parameters of each disulfide bond in the protein (Table 2, Fig. 5).

5 Notes

The disulfide bond analysis tool makes use of the information of protein structures as they are recorded in the PDB. As with any database, mistakes can be made in the annotation. To eliminate poorly defined or erroneous bonds, criteria of a maximum structure resolution of 2.5 Å and sulfur–sulfur distances <10% from the disulfide bond equilibrium length of 2.038 Å [14] could be applied.

References

1. Chiu J, Hogg PJ (2019) Allosteric disulphides: sophisticated molecular structures enabling flexible protein regulation. J Biol Chem 294(8):2949–5908. https://doi.org/10.1074/jbc.REV118.005604
2. Hogg PJ (2003) Disulphide bonds as switches for protein function. Trends Biochem Sci 28(4):210–214
3. Schmidt B, Ho L, Hogg PJ (2006) Allosteric disulphide bonds. Biochemistry 45(24):7429–7433
4. Harrison PM, Sternberg MJ (1996) The disulphide beta-cross: from cystine geometry and clustering to classification of small disulphide-rich protein folds. J Mol Biol 264(3):603–623
5. Hutchinson EG, Thornton JM (1996) PROMOTIF—a program to identify and analyze structural motifs in proteins. Protein Sci 5(2):212–220. https://doi.org/10.1002/pro.5560050204
6. Ozhogina OA, Bominaar EL (2009) Characterization of the kringle fold and identification of a ubiquitous new class of disulphide rotamers. J Struct Biol 168(2):223–233. https://doi.org/10.1016/j.jsb.2009.06.003
7. Pijning AE, Chiu J, Yeo RX, Wong JWH, Hogg PJ (2018) Identification of allosteric disulphides from labile bonds in X-ray structures. R Soc Open Sci 5(2):171058. https://doi.org/10.1098/rsos.171058
8. Marques JR, da Fonseca RR, Drury B, Melo A (2010) Conformational characterization of disulphide bonds: a tool for protein

classification. J Theor Biol 267(3):388–395. https://doi.org/10.1016/j.jtbi.2010.09.012
9. Cook KM, Hogg PJ (2013) Post-translational control of protein function by disulphide bond cleavage. Antioxid Redox Signal 18(15):1987–2015. https://doi.org/10.1089/ars.2012.4807
10. Wells JA, Powers DB (1986) In vivo formation and stability of engineered disulphide bonds in subtilisin. J Biol Chem 261(14):6564–6570
11. Kuwajima K, Ikeguchi M, Sugawara T, Hiraoka Y, Sugai S (1990) Kinetics of disulphide bond reduction in alpha-lactalbumin by dithiothreitol and molecular basis of superreactivity of the Cys6-Cys120 disulphide bond. Biochemistry 29(36):8240–8249
12. Wetzel R, Perry LJ, Baase WA, Becktel WJ (1988) Disulphide bonds and thermal stability in T4 lysozyme. Proc Natl Acad Sci U S A 85(2):401–405
13. Pjura PE, Matsumura M, Wozniak JA, Matthews BW (1990) Structure of a thermostable disulphide-bridge mutant of phage T4 lysozyme shows that an engineered cross-link in a flexible region does not increase the rigidity of the folded protein. Biochemistry 29(10):2592–2598
14. Zhou B, Baldus IB, Li W, Edwards SA, Grater F (2014) Identification of allosteric disulphides from prestress analysis. Biophys J 107(3):672–681. https://doi.org/10.1016/j.bpj.2014.06.025
15. Baldus IB, Grater F (2012) Mechanical force can fine-tune redox potentials of disulphide bonds. Biophys J 102(3):622–629. https://doi.org/10.1016/j.bpj.2011.12.039
16. Li W, Grater F (2010) Atomistic evidence of how force dynamically regulates thiol/disulphide exchange. J Am Chem Soc 132(47):16790–16795. https://doi.org/10.1021/ja104763q
17. Wiita AP, Ainavarapu SR, Huang HH, Fernandez JM (2006) Force-dependent chemical kinetics of disulphide bond reduction observed with single-molecule techniques. Proc Natl Acad Sci USA 103(19):7222–7227. https://doi.org/10.1073/pnas.0511035103
18. Wiita AP, Perez-Jimenez R, Walther KA, Grater F, Berne BJ, Holmgren A, Sanchez-Ruiz JM, Fernandez JM (2007) Probing the chemistry of thioredoxin catalysis with force. Nature 450(7166):124–127
19. Schmidt B, Hogg PJ (2007) Search for allosteric disulphide bonds in NMR structures. BMC Struct Biol 7:49. https://doi.org/10.1186/1472-6807-7-49
20. Zhou B, Hogg PJ, Grater F (2017) One-way allosteric communication between the two disulphide bonds in tissue factor. Biophys J 112(1):78–86. https://doi.org/10.1016/j.bpj.2016.12.003
21. Chen VM, Ahamed J, Versteeg HH, Berndt MC, Ruf W, Hogg PJ (2006) Evidence for activation of tissue factor by an allosteric disulphide bond. Biochemistry 45(39):12020–12028
22. Matthias LJ, Yam PT, Jiang XM, Vandegraaff N, Li P, Poumbourios P, Donoghue N, Hogg PJ (2002) Disulphide exchange in domain 2 of CD4 is required for entry of HIV-1. Nat Immunol 3(8):727–732
23. Butera D, Passam F, Ju L, Cook KM, Woon H, Aponte-Santamaria C, Gardiner E, Davis AK, Murphy DA, Bronowska A, Luken BM, Baldauf C, Jackson S, Andrews R, Grater F, Hogg PJ (2018) Autoregulation of von Willebrand factor function by a disulphide bond switch. Sci Adv 4(2):eaaq1477. https://doi.org/10.1126/sciadv.aaq1477
24. Passam F, Chiu J, Ju L, Pijning A, Jahan Z, Mor-Cohen R, Yeheskel A, Kolsek K, Tharichen L, Aponte-Santamaria C, Grater F, Hogg PJ (2018) Mechano-redox control of integrin de-adhesion. eLife 7:e34843. https://doi.org/10.7554/eLife.34843
25. Sun MA, Wang Y, Zhang Q, Xia Y, Ge W, Guo D (2017) Prediction of reversible disulphide based on features from local structural signatures. BMC Genomics 18(1):279. https://doi.org/10.1186/s12864-017-3668-8
26. Wong JW, Hogg PJ (2010) Analysis of disulphide bonds in protein structures. J Thromb Haemost 8(10):2345. https://doi.org/10.1111/j.1538-7836.2010.03894.x
27. Ioannou Y, Zhang JY, Passam FH, Rahgozar S, Qi JC, Giannakopoulos B, Qi M, Yu P, Yu DM, Hogg PJ, Krilis SA (2010) Naturally occurring free thiols within beta 2-glycoprotein I in vivo: nitrosylation, redox modification by endothelial cells, and regulation of oxidative stress-induced cell injury. Blood 116(11):1961–1970. https://doi.org/10.1182/blood-2009-04-215335
28. Kumar S, Chinnaraj M, Planer W, Zuo X, Macor P, Tedesco F, Pozzi N (2021) An allosteric redox switch in domain V of beta(2)-glycoprotein I controls membrane binding and anti-domain I autoantibody recognition. J Biol Chem 297(2):100890. https://doi.org/10.1016/j.jbc.2021.100890
29. Dupuy A, Aponte-Santamaria C, Yeheskel A, Hortle E, Oehlers SH, Grater F, Hogg PJ, Passam FH, Chiu J (2023) Mechano-redox control of Mac-1 de-adhesion by PDI promotes directional movement under flow. Circ Res 132(9):e151–e168. https://doi.org/10.1161/CIRCRESAHA.122.321926
30. Li Y, Xu X, Wang HJ, Chen YC, Chen Y, Chiu J, Li L, Wang L, Wang J, Tang Z, Ren L, Li H, Wang X, Jin S, Wu Y, Huang M, Ju LA, Fang C (2024) Endoplasmic reticulum protein 72 regulates integrin Mac-1 activity to influence neutrophil recruitment. Arterioscler

Thromb Vasc Biol 44(3):e82–e98. https://doi.org/10.1161/ATVBAHA.123.319771
31. Auwerx J, Isacsson O, Soderlund J, Balzarini J, Johansson M, Lundberg M (2009) Human glutaredoxin-1 catalyzes the reduction of HIV-1 gp120 and CD4 disulphides and its inhibition reduces HIV-1 replication. Int J Biochem Cell Biol 41(6):1269–1275
32. Barbouche R, Miquelis R, Jones IM, Fenouillet E (2003) Protein-disulphide isomerase-mediated reduction of two disulphide bonds of HIV envelope glycoprotein 120 occurs post-CXCR4 binding and is required for fusion. J Biol Chem 278(5):3131–3136
33. Gallina A, Hanley TM, Mandel R, Trahey M, Broder CC, Viglianti GA, Ryser HJ (2002) Inhibitors of protein-disulphide isomerase prevent cleavage of disulphide bonds in receptor-bound glycoprotein 120 and prevent HIV-1 entry. J Biol Chem 277(52):50579–50588. https://doi.org/10.1074/jbc.M204547200
34. Ou W, Silver J (2006) Role of protein disulphide isomerase and other thiol-reactive proteins in HIV-1 envelope protein-mediated fusion. Virology 350(2):406–417
35. Reiser K, Francois KO, Schols D, Bergman T, Jornvall H, Balzarini J, Karlsson A, Lundberg M (2012) Thioredoxin-1 and protein disulphide isomerase catalyze the reduction of similar disulphides in HIV gp120. Int J Biochem Cell Biol 44(3):556–562. https://doi.org/10.1016/j.biocel.2011.12.015
36. Azimi I, Matthias LJ, Center RJ, Wong JW, Hogg PJ (2010) Disulphide bond that constrains the HIV-1 gp120 V3 domain is cleaved by thioredoxin. J Biol Chem 285(51):40072–40080. https://doi.org/10.1074/jbc.M110.185371
37. Ang CG, Hyatt NL, Le Minh G, Gupta M, Kadam M, Hogg PJ, Smith 3rd AB, Chaiken IM (2025) Conformational activation and disulphide exchange in HIV-1 Env induce cell-free lytic/fusogenic transformation and enhance infection. J Virol 99(3):e0147124. https://doi.org/10.1128/jvi.01471-24
38. Kellett-Clarke H, Stegmann M, Barclay AN, Metcalfe C (2015) CD44 binding to hyaluronic acid is redox regulated by a labile disulphide bond in the hyaluronic acid binding site. PLoS One 10(9):e0138137. https://doi.org/10.1371/journal.pone.0138137
39. Li J, Kim K, Jeong S, Chiu J, Xiong B, Petukhov PA, Dai X, Li X, Andrews RK, Du X, Hogg PJ, Cho J (2019) Platelet protein disulphide isomerase promotes glycoprotein Ibα-mediated platelet-neutrophil interactions under thromboinflammatory conditions. Circulation 139(10):1300–1319. https://www.ahajournals.org/doi/10.1161/CIRCULATIONAHA.118.036323
40. Plugis NM, Weng N, Zhao Q, Palanski BA, Maecker HT, Habtezion A, Khosla C (2018) Interleukin 4 is inactivated via selective disulphide-bond reduction by extracellular thioredoxin. Proc Natl Acad Sci USA 115(35):8781–8786. https://doi.org/10.1073/pnas.1805288115
41. Metcalfe C, Cresswell P, Barclay AN (2012) Interleukin-2 signalling is modulated by a labile disulphide bond in the CD132 chain of its receptor. Open Biol 2(1):110036. https://doi.org/10.1098/rsob.110036
42. Ahamed J, Versteeg HH, Kerver M, Chen VM, Mueller BM, Hogg PJ, Ruf W (2006) Disulphide isomerization switches tissue factor from coagulation to cell signaling. Proc Natl Acad Sci USA 103(38):13932–13937
43. Doello S, Shvarev D, Theune M, Sauerwein J, Klon A, Keskin E, Boehm M, Gutekunst K, Forchhammer K (2024) Structural basis of the allosteric regulation of cyanobacterial glucose-6-phosphate dehydrogenase by the redox sensor OpcA. Proc Natl Acad Sci USA 121(50):e2411604121. https://doi.org/10.1073/pnas.2411604121
44. Perez-Vargas J, Teppa E, Amirache F, Boson B, Pereira de Oliveira R, Combet C, Bockmann A, Fusil F, Freitas N, Carbone A, Cosset FL (2021) A fusion peptide in preS1 and the human protein disulphide isomerase ERp57 are involved in hepatitis B virus membrane fusion process. eLife 10:e64507. https://doi.org/10.7554/eLife.64507
45. Chu Y, Zhang H, Yang M, Yu R (2023) Molecular dynamic simulations reveal the activation mechanisms of oxidation-induced TRPV1. Int J Mol Sci 24(11):9553. https://doi.org/10.3390/ijms24119553
46. Ushioda R, Miyamoto A, Inoue M, Watanabe S, Okumura M, Maegawa KI, Uegaki K, Fujii S, Fukuda Y, Umitsu M, Takagi J, Inaba K, Mikoshiba K, Nagata K (2016) Redox-assisted regulation of Ca2+ homeostasis in the endoplasmic reticulum by disulphide reductase ERdj5. Proc Natl Acad Sci USA 113(41):E6055–E6063. https://doi.org/10.1073/pnas.1605818113
47. Jha V, Xiong B, Kumari T, Brown G, Wang J, Kim K, Lee J, Asquith N, Gallagher J, Asherman L, Lambert T, Bai Y, Du X, Min JK, Sah R, Javaheri A, Razani B, Lee JM, Italiano JE, Cho J (2023) A critical role for ERO1alpha in arterial thrombosis and ischemic stroke. Circ Res 132(11):e206–e222. https://doi.org/10.1161/CIRCRESAHA.122.322473
48. Pimanda JE, Annis DS, Raftery M, Mosher DF, Chesterman CN, Hogg PJ (2002) The von

Willebrand factor-reducing activity of thrombospondin-1 is located in the calcium-binding/C-terminal sequence and requires a free thiol at position 974. Blood 100(8):2832–2838
49. Hotchkiss KA, Matthias LJ, Hogg PJ (1998) Exposure of the cryptic Arg-Gly-Asp sequence in thrombospondin-1 by protein disulphide isomerase. Biochim Biophys Acta 1388(2):478–488
50. Hotchkiss KA, Chesterman CN, Hogg PJ (1996) Catalysis of disulphide isomerization in thrombospondin 1 by protein disulphide isomerase. Biochemistry 35(30):9761–9767. https://doi.org/10.1021/bi9603938
51. Bowley SR, Fang C, Merrill-Skoloff G, Furie BC, Furie B (2017) Protein disulphide isomerase secretion following vascular injury initiates a regulatory pathway for thrombus formation. Nat Commun 8:14151. https://doi.org/10.1038/ncomms14151
52. Lv K, Chen S, Xu X, Chiu J, Wang HJ, Han Y, Yang X, Bowley SR, Wang H, Tang Z, Tang N, Yang A, Yang S, Wang J, Jin S, Wu Y, Schmaier AH, Ju LA, Hogg PJ, Fang C (2024) Protein disulphide isomerase cleaves allosteric disulphides in histidine-rich glycoprotein to regulate thrombosis. Nat Commun 15(1):3129. https://doi.org/10.1038/s41467-024-47493-0
53. Malojcic G, Glockshuber R (2010) The PAPS-independent aryl sulfotransferase and the alternative disulphide bond formation system in pathogenic bacteria. Antioxid Redox Signal 13(8):1247–1259. https://doi.org/10.1089/ars.2010.3119
54. Swaminathan S, Eswaramoorthy S (2000) Structural analysis of the catalytic and binding sites of *Clostridium botulinum* neurotoxin B. Nat Struct Biol 7(8):693–699
55. Mohammad A, Bon Ramos A, Lee BW, Cohen SW, Kiani MK, Iwata-Reuyl D, Stec B, Swairjo MA (2017) Protection of the queuosine biosynthesis enzyme QueF from irreversible oxidation by a conserved intramolecular disulphide. Biomolecules 7(1):30. https://doi.org/10.3390/biom7010030
56. Kim S, Jang YS, Ha SC, Ahn JW, Kim EJ, Lim JH, Cho C, Ryu YS, Lee SK, Lee SY, Kim KJ (2015) Redox-switch regulatory mechanism of thiolase from *Clostridium acetobutylicum*. Nat Commun 6:8410. https://doi.org/10.1038/ncomms9410
57. Cook KM, McNeil HP, Hogg PJ (2013) Allosteric control of bII-tryptase by a redox active disulphide bond. J Biol Chem 288(48):34920–34929. https://doi.org/10.1074/jbc.M113.523506
58. Chiu J, Wong JW, Hogg PJ (2014) Redox regulation of methionine aminopeptidase 2 activity. J Biol Chem 289(21):15035–15043. https://doi.org/10.1074/jbc.M114.554253
59. Gopalan G, He Z, Balmer Y, Romano P, Gupta R, Heroux A, Buchanan BB, Swaminathan K, Luan S (2004) Structural analysis uncovers a role for redox in regulating FKBP13, an immunophilin of the chloroplast thylakoid lumen. Proc Natl Acad Sci USA 101(38):13945–13950
60. Pinkas DM, Strop P, Brunger AT, Khosla C (2007) Transglutaminase 2 undergoes a large conformational change upon activation. PLoS Biol 5(12):e327. https://doi.org/10.1371/journal.pbio.0050327
61. Melkonian AV, Loppinet E, Martin R, Porteus M, Khosla C (2021) An unusual "OR" gate for allosteric regulation of mammalian transglutaminase 2 in the extracellular matrix. J Am Chem Soc 143(28):10537–10540. https://doi.org/10.1021/jacs.1c04616
62. Nishii W, Kukimoto-Niino M, Terada T, Shirouzu M, Muramatsu T, Kojima M, Kihara H, Yokoyama S (2015) A redox switch shapes the Lon protease exit pore to facultatively regulate proteolysis. Nat Chem Biol 11(1):46–51. https://doi.org/10.1038/nchembio.1688
63. Sagong HY, Kim KJ (2017) Structural basis for redox sensitivity in *Corynebacterium glutamicum* diaminopimelate epimerase: an enzyme involved in L-lysine biosynthesis. Sci Rep 7:42318. https://doi.org/10.1038/srep42318
64. Mauney CH, Rogers LC, Harris RS, Daniel LW, Devarie-Baez NO, Wu H, Furdui CM, Poole LB, Perrino FW, Hollis T (2017) The SAMHD1 dNTP triphosphohydrolase is controlled by a redox switch. Antioxid Redox Signal 27(16):1317–1331. https://doi.org/10.1089/ars.2016.6888
65. Oliveira AR, Mota C, Vilela-Alves G, Manuel RR, Pedrosa N, Fourmond V, Klymanska K, Leger C, Guigliarelli B, Romao MJ, Cardoso Pereira IA (2024) An allosteric redox switch involved in oxygen protection in a CO(2) reductase. Nat Chem Biol 20(1):111–119. https://doi.org/10.1038/s41589-023-01484-2
66. Ravilious GE, Nguyen A, Francois JA, Jez JM (2012) Structural basis and evolution of redox regulation in plant adenosine-5'-phosphosulfate kinase. Proc Natl Acad Sci USA 109(1):309–314. https://doi.org/10.1073/pnas.1115772108
67. Zhou A, Carrell RW, Murphy MP, Wei Z, Yan Y, Stanley PLD, Stein PE, Pipkin FB, Read RJ (2010) A redox switch in angiotensinogen modulates angiotensin release. Nature 468(7320):108–111. http://www.nature.com/nature/journal/v468/n7320/abs/nature09505.html#supplementary-information
68. Zhou A, Carrell RW, Murphy MP, Wei Z, Yan Y, Stanley PLD, Stein PE, Pipkin FB, Read RJ (2010) A redox switch in angiotensinogen modulates angiotensin release. Nature 468(7320):108–111. http://www.nature.com/nature/

journal/v468/n7320/abs/nature09505.html#supplementary-information
69. Giannakopoulos B, Gao L, Qi M, Wong JW, Yu DM, Vlachoyiannopoulos PG, Moutsopoulos HM, Atsumi T, Koike T, Hogg PJ, Qi JC, Krilis SA (2012) Factor XI is a substrate for oxidoreductases: enhanced activation of reduced FXI and its role in antiphospholipid syndrome thrombosis. J Autoimmun 39(3):121–129. https://www.sciencedirect.com/science/article/pii/S0896841112000625?via%3Dihub
70. Kaiser BK, Yim D, Chow IT, Gonzalez S, Dai Z, Mann HH, Strong RK, Groh V, Spies T (2007) Disulphide-isomerase-enabled shedding of tumour-associated NKG2D ligands. Nature 447(7143):482–486
71. Butera D, Wind T, Lay AJ, Beck J, Castellino FJ, Hogg PJ (2014) Characterization of a reduced form of plasma plasminogen as the precursor for angiostatin formation. J Biol Chem 289(5):2992–3000. https://doi.org/10.1074/jbc.M113.539924
72. Pinjusic K, Bulliard M, Rothe B, Ansaryan S, Liu YC, Ginefra P, Schmuziger C, Altug H, Constam DB (2024) Stepwise release of activin-A from its inhibitory prodomain is modulated by cysteines and requires furin coexpression to promote melanoma growth. Commun Biol 7(1):1383. https://doi.org/10.1038/s42003-024-07053-0
73. Miller WP, Sha CM, Sunilkumar S, Toro AL, VanCleave AM, Kimball SR, Dokholyan NV, Dennis MD (2022) Activation of disulphide redox switch in REDD1 promotes oxidative stress under hyperglycemic conditions. Diabetes 71(12):2764–2776. https://doi.org/10.2337/db22-0355
74. Chiu J, Wong JW, Gerometta M, Hogg PJ (2014) Mechanism of dimerization of a recombinant mature vascular endothelial growth factor C. Biochemistry 53(1):7–9. https://doi.org/10.1021/bi401518b
75. Banerji S, Lawrance W, Metcalfe C, Briggs DC, Yamauchi A, Dushek O, van der Merwe PA, Day AJ, Jackson DG (2016) Homodimerization of the lymph vessel endothelial receptor LYVE-1 through a redox-labile disulphide is critical for hyaluronan binding in lymphatic endothelium. J Biol Chem 291(48):25004–25018. https://doi.org/10.1074/jbc.M116.736926

Check for updates

Chapter 2

Structural Analysis of Disulfide Bonds in the RCSB Protein Data Bank Using proteusPy

Eric G. Suchanek

Abstract

A structural analysis of disulfide bonds extracted from the RCSB Protein Data Bank (PDB) is presented. This analysis utilized the **proteusPy** Python package and provides a statistical basis for yielding a dataset of geometrically reasonable disulfide bond structures. Additionally, a method for quantization of dihedral angles to reduce structural complexity and create disulfide structural classes is described. This method allows for the segmentation of the disulfide bonds into structural families. The results of this analysis provide a framework for understanding the geometric and structural families of disulfide bonds in the RCSB databank.

Keywords Disulfide bond, Configuration, Conformation, Geometry, Structure, ProteusPy, Strain, Dihedral angle

1 Introduction

Disulfide bonds are formed when two Cysteine residues are oxidized resulting in a sulfur–sulfur covalent bond. A schematic of a modeled disulfide bond is shown in Fig. 1. The proximal and distal residues are "1," the chain is "A" and the pdbid is "1egs." This figure represents an illustration of disulfide bond connectivity and torsions.

These structural elements play pivotal roles in structural stabilization within and between protein subunits. Moreover, they participate in enzymatic catalysis, regulate protein activities, and offer protection against oxidative stress. Disulfide bonds also play an important role in the development of neurodegenerative diseases like Alzheimer's since they are intimately involved in tau protein aggregation [1–3].

The Python package **proteusPy** has been developed to facilitate analysis and characterization of disulfide bonds. This package is available on the GitHub repository [4] and contains the disulfide bond structural database extracted from the RCSB Protein Data Bank: https://www.rcsb.org/. **proteusPy** has a number of features including:

Philip J. Hogg (ed.), *Functional Disulphide Bonds*, Methods in Molecular Biology, vol. 3016,
https://doi.org/10.1007/978-1-0716-5158-2_2,

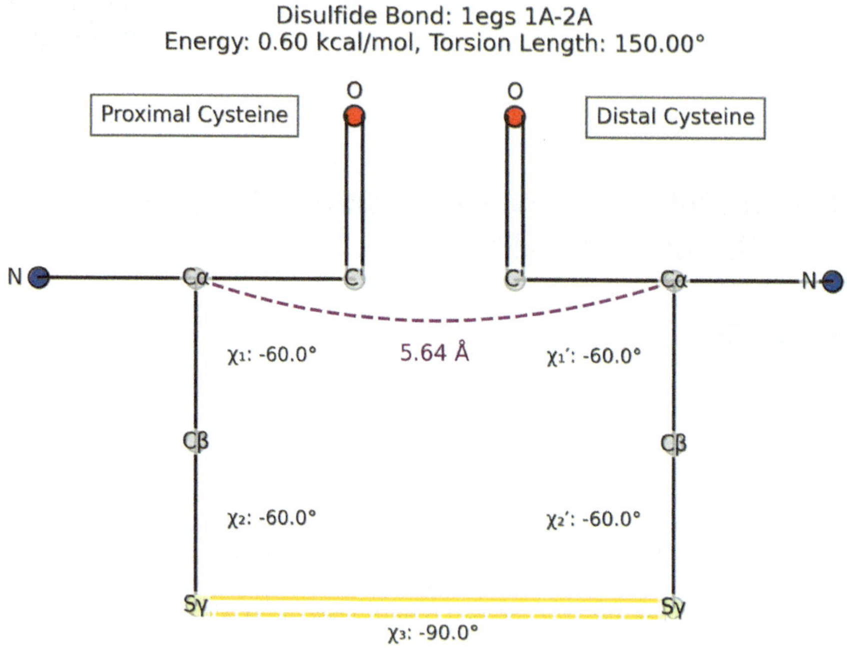

Fig. 1 Schematic of a modeled disulfide bond

- Interactively display disulfides contained in the RCSB in a variety of display styles.
- Calculate geometric and energetic properties about these disulfides.
- Create binary and octant structural classes by quantizing the disulfide torsional angles.
- Build idealized disulfide bonds from dihedral angle input.
- Find disulfide neighbors based on dihedral angle input.
- Overlap disulfides onto a common frame of reference for display.
- Build protein backbones from backbone phi, psi dihedral angle templates.

Refer to the **proteusPy** documentation for more information on the package features and usage. This is available at https://suchanek.github.io/proteusPy. The package is available for installation via PyPi, and the source code is available on GitHub at https://github.com/suchanek/proteusPy.

2 Methods

The general process employed to create the database involves the following steps:

1. Extraction of disulfide bonds from the RCSB Protein Data Bank. This required the development of a parser to extract the disulfide bond information. This seemingly simple task was complicated by the fact that there are many errors in the PDB files, and the parser had to be able to manage them. This led to the development of a robust logging system to track the parsing and ensure that the parser could handle the anomalies without crashing.
2. Construction of a master list of disulfide bonds using the **DisulfideClassManager** class. This list is stored on Google Drive and is downloaded automatically by the proteusPy package upon installation.
3. Filtering based on geometric criteria to ensure structural validity for analysis purposes.
4. Quantization of dihedral angles for structural classification.
5. Visualization of disulfide bonds via the rcsbviewer.py program.

2.1 Disulfide Database Creation

A query to identify proteins containing one or more disulfide bonds using the query builder on https://www.rcsb.org was used. This yielded 38,214 entries. These were downloaded via a Python script **DisulfideDownloader.py**. Once downloaded, the structures were extracted with the **proteusPy.DisulfideExtractor** entry point. This multiprocessing program can extract over 36,000 PDB structure files in just under 3 min using 14 cores on a M3 Max Macbook Pro. Structures that were nonparsable were discarded. The remaining 36,968 were used as the basis for the geometric analysis.

2.2 Disulfide Bond Distance Analysis

The structures used are the product of X-ray crystallography, and as such all models have errors associated with their atomic positions. Structure quality is generally tied to the **resolution** structural parameter. Resolutions <2.0°A are considered high quality with sufficient detail to easily distinguish individual atoms.

To ensure that the structures used in the analysis are geometrically reasonable, structures were filtered based on the criteria indicated above:

- The distance between the two sulfur atoms in the disulfide bond. This should be close to the covalent bond length of 2.04°A.
- The distance between the two C_α atoms. This distance should not exceed 8°A.
- Bond length deviations from ideality.
- Bond angle deviations from ideality.

Since the S_γ–S_γ bond is covalent, the interatomic distance should be close to the covalent bond length of 2.04°A. The overall S_γ–S_γ and C_α–C_α distance distributions are shown in Table 1.The mean distance

Table 1
Disulfide bond distances, full database

Distance (°A)	Mean (°A)	Std
$S\gamma$–$S\gamma$	2.12	1.95
$C\alpha$–$C\alpha$	5.67	1.87

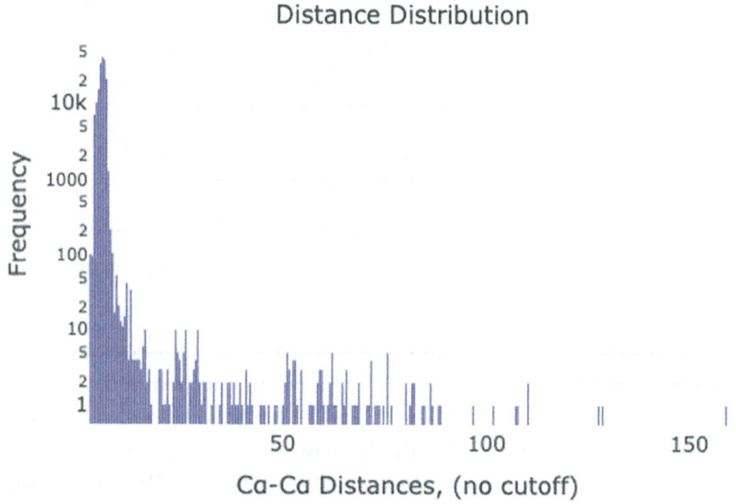

Fig. 2 Histogram of unfiltered disulfide bond C_α–C_α distances

is 2.12°A, with a standard deviation of 1.95 for the entire database. The distance between the proximal and distal residue C_α atoms is another useful screening property. This distance cannot exceed 8°A, since this would yield a structure with physically impossible geometry. The C_α–C_α distance distribution before any filtering is shown in Fig. 2. The mean distance is 5.67°A, with a standard deviation of 1.87. This is a wide distribution, indicating that the quality of the structures varies significantly. There are also a number of structures with C_α–C_α distances well above 8°A, indicating physically impossible geometry.

The longest disulfide bond in the database is not shown, since the C_α–C_α bond distance is over 100°A. For interested readers, the ID is: 6zz8-178C-178D. This structure is clearly an outlier and would be discarded in the filtering process. Conversely, the shortest disulfide has a C_α–C_α distance of 2.71°A, which is extremely short and well below the average. This is illustrated in Fig. 3. The structure exhibits a highly strained configuration with a high torsional energy and practically *cis* S_γ–S_γ bond (Fig. 4).

Statistics (mean, standard deviation, and confidence intervals) were calculated for the bond lengths and angles in order to filter them for reasonable geometry (Figs. 5 and 6).

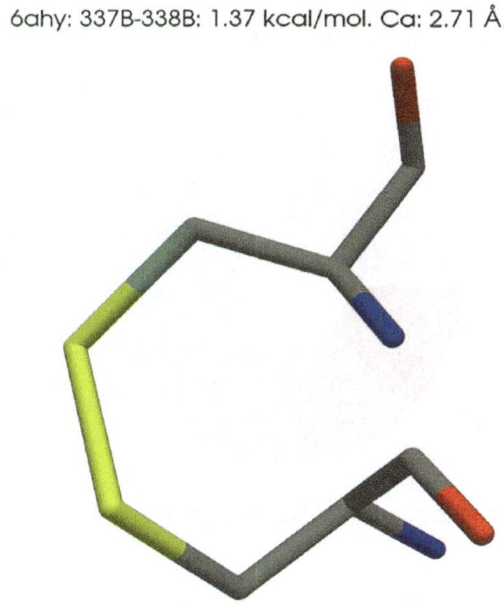

Fig. 3 The shortest disulfide in RCSB

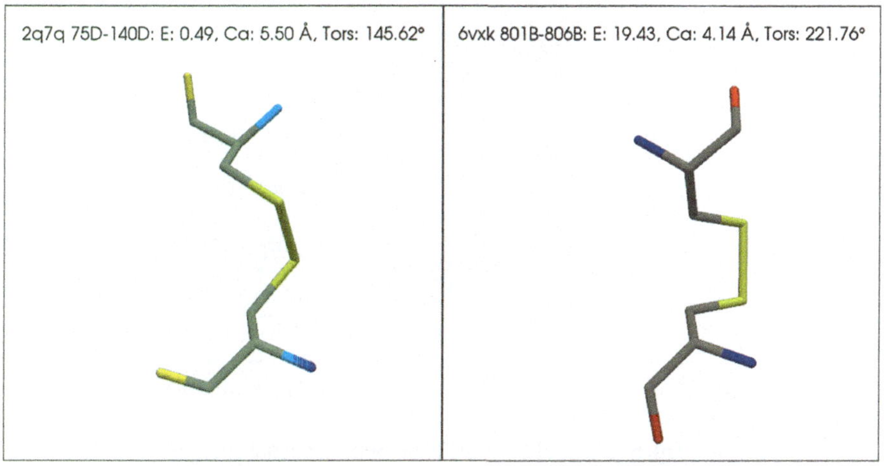

Fig. 4 Minimum and maximum torsional energy structures

The cutoffs for the C_α–C_α and S_γ–S_γ bond distances are shown in Table 2.

The application of the two filters resulted in the loss of 16,312 structures, leaving a total of 35,367 structures containing 158,965 disulfide bonds. The overall C_α–C_α distance distribution after filtering is shown in Fig. 5. The S_γ–S_γ distance distribution after filtering is shown in Fig. 6.

2.3 Disulfide Bond Strain Energy Analysis

The approximate bond torsional energy (kcal/mol) of a given disulfide bond is calculated using the following Equation. [5]:

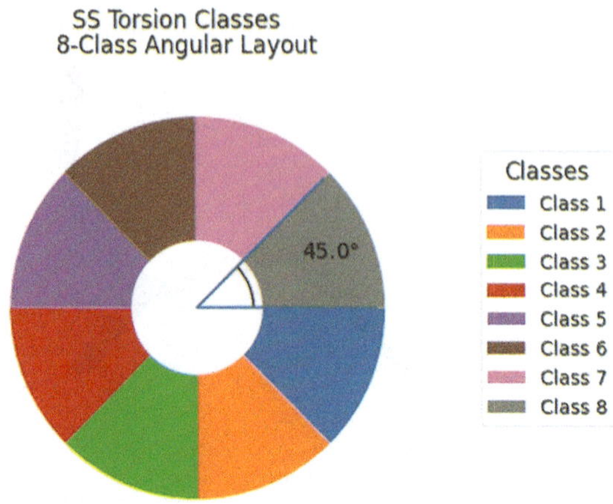

Fig. 5 Disulfide bond octant classification

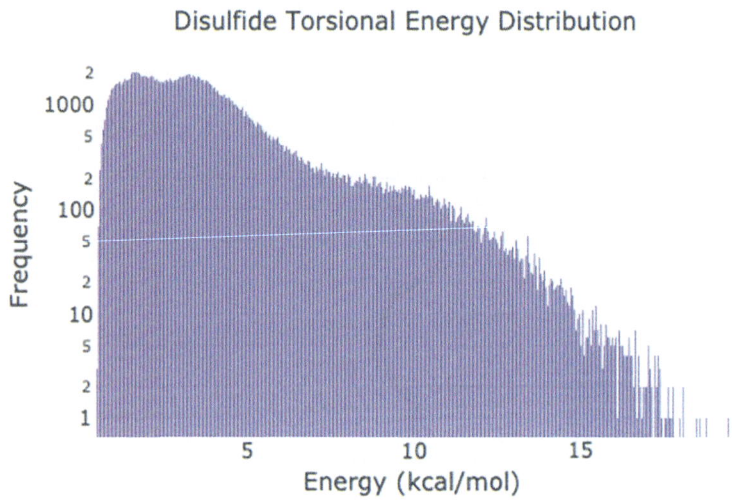

Fig. 6 Distribution of Disulfide Bond Torsional Energies

Table 2
Disulfide bond distance cutoffs for 95th percentile

Bond	Distance (°A)
$S\gamma$–$S\gamma$	2.52
$C\alpha$–$C\alpha$	6.71

$$E_{kcal\ mol^{-1}} \approx 2.0 \cdot \cos(3.0 \cdot \chi_1) + \cos(3.0 \cdot \chi_{1'}) + \cos(3.0 \cdot \chi_2) + \cos(3.0 \cdot \chi_{2'}) + 3.5 \cdot \cos(2.0 \cdot \chi_3) + 0.6 \cdot \cos(3.0 \cdot \chi_3) + 10.1 \quad (1)$$

The equation embodies the typical three-fold rotation barriers associated with single bonds ($\chi_1, \chi_1', \chi_2, \chi_2'$), and a high two-fold barrier for χ_3, resulting from the partial double bond character of the S_γ–S_γ bond. This property leads to two major disulfide families, characterized by the sign of χ_3. *Left-handed* disulfides have $\chi_3 < 0°$ and *right-handed* disulfides have $\chi_3 > 0°$. The overall distribution of disulfide bond torsional energies after filtering is shown in Fig. 7[1].

2.4 Dihedral Angle Quantization

The overall conformation of a particular disulfide bond is characterized by five rotatable bonds. These are denoted $\chi_1, \chi_2, \chi_3, \chi_2', \chi_1'$, and reflect the overall "twist" of the two residues between the first amino acid (proximal residue) and second amino acid (distal residue). The "twist" between each of these sidechain bonds is known as the *torsion* or *dihedral* angles. Schmidt et al. [6] proposed a method for quantizing these angles into binary classes by calling a dihedral "+" if it is greater than 0 and "−" if it is less than 0. This method yielded $2^5 = 32$ possible classes. The proteusPy package uses:

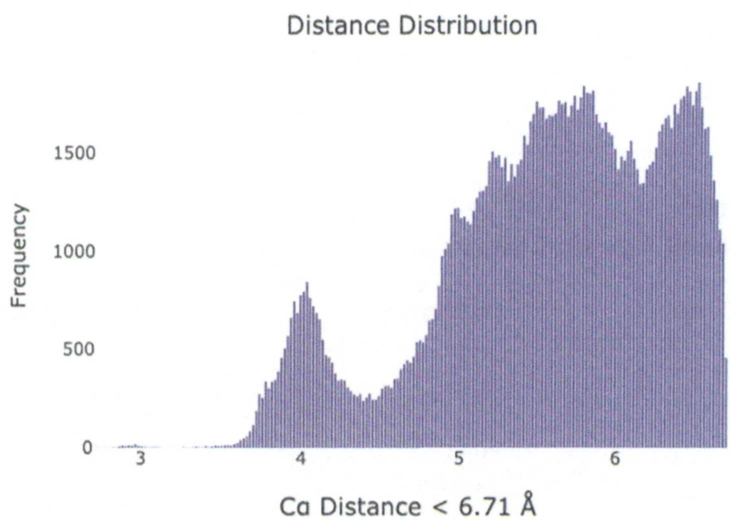

Fig. 7 Distribution of disulfide bond C_α–C_α distances after filtering

[1] The distribution prior to filtering was similar

- "0" represents a negative angle
- "2" represents a positive angle

For example, the class "00000" represents a disulfide bond where all five dihedral angles are negative, known as a "–LHSpiral" (Left-Handed Spiral).

2.4.1 Octant Quantization

The octant quantization system divides the 360° range of each dihedral angle into eight equal segments of 45° each, numbered from 1 to 8 clockwise starting from 0° as shown in Table 3. This is illustrated in Fig. 8. This creates $8^5 = 32{,}768$ possible classes, providing much finer granularity in classifying disulfide conformations.

2.4.2 Mapping Between Systems

The binary and octant systems are related as follows:

- Binary "0" (negative angle) maps to octants 1, 2, 3, and 4
- Binary "2" (positive angle) maps to octants 5, 6, 7, and 8

Since there are 32,768 possible octant classes, and only 32 binary classes, each binary class maps to 1024 octant classes. **proteusPy** provides a method (proteusPy.DisulfideLoader.toclass()) to map from binary to octant classes.

2.4.3 Example: Mapping Dihedrals to a Class String

Let us consider a specific disulfide bond with the following dihedral angles:

$\chi 1 = -60^\circ$
$\chi 2 = -60^\circ$
$\chi 3 = -90^\circ$
$\chi_2' = -60^\circ$
$\chi 1_{\prime} = -60^\circ$

Table 3
Dihedral angle octant quantization

Octant	Positive range	Negative equivalent
1	315°–360°	–45° to 0°
2	270°–315°	–90° to –45°
3	225°–270°	–135° to –90°
4	180°–225°	–180° to –135°
5	135°–180°	–
6	90°–135°	–
7	45°–90°	–
8	0°–45°	–

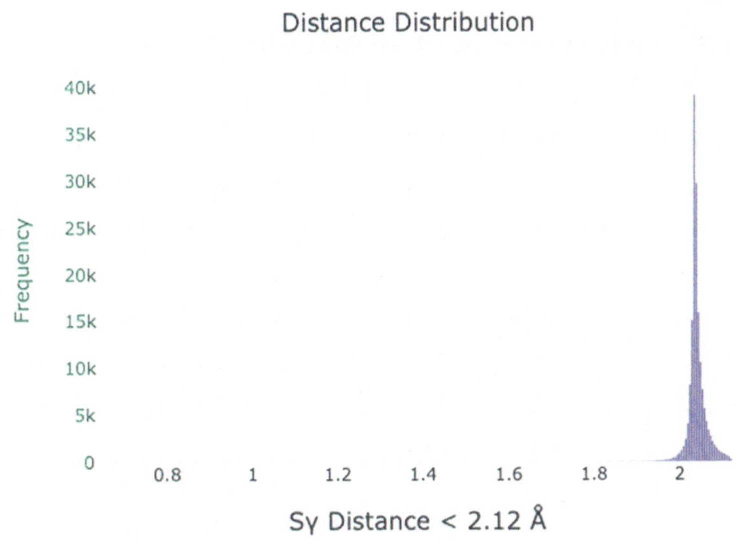

Fig. 8 Distribution of disulfide bond S_γ–S_γ distances after filtering

Binary Classification:

Since all angles are negative, they all map to "0" in the binary system:

- Binary class string: "00000" (the "–LHSpiral" class)
  Octant Classification:
- $\chi_1 = -60°$ falls in the range $-90°$ to $-45° \rightarrow$ Octant $2\chi_2 = -60°$ falls in the range $-90°$ to $-45° \rightarrow$ Octant 2
- $\chi_3 = -90°$ falls in the range $-90°$ to $-45° \rightarrow$ Octant 2
- $\chi_2' = -60°$ falls in the range $-90°$ to $-45° \rightarrow$ Octant 2
- $\chi_1' = -60°$ falls in the range $-90°$ to $-45° \rightarrow$ Octant 2

Thus, the octant class string is "22222." This example demonstrates how the same disulfide bond is classified as "00000" in the binary system and "22222" in the octant system. The octant system then provides a structural filter that reduces the possible number of structures within a structural family.

2.5 Disulfide Bond Classification

2.5.1 Binary Classes

The binary class results are shown in Table 4. The class names are per Schmidt et al. [6]. The most common class is 00000b or "—." representing a "–LHSpiral" conformation. This class represents 23.36% of the total disulfide bonds in the database. The least common class is 20,202, representing a "+RHStaple" conformation, which represents 0.31% of the total disulfide bonds in the database.

Table 4
Disulfide binary classes in the filtered (95%) RCSB database

$\chi1,\chi2,\chi3,\chi_2',\chi_1'$	Count	Incidence (%)	Energy (kcal mol^{-1})	C_α–C_α(°A)	Classname
–,–,–,–,–	39,140	24.63	2.68 (0.84–4.53)	5.74 (5.37–6.11)	–LHSpiral
–,–,–,–,+	7489	4.71	3.25 (1.39–5.11)	6.04 (5.63–6.45)	–,–,–,–,+
–,–,–,+,–	4397	2.77	4.79 (1.84–7.73)	5.30 (4.55–6.05)	–LHHook
–,–,–,+,+	2189	1.38	3.90 (1.42–6.38)	5.37 (4.76–5.98)	–,–,–,+,+
–,–,+,–,–	15,390	9.68	4.47 (2.98–5.96)	4.20 (3.75–4.66)	–RHStaple
–,–,+,–,+	1231	0.78	4.67 (1.86–7.48)	5.08 (4.48–5.69)	–,–,+,–,+
–,–,+,+,–	6638	4.18	4.19 (1.67–6.70)	5.33 (4.76–5.91)	–,–,+,+,–
–,–,+,+,+	6313	3.97	3.45 (1.08–5.82)	5.21 (4.74–5.68)	–,–,+,+,+
–,+,–,–,–	6512	4.10	3.93 (1.01–6.85)	5.71 (4.94–6.48)	–,+,–,–,–
–,+,–,–,+	7154	4.50	3.22 (0.73–5.70)	6.18 (5.63–6.74)	–,+,–,–,+
–,+,–,+,–	2910	1.83	4.64 (0.98–8.30)	5.42 (4.40–6.44)	–LHStaple
–,+,–,+,+	1046	0.66	4.22 (1.18–7.25)	5.52 (4.66–6.38)	–,+,–,+,+
–,+,+,–,–	6470	4.07	3.76 (1.03–6.49)	5.38 (4.88–5.88)	–RHHook
–,+,+,–,+	813	0.51	5.89 (3.10–8.68)	5.86 (5.38–6.33)	–,+,+,–,+
–,+,+,+,–	7910	4.98	3.20 (0.92–5.47)	5.81 (5.40–6.21)	–RHSpiral
–,+,+,+,+	6647	4.18	4.00 (2.20–5.80)	6.25 (5.87–6.63)	–,+,+,+,+
+,–,–,–,–	4211	2.65	3.32 (1.33–5.32)	6.03 (5.66–6.41)	±LHSpiral
+,–,–,–,+	1256	0.79	3.86 (1.08–6.64)	6.26 (5.92–6.61)	+LHSpiral
+,–,–,+,–	1043	0.66	5.65 (2.52–8.78)	5.75 (5.15–6.36)	±LHHook
+,–,–,+,+	703	0.44	4.45 (1.92–6.97)	5.67 (5.08–6.26)	+LHHook
+,–,+,–,–	629	0.34	5.86 (2.93–8.79)	5.29 (4.58–6.00)	±RHStaple
+,–,+,–,+	317	0.20	7.72 (4.51–10.9)	5.57 (4.76–6.38)	+RHStaple
+,–,+,+,–	1593	1.00	4.53 (1.82–7.23)	5.59 (5.10–6.08)	±RHHook
+,–,+,+,+	639	0.40	6.52 (3.44–9.60)	6.09 (5.53–6.65)	+,–,+,+,+
+,+,–,–,–	3401	2.14	3.27 (0.72–5.83)	5.52 (4.79–6.24)	–/+LHHook
+,+,–,–,+	1866	1.17	3.69 (0.78–6.60)	6.03 (5.55–6.52)	+,+,–,–,+
+,+,–,+,–	1632	1.03	4.58 (1.40–7.76)	5.12 (4.05–6.19)	±LHStaple
+,+,–,+,+	500	0.32	4.87 (1.25–8.49)	5.65 (4.92–6.37)	+LHStaple
+,+,+,–,–	7765	4.89	3.04 (0.89–5.20)	5.13 (4.72–5.54)	–/+RHHook
+,+,+,–,+	654	0.41	5.03 (2.13–7.93)	5.70 (5.17–6.23)	+RHHook
+,+,+,+,–	3260	2.05	3.43 (1.13–5.72)	5.94 (5.58–6.29)	±RHSpiral
+,+,+,+,+	7220	4.54	3.70 (2.20–5.21)	6.36 (6.06–6.67)	+RHSpiral

2.5.2 Octant Classes

The octant classification strategy was used since the 45° octants span the normal tri-fold dihedral minima of ±60°, and the number of possible classes (8^5 or 32,768) is manageable with current computational resources. These classes are constructed at runtime when the **DisulfideDownloader** class is instantiated.

In order to map from a binary class string to all possible octant class strings the **itertools** python library is used. This allows for the generation of all permutations of the class strings efficiently. The octant classification for the binary class 00000b is shown in Fig. 9.

The **DisulfideClassAnalysis.py** program was used to generate the binary and octant class distributions and statistics. This program enumerates class membership for these classes and calculates and displays dihedral angle and distance statistics for each class. Given the large number of octant classes present (over 9000), a cutoff factor was introduced to reduce the number of classes displayed to a manageable number. The typical cutoff factor used was 0.04%, which yields 325 described classes. This means that only classes with an incidence of 0.04% or greater are described.

Consider binary class 00000b (the "–LHSpiral" structural family, which has the highest representation in the entire database at 24.63%) once more (Fig. 9); the two most prominent peaks correspond to octant classes 22222o and 22322o, respectively. These differ only by the movement of χ_3 into octant class 3 (−135° to −90°). Only 617/1024 possible classes are present in this class. The dihedral

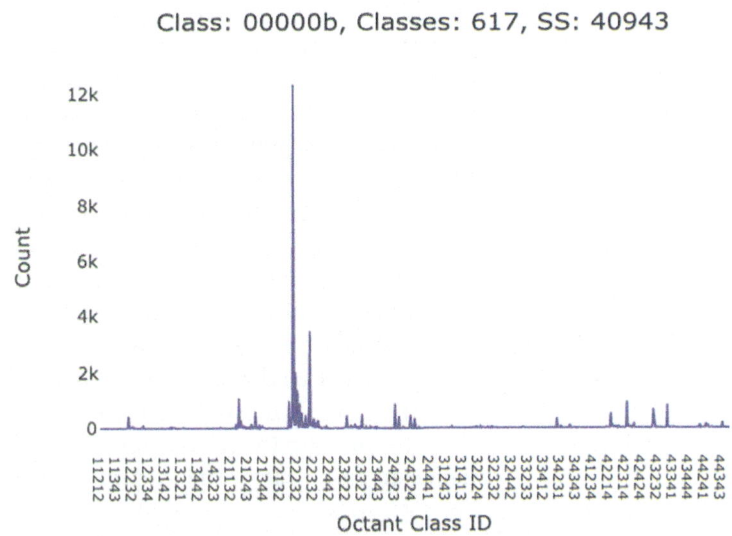

Fig. 9 Octant classes for binary class 00000b

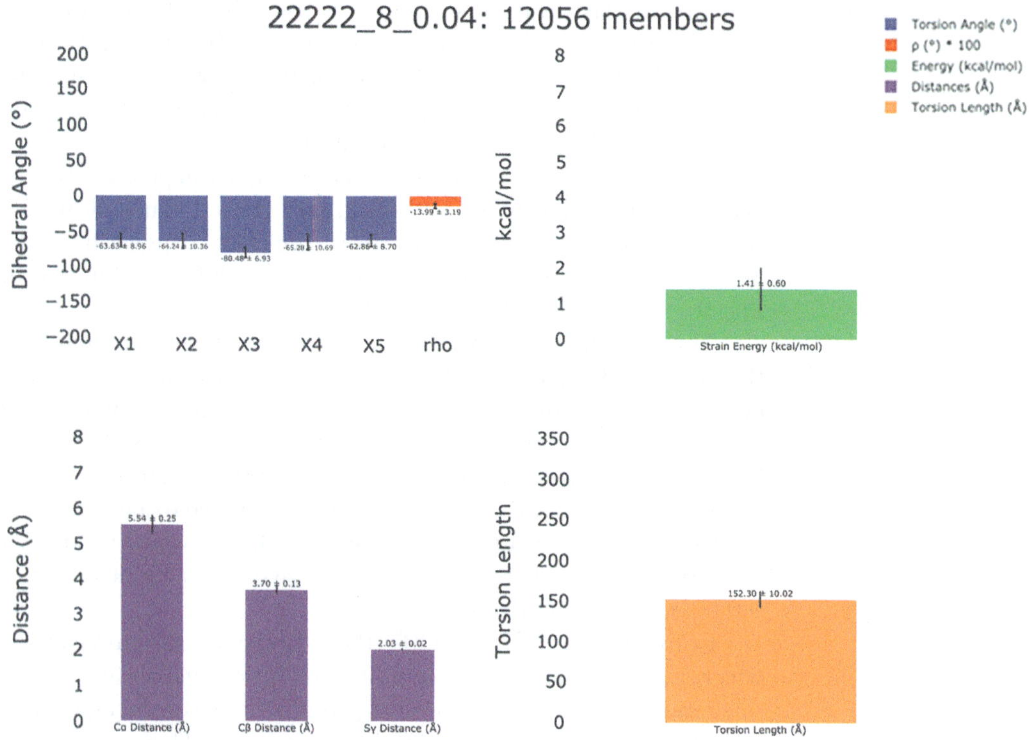

Fig. 10 Class 22222o dihedral angle details

angle statistics for the two classes are shown in Figs. 10 and 11. The error bars display relatively small values, suggesting that the class structures are structurally similar. Boxplots for the binary and octant classes are presented in Figs. 12–16. The boxplot for binary classes reveals a broad range of energy variation across classes, whereas the octant classes exhibit a narrower distribution. The tighter clustering of octant classes reflects the 1024× better resolution than a binary classification. The figures illustrate a wide range of torsional energies, and there are several classes that occupy low energy regions of the space. The lowest energy disulfide in the entire database is from pdbid *2q7q75D–140D*, with an approximate torsional energy of 0.49 kcal mol^{-1}. This is virtually identical to the theoretical minimum energy calculated through the **scikit-learn**[2]. The modeled and actual structures are shown in Fig. 17.

The **proteusPy.DisulfideClassGenerator** class is used to visualize an ensemble of the average structure for each of these peaks.

[2] Data not shown but available in the proteusPy repository.

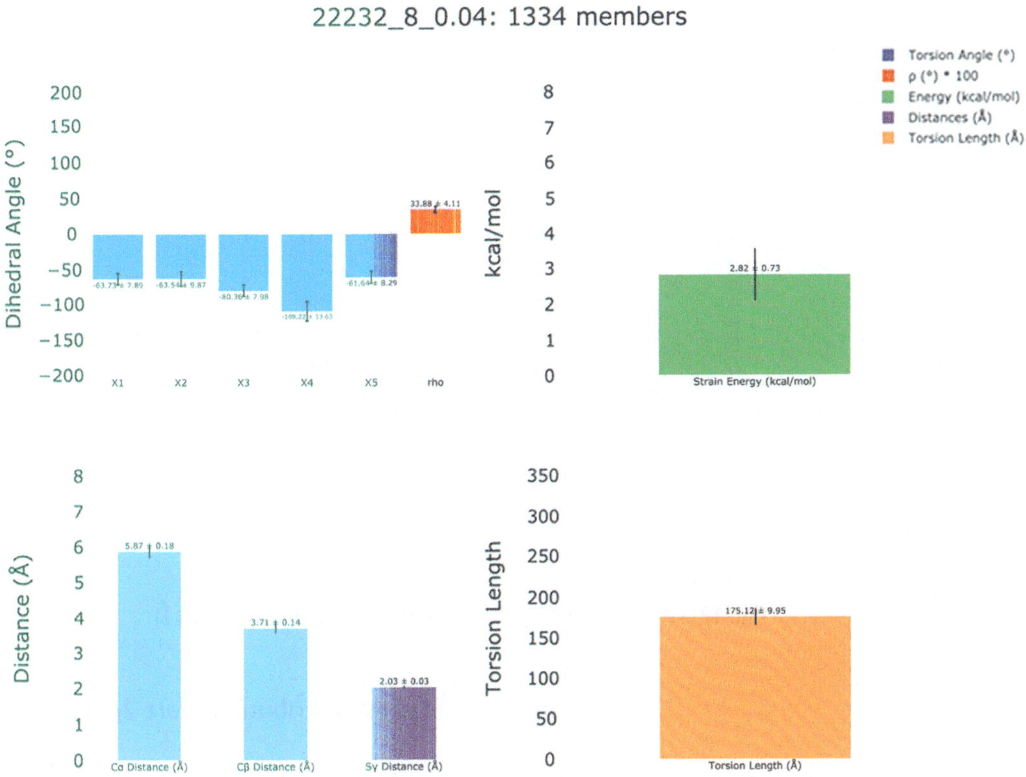

Fig. 11 Class 22322o dihedral angle details

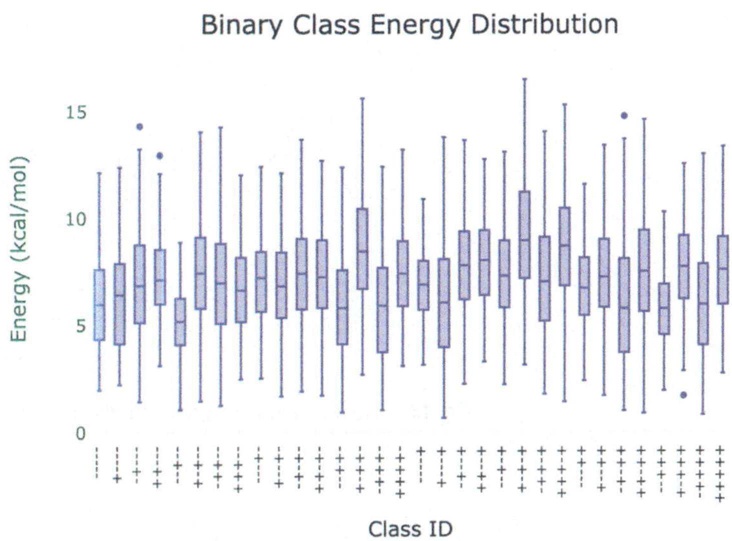

Fig. 12 Disulfide bond torsional energies by binary class

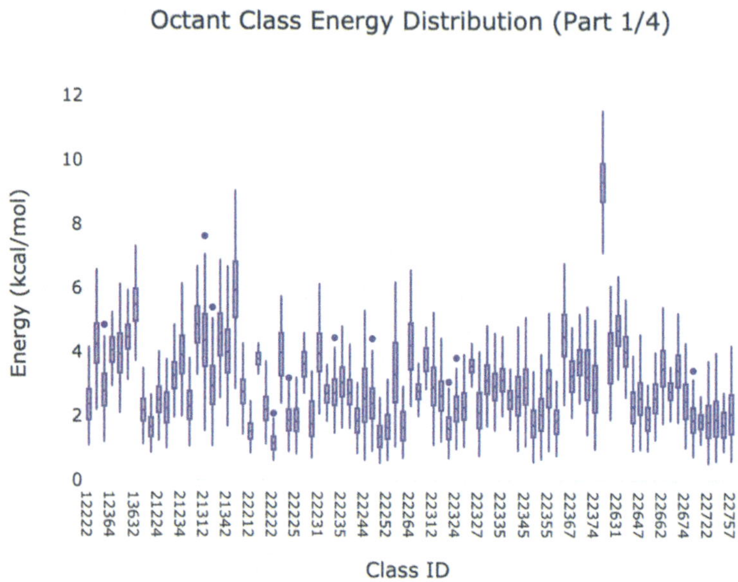

Fig. 13 Disulfide bond torsional energies by octant class (part 1)

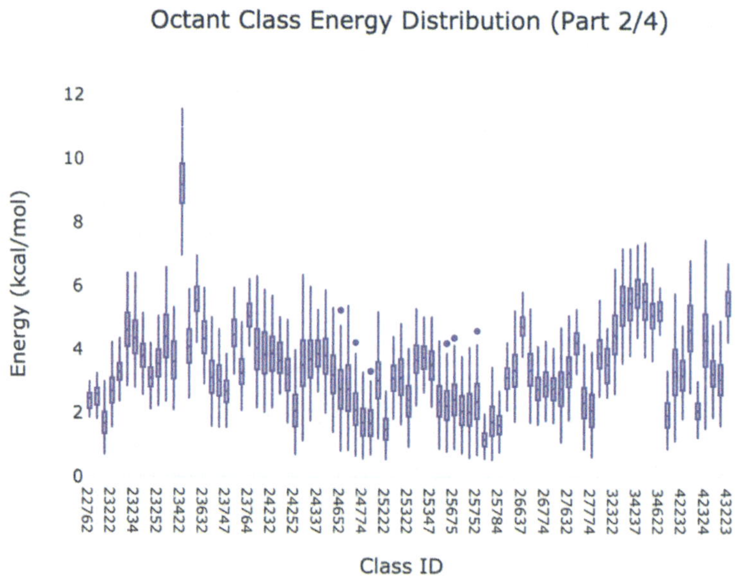

Fig. 14 Disulfide bond torsional energies by octant class (part 2)

This class takes the mean and standard deviations for the dihedral angles for each octant class and generates an idealized disulfide bond structure ensemble. These represent the average structure for each class with their respective dihedral angle standard deviations.

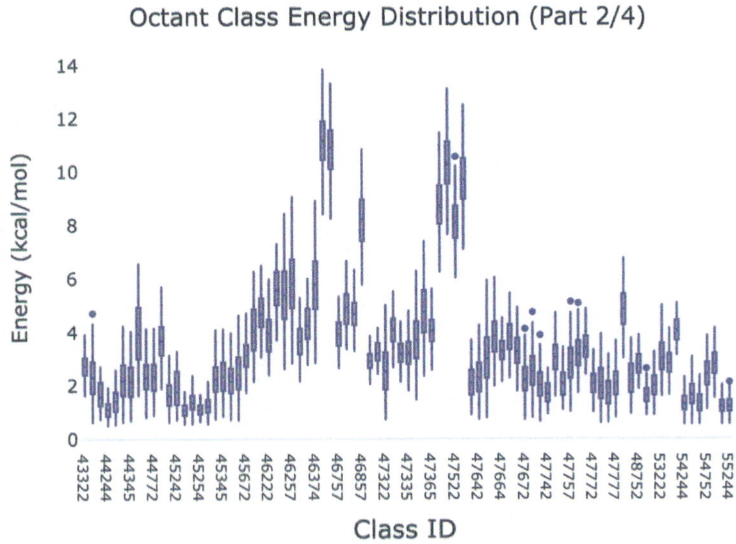

Fig. 15 Disulfide bond torsional energies by octant class (part 3)

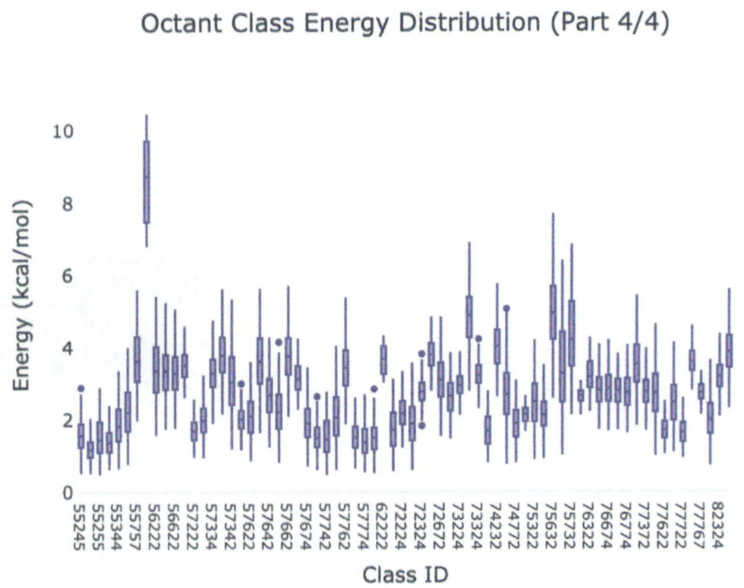

Fig. 16 Disulfide bond torsional energies by octant class (part 4)

Each dihedral angle, therefore, has three possible states, (mean–std, mean, mean+std). This generates 3^5 or 243 structures. An example is shown in Figs. 18 and 19. The figures are unable to capture the real differences between the ensemble structures, but the areas of space

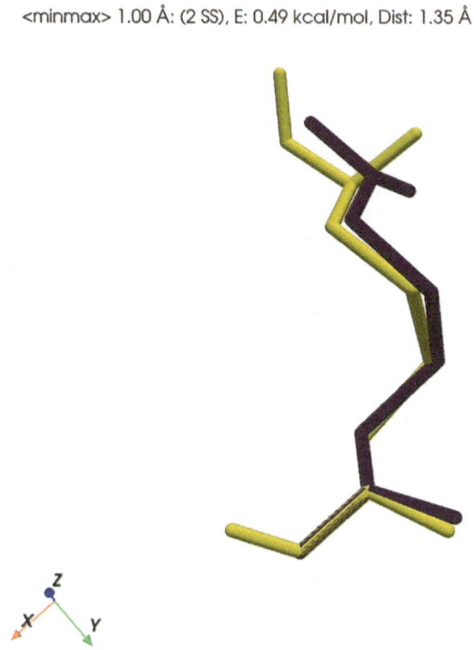

Fig. 17 Overlay of modeled and actual 2q7q disulfide structure

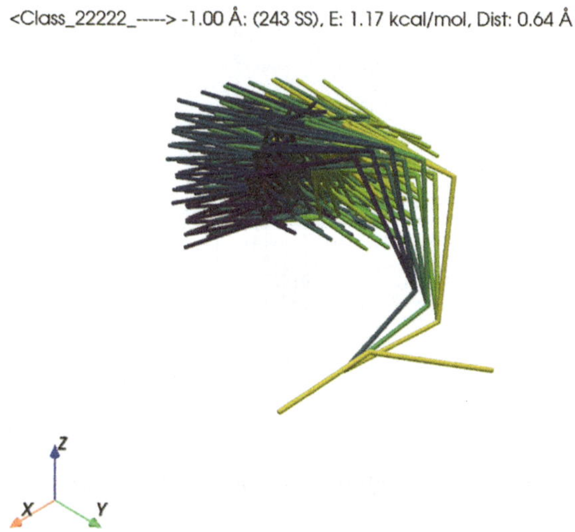

Fig. 18 Class 22222o ensemble

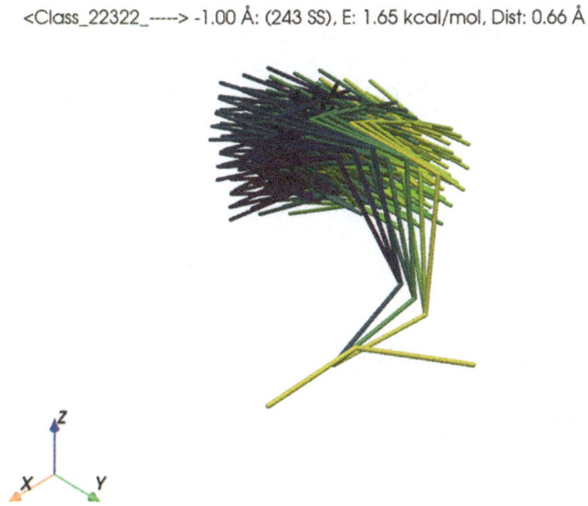

Fig. 19 Class 22322o ensemble

covered and accessible to the subclasses are readily apparent when viewed in 3D (Figs. 13, 14, 15, 16, 17, 18 and 19).

3 Discussion

The analysis has yielded a high-quality dataset of disulfide bonds segmented into structural families based on eight-fold (octant) dihedral angle quantization. Notable observations include:

- Disulfide bond model quality varies significantly in the RCSB database. Filtering at 95% confidence removes over 16,000 bonds based on C_α–C_α and S_γ–S_γ distances alone.
- The disulfide structural classes present represent ≈28% of the total possible classes (≈8500/32,768), which is a significant reduction in structural complexity. This indicates that the structural space of disulfide bonds is constrained heavily; local non-bonded interactions and steric hindrance are likely to be the primary factors.
- The octant approach allows for the further elucidation of structural families within the binary classification.
- There are disulfides whose conformation is close to the global energy minimum predicted through scikit-learn.
- Ensemble structures for disulfide consensus classes can be used to predict sites in proteins of known structure capable of accommodating those bonds.

Acknowledgments

The author would like to acknowledge the wonderful resource provided by http://www.rcsb.org for providing the structural data used in this study. In addition, I would like to thank Philip Hogg for his insight and counsel.

References

1. Meraz-Ríos MA, León KIL-De (2010) Tau oligomers and aggregation in Alzheimer's disease. J Neurochem. https://onlinelibrary.wiley.com/doi/abs/10.1111/j.1471-4159.2009.06511.x
2. Saito T, Chiku T, Oka M, Wada-Kakuda S et al (2021) Hum Mol Genet 30(21):1955. https://academic.oup.com/hmg/article-abstract/30/21/1955/6302471
3. Mossuto MF (2013) Int J Cell Biol. https://onlinelibrary.wiley.com/doi/abs/10.1155/2013/318319
4. Suchanek EG (2024) proteuspy: A package for modeling and analyzing proteins of known structure. GitHub repository. https://github.com/suchanek/proteusPy
5. Pabo CO, Suchanek EG (1986) Biochemistry 25(20):5987. https://doi.org/10.1021/bi00368a023
6. Schmidt B (2006) Biochemistry 45(24):7429. https://doi.org/10.1021/bi0603064

Chapter 3

Assessing the Evolutionary Conservation of Protein Disulfide Bonds

Qiuyu Jing and Jason W. H. Wong

Abstract

Studying the evolutionary conservation of proteins can be a valuable tool for understanding its function. At the sequence level, the conservation of each residue can be used to infer the importance of the particular regions of proteins. In the case of protein disulfide bonds, the conservation of the cysteines involved can be used to infer the conservation of the disulfide bond itself. In this chapter, bioinformatics methods are described that can be used to assess the conservation of a protein disulfide bond with a focus on the study of human proteins. Conservation will be assessed at the species and at the human population levels. The methods described make use of publicly available databases and can be applied by any researcher using a standard desktop computer with internet access.

Keywords Disulfide bonds, Cysteine, Evolution, Conservation, Bioinformatics, SNP, Population genetics

1 Introduction

Cysteine is one of the most chemically reactive amino acids. In proteins, it plays a major role in posttranslational modifications including disulfide bond formation, S-nitrosylation, S-glutathionylation, S-prenylation, as well as metal ion binding [1]. Among these, disulfide bond formation is most common. The formation of disulfide bonds is a key posttranslational modification for the maintenance of protein structure and function. The correct formation of disulfide bonds is often vital for the stability of the final structure of a protein [2] and has also been shown to control protein function through enzymatically controlled reduction and oxidation of a disulfide bond [3].

Considering the importance of disulfide bonds in the maintenance of protein structure and function, it has been shown that disulfide bonds are generally highly conserved through evolution. In fact, not only are disulfide bonds highly conserved, but they have also accrued in homologous classes of proteins through eukaryotic

Philip J. Hogg (ed.), *Functional Disulphide Bonds*, Methods in Molecular Biology, vol. 3016,
https://doi.org/10.1007/978-1-0716-5158-2_3,

evolution [4]. The acquisition of disulfide bonds may reflect the demand for greater sophistication in protein function in complex multicellular organisms. Evidence of gain of protein function in the course of eukaryotic evolution as a result of disulfide bond acquisition includes CD4 co-receptor on helper T-cells, where the acquisition of a cysteine in domain 2 of the protein enables regulated formation of homodimers on the cell surface [5].

Although many disulfide bonds are highly conserved [4], the study of the conservation of individual disulfide bonds can be useful in inferring disulfide and protein function. Conservation can be studied at both the species and at the population levels. It can also be used to assess across paralogous proteins to determine whether disulfide bonds play a role in the loss or acquisition of specific functions in related proteins. At the species level, disulfide bond conservation can often be analyzed based on protein sequence by observing the conservation of the cysteines involved. The general interpretation is that if a disulfide bond is highly conserved across a wide range of distantly related species, then the bond is likely indispensable to the function of the protein. Conversely, if it is only conserved in a subset of species, then the bond may be dispensable or the presence/absence of the bond alters the function of the protein in a manner that provides a selective advantage of the respective species.

At the population level, conservation is defined by the frequency of an alternative allele observed in the nucleotides of codons. In the case of cysteine, since it is encoded by only two codons (namely TGT and TGC), all alternative alleles in the codon (commonly called single nucleotide polymorphisms or SNPs), except a T>C variant at the degenerate third position in the codon, would result in the mutation of cysteine into another amino acid or stop codon and thus leading to the loss of a disulfide bond. At the population level, generally, invariant sites, where there are no SNPs, are considered essential for protein function, as no variants can be observed in the human population. If a SNP is frequently observed in the human population, it would suggest that the site is not important in terms of human population fitness, but it could still nevertheless alter protein function that underlie nonfitness dependent phenotype. "Common" SNPs are generally defined as those present in greater than 5% of the human population [6]. For "rare" SNPs (i.e., those present in less than 5% of the population), they are considered to have some impact on population fitness and their frequency reflects their negative selection. Extremely "rare" SNPs (i.e., those present in less than 0.01% of the population) are often of interest as they can be associated with rare diseases and genetic disorders. In the case of disulfide bonds, those that overlap with rare SNPs are potentially interesting, as it may suggest that loss of a disulfide bond could underlie a change in protein function and human disease.

In this chapter, bioinformatics methods will be outlined to study the conservation of disulfide bonds in proteins both at the species and the human population levels. The methods will make exclusive use of publicly available resources and will be accessible for any researchers interested in assessing the conservation of disulfide bonds in their proteins of interest.

2 Materials

This chapter will use two proteins, CD4 and EPCAM, as examples for assessing the conservation of disulfide bonds across species and in the human population, respectively. Based on the methods described, the reader should be able to apply it to their own proteins/disulfide bonds of interest. All analyses described can be carried out on a standard desktop computer with a web browser. Below is a list of free-for-academic-use databases and software required for the data analysis:

- Uniprot [7] (www.uniprot.org)—The Universal Protein Resource used for standardized protein annotation and linkage to other external resources.
- OMA browser [8] (https://omabrowser.org/)—"Orthologous Matrix" is a database containing sets of 1:1 orthologous proteins.
- Jalview [9] (www.jalview.org/) – Jalview is a Java-based application for the visualization and analysis of protein sequences.
- gnomAD browser [10] (https://gnomad.broadinstitute.org/)—The Genome Aggregation Database (gnomAD, the v4 dataset based on GRCh38/hg38) with allele frequencies spanning 730,947 exome sequences and 76,215 whole-genome sequences.
- UCSC table browser [11] (https://genome.ucsc.edu/cgi-bin/hgTables)—Web interface to extract genomic annotations for a wide range of genome-related data sources.

3 Methods

A. **Conservation Across Species**

1. Use the search function on the UniProt database to locate the entry for the protein of interest. In this example, the CD4 protein will be used. The CD4 protein has three disulfide bonds. Under the "PTM/Processing" section

of the UniProt page for the human CD4 protein, these disulfide bonds are shown as linking Cys41–Cys109, Cys155–Cys184, and Cys328–Cys370.

2. Scroll down to the "Family & Domains" section and click on the link to the OMA entry "KTCQCSH". A new window will open to the OMA browser (*see* **Notes 1 and 2**) (Fig. 1a).
3. Click on the link under the "Protein ID" header for *Homo sapiens*—HUMAN10779. The page will show that there are 143 Orthologs for the Human CD4 protein (Fig. 1b).
4. To establish whether the disulfide bonds of interest are conserved across the species with the orthologous protein, the sequences are downloaded in FASTA format for multiple sequence alignment (Fig. 1b). The file will be automatically named as "HUMAN10779_orthologs.txt". Note that the HUMAN10779 sequence is not included in the downloaded file. To download it separately, scroll down to the "Sequences/Isoforms" section and download the

Fig. 1 Identifying orthologous proteins. (**a**) Phylogenomic databases listed for human CD4 (P01730) in UniProt. The entry for the OMA browser is boxed in red. (**b**). OMA browser entry showing orthologs of human CD4. The number of orthologs and the link to download the sequences in FASTA format are boxed in red. (**c**). OMA browser entry showing different isoforms of human CD4. The link to download human CD4 amino acid sequence in FASTA format is boxed in red

protein sequence in FASTA format for reference isoform HUMAN10779 (Fig. 1c). Copy the sequence into the "HUMAN10779_orthologs.txt" file.

5. For multiple sequence alignment and visualization, we use the JalView tool. Startup JalView following the instructions for your operating system here: http://www.jalview.org/Download.
6. Load up the FASTA file—"HUMAN10779_orthologs.txt" through File→ Input Alignment → From File.
7. Perform multiple sequence alignment within JalView using the Muscle alignment algorithm [12] through Web Service →Alignment → Muscle with defaults (*see* **Note 3**).
8. The alignments will appear in a new window. In order to locate the disulfide bonded cysteines (i.e., Cys41–Cys109, Cys155–Cys184, and Cys328–Cys370), find the HUMAN CD4 sequence label (HUMAN10779) through Select → Find option, right click on the label → HUMAN10779 → Set as Reference. The residue ruler at the top of the window will now be set to match the HUMAN CD4 sequence. To make the Cysteine residues easier to see, additionally highlight them by selecting the Color Menu → User Defined and then set the color of "Cys" to yellow (Fig. 2a).
9. The conservation of each of the disulfide bonded cysteines can now be easily visualized by scrolling to the appropriate Human CD4 residue indicated by the ruler. To determine the conservation in relation to the phylogenetic tree based on the protein, a phylogenetic tree can be generated through Calculate → Calculate Tree or PCA using default parameters (Fig. 2b) (*see* **Note 4**).

B. **Conservation Within Humans**

1. SNPs are defined by genomic coordinates; thus, the first step in making use of SNP databases such as gnomAD [10] is to identify the genomic coordinates corresponding to the disulfide bonded cysteine residues. To do this, we use the UCSC table browser (https://genome.ucsc.edu/cgi-bin/hgTables). First ensure that the assembly is "Dec. 2013 (GRCh38/hg38)" (*see* **Note 5**), group is "Gene and Gene Predictions", and track is "UniProt" and table is "Disulf. Bonds (unipDisulfBond)". For this example, we will use the protein EPCAM. To specifically obtain the genomic coordinates for the EPCAM disulfide bonds, click on the "Create" button next to filter" and enter the Uniprot ID for EPCAM—"P16422" into the "uniProtId" row (*see* **Note 6**) (Fig. 3a).
2. View the disulfide bond entries by selecting "Output format" as "All fields from selected table." To download the

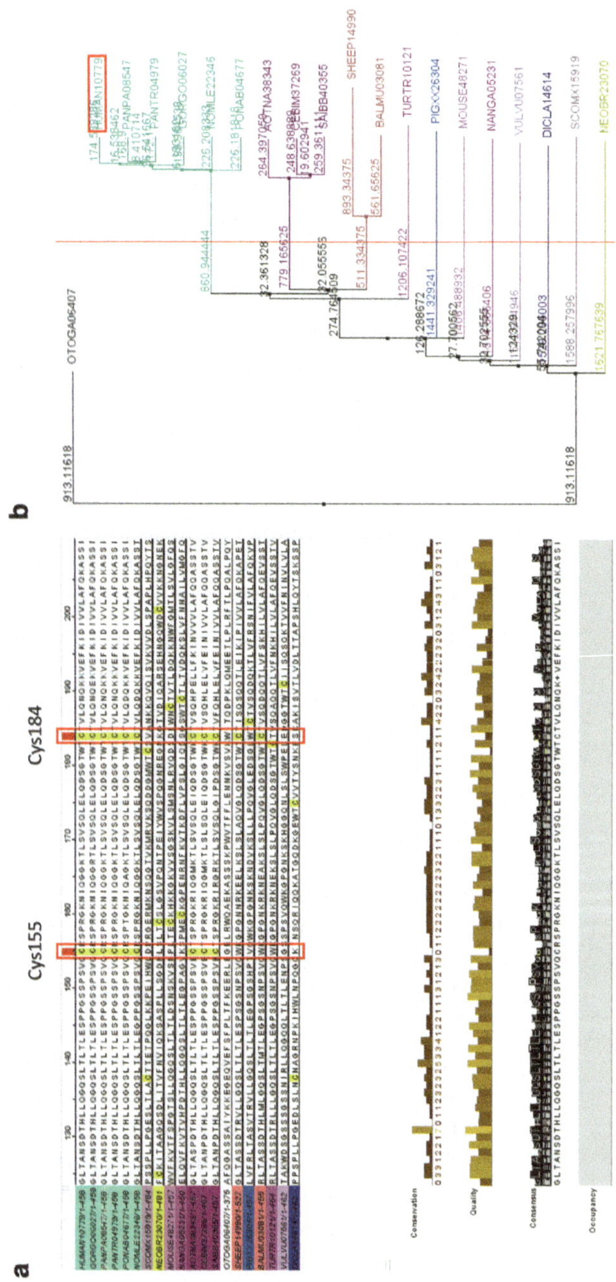

Fig. 2 Multiple sequence alignment of human CD4 with its orthologs. (**a**) Multiple sequence alignment showing the disulfide bond Cys155–Cys184 (boxed in red) based on 20 orthologs as demonstration. (**b**) Phylogenetic tree of the selected 20 orthologs. Orthologs from the same branch rooted from whether the red bar locations are showing in the same color

a

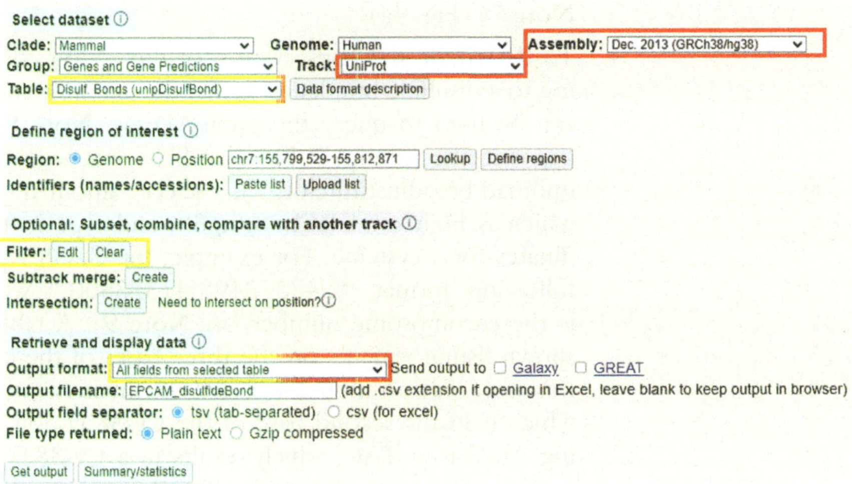

b

#chrom	chromStart	chromEnd	name	score	strand	thickStart	thickEnd	...
# Filtering on 1 columns								
chr2	47373464	47373524	disulf bond	1000	+	47373464	47373524	...
chr2	47373470	47373563	disulf bond	1000	+	47373470	47373563	...
chr2	47373497	47373530	disulf bond	1000	+	47373497	47373530	...
chr2	47373818	47373920	disulf bond	1000	+	47373818	47373920	...
chr2	47373950	47373971	disulf bond	1000	+	47373950	47373971	...
chr2	47373974	47374028	disulf bond	1000	+	47373974	47374028	...

Fig. 3 Extraction of genomic coordinates for disulfide bonds. (**a**) Interface of the UCSC table browser. The parameters that need to be changed to extract disulfide bond genomic coordinates are boxed in red. (**b**) Output table of disulfide bond genomic coordinates for the example protein EPCAM

table, enter a filename into the "Output filename" box. From the table, it can be seen that Cys27–Cys46 corresponds to "chr2 47373464 47373524," Cys29–Cys59 corresponds to "chr2 47373470 47373563," Cys38–Cys48 corresponds to "chr2 47373497 47373530," and so on (Fig. 3b).

3. The genomic coordinates listed are for the genomic range between the two disulfide bonded cysteine. The precise genomic coordinates of the disulfide bonded cysteine codons can be calculated as the first and last three bases within the genomic range. Using Cys38–Cys48 as an example, for a positive stranded gene such as EPCAM (indicated by the strand column in the table), the genomic coordinate for the codon corresponding to Cys38 is chr2 47373498

47373500 while Cys48 is chr2 47373528 47373530 (*see* **Note 7**) (Fig. 4a).

4. To determine the conservation of the codon corresponding to a disulfide bonded cysteine, the genomic coordinates can be used to query the gnomAD (*see* **Note 8**). To do so, visit the gnomAD browser [10] webpage (https://gnomad.broadinstitute.org/), select gnomAD v4.1.0, which is built on GRCh38/hg38, and type in the coordinates for a cysteine. For example, for Cys38 type in the following format "2:47373498-47373500," where "2" is the chromosome number (*see* **Note 9**). A table will be shown listing variants on the three bases of the codon. In this case, only position 2:47373499 contains a G>A variant which is in the second base of the Cys38 codon, converting TGC to a TAC, which results in a Cys38Tyr protein variant (Fig. 4b). The table also shows that the variant has an allele count of 2 meaning that it is only present in 2/1613384 alleles with an allele frequency of 1.24e-6, suggesting that this loss of the cysteine and disulfide bond is extremely rare in the human population. For the other two

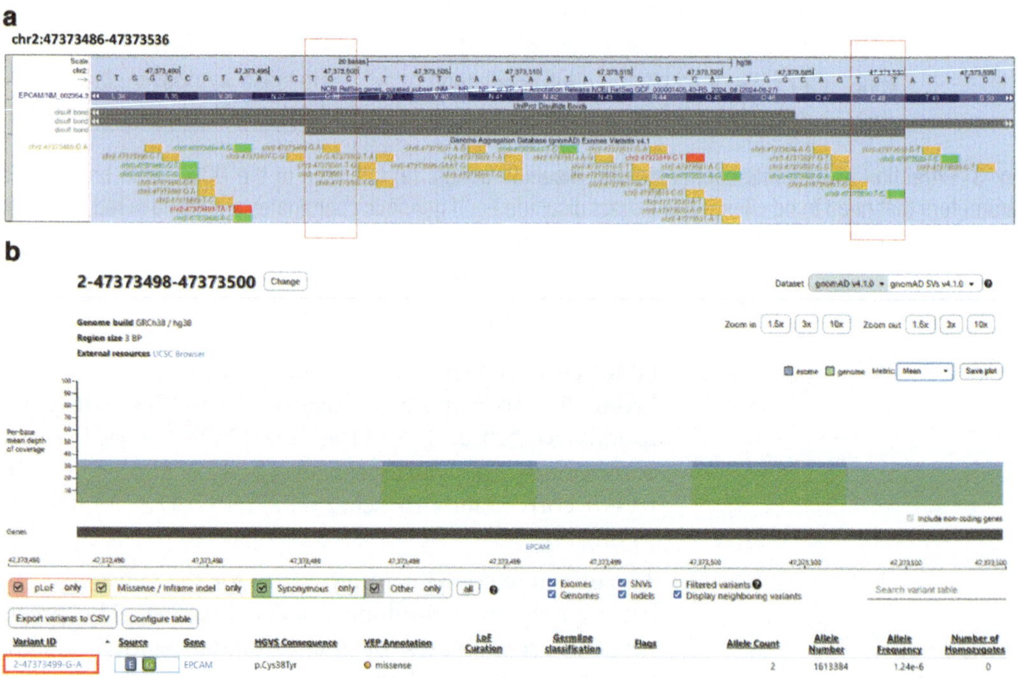

Fig. 4 Single nucleotide polymorphisms (SNPs) at disulfide bonded cysteines. (**a**) UCSC genome browser tracks showing the location of the disulfide bonds in EPCAM, the specific DNA and protein sequence including the disulfide bond Cys38–Cys48 (boxed in red), and a track of gnomAD SNPs profiled in exome sequencing data. (**b**) The interface of gnomAD browser showing the variants within the codon Cys38 from EPCAM (i.e., chr2:47373498-47373500). The single variant in the region at chr2:47373499 is boxed in red

positions in the codon, there are no variant alleles present; therefore these are not shown in the gnomAD Browser (*see* **Note 10**).

5. Clicking on the variant in the table will open a new window which gives more depth of information regarding the variant including a population level breakdown of allele frequency and additional sequencing quality control characteristics of the site.

4 Notes

1. For the analysis of species conservation, orthologs are generally used as the basis of comparison. Orthologs are genes in different species that evolved from a common ancestral gene by speciation. In the ideal case, orthologs between two species have a one-to-one homology relationship; however it is possible that an ortholog in another species has undergone duplication after speciation, resulting in a one-to-many relationships. Many orthologs retain the same function in the course of evolution, but in the latter case, it is possible that some orthologs partially lose or gain additional functions. The OMA browser annotates the relationship between orthologs. It is important to take this information into consideration when interpreting disulfide conservation.
2. A link to the OMA browser may not be available to all proteins. Uniprot also provides links to other protein orthology databases. If other databases are used, so long as a FASTA file of orthologous proteins can be obtained, it will be possible to proceed to Step 4.
3. There is a range of multiple sequence alignment algorithms available with differences in speed and accuracy. Muscle is chosen here for its speed and generally accurate performance. Similarity matrices can also be an important consideration. For a comprehensive review on multiple sequence alignment, see [13].
4. While it is possible to construct a phylogenetic tree based on the protein being analysis, this may not be ideal for the representation of the true phylogenetic relationship between the species being analyzed. Therefore, if the conservation of a disulfide bond needs to be analyzed in the context of the true phylogenetic relationship between species, published phylogenetic relationships should be used [14].
5. The human genome is available in different assemblies—currently hg16 to hg38 and T2T CHM 13v2.0/hs1 on the UCSC genome browser. The latest version is CHM13v2.0

(Jan 2022), which is the first complete, gapless human genome build. However, the full gene annotation and protein tracks are not completely available yet. It is important to note that the genome coordinates for genes vary across each genome assembly. gnomAD provides resources based on both hg19 and hg38 builds. It is essential that the matched genomic coordinates (i.e., hg38) are used for the disulfide bonded cysteines.

6. The UCSC genome browser is an online portal for the visualization of genomic data and genome annotations. The UCSC table browser provides an interface to access the underlying data in the form of a table in text format. The UniProt tables provide genomic coordinates for protein annotations in the UniProt database. To find the protein of interest, use the UniProt ID which can be found from the UniProt database.

7. When converting the disulfide bond genomic coordinates to cysteine codon coordinates, there are two points to note. First, if a gene is on the positive strand (denoted by "+" in the strand column of the table), the N-terminal cysteine will be at the 5′ end of the disulfide genomic range (noting that as the genomic coordinates are 0-based, the first position of the codon corresponds to the second base in the genomic range), while the C-terminal cysteine will be at the 3′ end of the disulfide genomic range (as in the example). In contrast, if a gene is on the negative strand (denoted by "−" in the strand column of the table), the codon assignment will be reversed and complemented such that the N-terminal cysteine will start at the 3′ end with the last based in the genomic range corresponding to the first position of the codon. The second point to note is that it is possible that a codon overlaps an exon boundary. In this case, the genomic coordinates of the codon would have to be identified accordingly. The simplest way to identify if a codon overlaps a boundary is to examine the "blockCount" column and the "blockSizes" column. If "blockCount" is greater than one, it means that the disulfide bond genomic coordinates spans exon boundaries. In this case, examine the corresponding comma delimited "blockSizes" entry. Each value in the comma delimited list denotes the number of bases (i.e., blockSize) covered over each relevant exon. If the first or last blockSize is less than 3, it means that the corresponding cysteine codon spans an exon boundary. In this case, the precise genomic coordinates of the spliced part of the codon must be carefully verified using the UCSC genome browser.

8. There are numerous human SNP databases that document the allele frequency of SNPs within the human population. For example, Exome Aggregation Consortium (ExAC) [10] contains SNPs of protein coding region from 60,671 humans

derived from exome-sequencing data based on hg19. Now the ExAC browser is directed to gnomAD browser (https://gnomad.broadinstitute.org/). The v4 data set (GRCh38) of gnomAD spans 730,947 exome sequences and 76,215 whole-genome sequences from unrelated individuals, of diverse ancestries, sequenced as part of various disease-specific and population genetic studies.

9. There are two standard annotations for chromosomes. UCSC genome browser prefixes all chromosomes with "chr" i.e., chr1, chr2, chrX, etc. In contrast, the gnomAD browser labels chromosomes without prefixes, i.e., 1, 2, X, etc. Therefore, the "chr" prefix has to be removed prior to searching.

10. As shown in Fig. 4a, the SNPs from gnomAD can also be visualized in the UCSC genome browser by adding the gnomAD track. Although the track in UCSC provides less information compared with the gnomAD browser, it is nevertheless useful for visual inspection that the correct disulfide/cysteine/codon is being analyzed.

Acknowledgments

This work was supported by the Centre for Oncology and Immunology under the Health@InnoHK Initiative, funded by the Innovation and Technology Commission, the Government of Hong Kong SAR, China.

References

1. Poole LB (2015) The basics of thiols and cysteines in redox biology and chemistry. Free Radic Biol Med 80:148–157. https://doi.org/10.1016/j.freeradbiomed.2014.11.013
2. Fass D (2012) Disulfide bonding in protein biophysics. Annu Rev Biophys 41:63–79. https://doi.org/10.1146/annurev-biophys-050511-102321
3. Butera D, Cook KM, Chiu J, Wong JW, Hogg PJ (2014) Control of blood proteins by functional disulfide bonds. Blood 123(13):2000–2007. https://doi.org/10.1182/blood-2014-01-549816
4. Wong JW, Ho SY, Hogg PJ (2011) Disulfide bond acquisition through eukaryotic protein evolution. Mol Biol Evol 28(1):327–334. https://doi.org/10.1093/molbev/msq194
5. Matthias LJ, Yam PT, Jiang XM, Vandegraaff N, Li P, Poumbourios P, Donoghue N, Hogg PJ (2002) Disulfide exchange in domain 2 of CD4 is required for entry of HIV-1. Nat Immunol 3(8):727–732
6. International HapMap C (2005) A haplotype map of the human genome. Nature 437(7063):1299–1320. https://doi.org/10.1038/nature04226
7. UniProt Consortium T (2018) UniProt: the universal protein knowledgebase. Nucleic Acids Res 46(5):2699. https://doi.org/10.1093/nar/gky092
8. Altenhoff AM, Glover NM, Train CM, Kaleb K, Warwick Vesztrocy A, Dylus D, de Farias TM, Zile K, Stevenson C, Long J, Redestig H, Gonnet GH, Dessimoz C (2018) The OMA orthology database in 2018: retrieving evolutionary relationships among all domains of life through richer web and programmatic interfaces. Nucleic Acids Res 46(D1):D477–D485. https://doi.org/10.1093/nar/gkx1019
9. Waterhouse AM, Procter JB, Martin DM, Clamp M, Barton GJ (2009) Jalview Version 2 – a multiple sequence alignment editor and analysis workbench. Bioinformatics 25(9):1189–1191. https://doi.org/10.1093/bioinformatics/btp033

10. Lek M, Karczewski KJ, Minikel EV, Samocha KE, Banks E, Fennell T, O'Donnell-Luria AH, Ware JS, Hill AJ, Cummings BB, Tukiainen T, Birnbaum DP, Kosmicki JA, Duncan LE, Estrada K, Zhao F, Zou J, Pierce-Hoffman E, Berghout J, Cooper DN, Deflaux N, DePristo M, Do R, Flannick J, Fromer M, Gauthier L, Goldstein J, Gupta N, Howrigan D, Kiezun A, Kurki MI, Moonshine AL, Natarajan P, Orozco L, Peloso GM, Poplin R, Rivas MA, Ruano-Rubio V, Rose SA, Ruderfer DM, Shakir K, Stenson PD, Stevens C, Thomas BP, Tiao G, Tusie-Luna MT, Weisburd B, Won HH, Yu D, Altshuler DM, Ardissino D, Boehnke M, Danesh J, Donnelly S, Elosua R, Florez JC, Gabriel SB, Getz G, Glatt SJ, Hultman CM, Kathiresan S, Laakso M, McCarroll S, McCarthy MI, McGovern D, McPherson R, Neale BM, Palotie A, Purcell SM, Saleheen D, Scharf JM, Sklar P, Sullivan PF, Tuomilehto J, Tsuang MT, Watkins HC, Wilson JG, Daly MJ, MacArthur DG, Exome Aggregation C (2016) Analysis of protein-coding genetic variation in 60,706 humans. Nature 536(7616):285–291. https://doi.org/10.1038/nature19057

11. Rosenbloom KR, Armstrong J, Barber GP, Casper J, Clawson H, Diekhans M, Dreszer TR, Fujita PA, Guruvadoo L, Haeussler M, Harte RA, Heitner S, Hickey G, Hinrichs AS, Hubley R, Karolchik D, Learned K, Lee BT, Li CH, Miga KH, Nguyen N, Paten B, Raney BJ, Smit AF, Speir ML, Zweig AS, Haussler D, Kuhn RM, Kent WJ (2015) The UCSC Genome Browser database: 2015 update. Nucleic Acids Res 43(Database issue): D670–D681. https://doi.org/10.1093/nar/gku1177

12. Edgar RC (2004) MUSCLE: multiple sequence alignment with highaccuracy and high throughput. Nucleic Acids Res 32(5):1792–1797. https://doi.org/10.1093/nar/gkh340

13. Thompson JD, Linard B, Lecompte O, Poch O (2011) A comprehensive benchmark study of multiple sequence alignment methods: current challenges and future perspectives. PLoS One 6(3):e18093. https://doi.org/10.1371/journal.pone.0018093

14. Letunic I, Bork P (2016) Interactive tree of life (iTOL) v3: an online tool for the display and annotation of phylogenetic and other trees. Nucleic Acids Res 44(W1):W242–W245. https://doi.org/10.1093/nar/gkw290

Chapter 4

Estimate of Numbers of Disulfide-Bonded Protein States

Philip J. Hogg

Abstract

Disulfide bonds form as proteins fold in the cell and formation was assumed to be complete when the mature protein emerges. This is not the situation for many proteins that are constitutively produced as multiple partially disulfide-bonded states. For some proteins, thousands of different states are predicted assuming no dependencies on disulfide bond formation. In this chapter, probabilities for disulfide bond formation are employed to estimate the number of disulfide-bonded states. Situations where disulfide bond formation is independent of, or dependent on, the state of other bonds in the protein are considered. Estimates of the number of disulfide-bonded protein states will assist with the conceptual and experimental challenges of studying this biology.

Keywords Disulfide bond, Cystine, Cysteine, Probability, Allosteric, Factor VIII

1 Introduction

Since the identification of protein disulfide bonds in the 1960s [1], it was long assumed that these bonds were fully formed and inert in the mature protein. Research over the last two decades has demonstrated that disulfide bonds are not all inert in native proteins. A subset of disulfide bonds called the allosteric disulfides [2–4] control the function of the native protein when cleaved or formed by different factors [5, 6]. Research over the last few years has unexpectedly demonstrated that disulfide bonds are also not fully formed in native proteins [7]. Some proteins are constitutively produced and function as multiple, sometimes hundreds, of partially disulfide-bonded states [8–11].

A protein containing n disulfide bonds, where the bonds are either formed or unformed, has 2^n possible disulfide-bonded states. For example, a protein containing eight disulfide bonds can exist in a maximum of 256 disulfide-bonded states. This estimate assumes that formation of any disulfide bond in the protein is independent of any other, which is very possibly not the case. Herein, I describe how probabilities for disulfide bond formation can be employed to estimate the number of protein disulfide-bonded states.

Philip J. Hogg (ed.), *Functional Disulphide Bonds*, Methods in Molecular Biology, vol. 3016,
https://doi.org/10.1007/978-1-0716-5158-2_4,

This information will help researchers better define scientific hypotheses and experimental strategies.

2 Multiple Partially Disulfide-Bonded Protein States

Hemophilia A is a chronic, life-threatening condition caused by a deficiency or dysfunction of plasma coagulation factor VIII (FVIII) due to mutations of the *F8* gene encoding this protein [12, 13]. The result is reduced thrombin generation and fibrin clot formation leading to bleeding episodes, chronic pain, joint deformity, reduced mobility, and increased mortality.

Eight disulfide bonds have been structurally defined in X-ray structures of FVIII (C172–C198 and C267–C348 in the A1 domain, C547–C573 and C649–C730 in the A2 domain, C1851–C1877 and C1918–C1922 in the A3 domain, C2040–C2188 in the C1 domain, and C2193–C2345 in the C2 domain) [14–16]. The redox state of the disulfide cysteine thiols in recombinant FVIII was determined using differential cysteine alkylation with a pair of isotopic alkylators and mass spectrometry [8, 9, 17]. All eight disulfide bonds were found to exist in unformed states in different fractions of the FVIII molecules (Fig. 1).

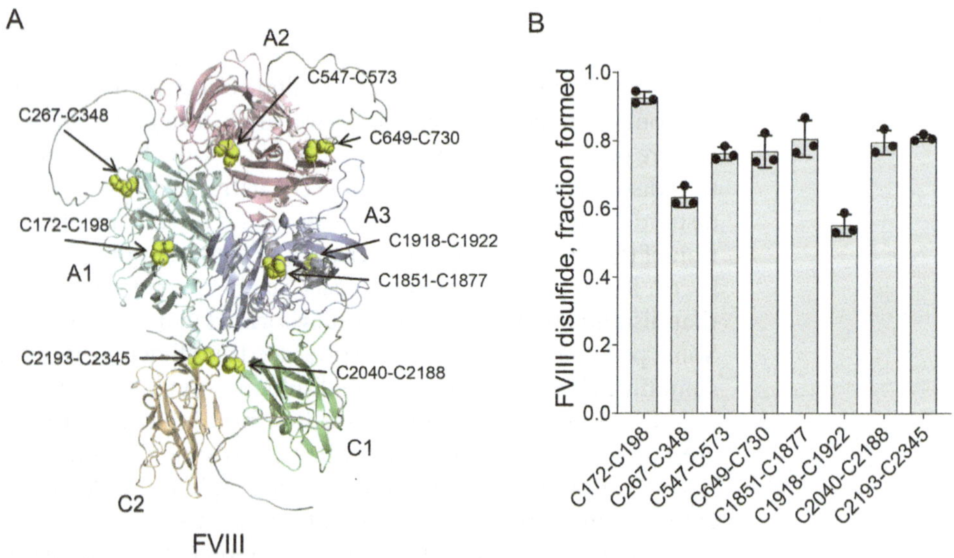

Fig. 1 FVIII is constitutively produced as multiple partially disulfide-bonded states. (**A**). Ribbon representation of the AlphaFold structure of human VIII lacking the unstructured B domain that does not contain disulfide bonds. The domains (A1, A2, A3, C1, and C2) are color coded, and the eight disulfide bonds are shown as yellow spheres, and residue numbers indicated (UniProt P00451 numbering). (**B**). Incidence of formed disulfide bonds in recombinant FVIII ($n = 3$ FVIII preparations). The approximate incidence of the formed disulfide bond in FVIII is indicated for each disulfide. For example, the A3 domain C1918–C1922 bond is formed in one in two molecules of the FVIII samples. The bars and errors are mean ± SD

Table 1

Conformation of the FVIII disulfide bonds and fraction of the bonds that are formed in protein populations of the recombinant protein. The disulfide conformations were determined using the Disulfide Bond Analysis tool [19] (accessed at ssp-main.fly.dev) and the PDB identifier 7k66 structure of FVIII [20]. The fraction formed is the mean values of three FVIII preparations (data from Fig. 1B) and is equivalent to the probability that the disulfide bond is formed in the population of FVIII molecules

FVIII disulfide	Conformation	Fraction formed
C172–C198	–LHhook	0.93
C267–C348	–LHhook	0.63
C547–C573	–RHstaple	0.76
C649–C730	+/–RHstaple	0.77
C1851–C1877	–LHstaple	0.81
C1918–C1922	n.d.[a]	0.55
C2040–C2188	+/–RHhook	0.80
C2193–C2345	+/–RHhook	0.81

[a]This bond is not present in the 7k66 FVIII structure

The eight FVIII disulfide bonds ranged from a mean fraction of 0.55–0.93 formed in the FVIII preparations (Table 1). Disulfide bonds can adopt 20 different conformations defined by the geometry of the five dihedral angles that describe the cystine residue [2]. Three of the 20 disulfide conformations, the –RHstaple, –/+RHhook, and –LHhook bonds, are often associated with allosteric disulfide bonds [2, 4]. It is noteworthy that three of the six FVIII disulfides have "allosteric" conformations (Table 1), which may contribute to disulfide control of the protein.

3 Probabilities of Disulfide Bond Formation

All FVIII disulfide bonds are unformed to some extent, which equates to maximum of 256 (2^8) possible disulfide-bonded states of the protein. Probability (P) is defined as the extent to which an event is likely to occur, measured by the ratio of the favorable cases to the whole number of cases possible. The event for disulfides is binary, that is, the bond is either formed or unformed and the favorable case for this analysis is the formed bond. Approximately 5 μg of plasma FVIII was analyzed in each experiment that equates to ~10^{12} molecules of the protein, which is an exhaustive sampling of possible cases. Therefore, the mean formed fraction for the individual FVIII disulfides represents the probability that the bond is formed in the population of molecules (Table 1).

To explore the variables involved in estimating types and numbers of FVIII states, a model polypeptide containing five disulfide bonds is simulated. A polypeptide containing five disulfide bonds, where the bonds are either formed or broken, can exist in 32 (2^5) possible disulfide-bonded states (Fig. 2). The probability a bond is formed is the ratio of the number of states to total number of states (32). For example, there are ten possible states where only two of the five bonds are formed and each state occurs in eight of the 32 possible isomers, so the probability of an isomer containing only two disulfide bonds is 8/32 or 0.25. For the five situations, that is, formation of one bond given the formation of 1, 2, 3, or 4 others, the following conditional probabilities apply:

$$\begin{aligned}
&P(1\ \text{bond}) = 0.5\\
&P(1\ \text{bond}|1\ \text{other}) = 0.25\\
&P(1\ \text{bond}|2\ \text{others}) = 0.125\\
&P(1\ \text{bond}|3\ \text{others}) = 0.0625\\
&P(1\ \text{bond}|4\ \text{others}) = 0.03125.
\end{aligned} \tag{1}$$

For the general case for any number of disulfides, the probability that a given number (n) of disulfide bond(s) will be formed in a protein, assuming there is no special dependence in formation of bonds, is:

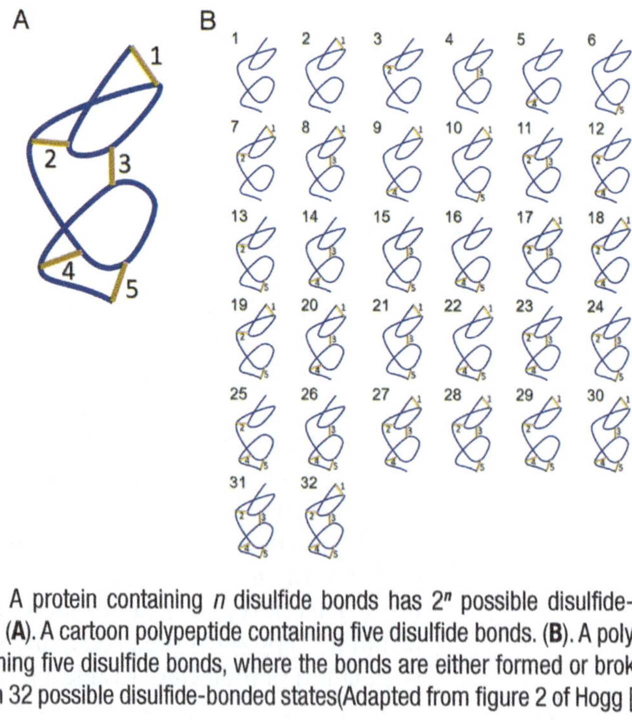

Fig. 2 A protein containing n disulfide bonds has 2^n possible disulfide-bonded states. (**A**). A cartoon polypeptide containing five disulfide bonds. (**B**). A polypeptide containing five disulfide bonds, where the bonds are either formed or broken, can exist in 32 possible disulfide-bonded states(Adapted from figure 2 of Hogg [11])

$$P(n) = 2^{-n}. \quad (2)$$

For example, for the eight disulfide bonds in FVIII, the probability that all bonds are formed, assuming no special dependencies, is very small at 0.004.

4 Conditional Formation of Disulfide Bonds

It is likely, particularly in complex proteins containing many disulfide bonds, that one or more bonds might be required to form before another bond can form. That is, disulfide bond formation in the protein is dependent on the state of other bonds. These situations

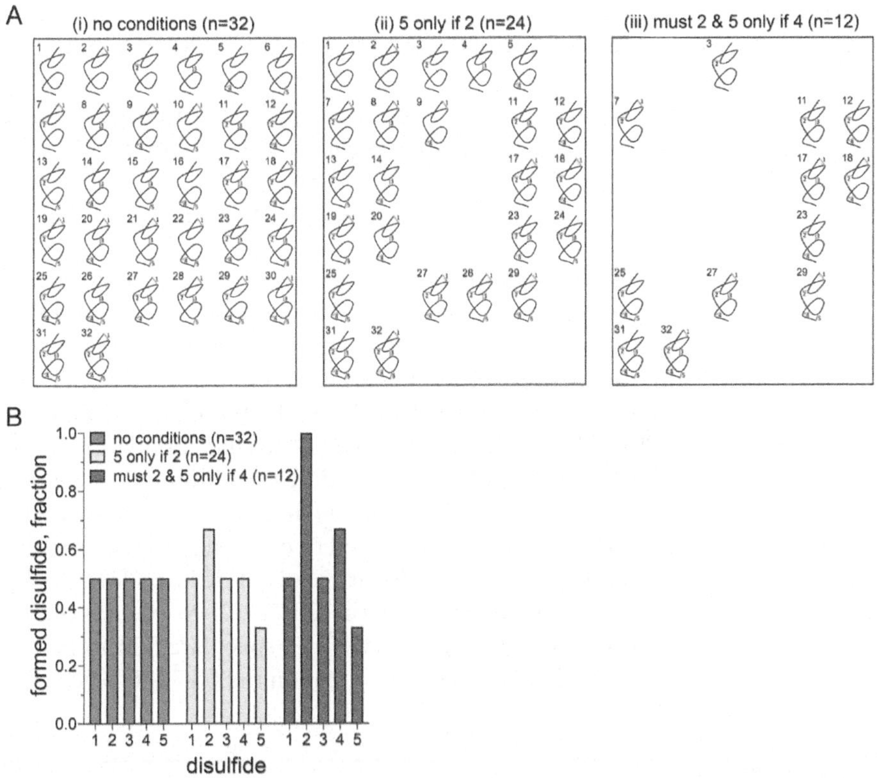

Fig. 3 Special conditional formation of disulfide bonds. (**A**). A polypeptide containing five disulfide bonds, where the bonds are either formed or broken, can exist in 32 possible disulfide states (part (i)). If the condition is applied that disulfide number 5 only forms if disulfide number 2 has formed, the total number of disulfide states reduces from 32 to 24 (part (ii)). If the condition is applied that disulfide number 2 must form and disulfide number 5 forms only if disulfide number 4 has formed, the total number of disulfide states reduces further to 12 (part (iii)). (**B**). Fraction of the five disulfide bonds that are formed in the scenarios in part A, which represents the probability that the disulfide bonds are formed in the population. This representation mirrors the expression of the experimental data shown in Fig. 1A (Adapted from figure 3 of Hogg [11]. Note that there is an error in the original figure that has been corrected here)

are illustrated using the model polypeptide and two different special conditions.

The first condition assumes that disulfide number 5 forms only if disulfide number 2 has formed. This results in a reduction of the total number of disulfide states from 32 to 24 (Fig. 3Aii). The second condition assumes that disulfide number 2 must form and disulfide number 5 forms only if disulfide number 4 has formed. In this situation, the number of disulfide states reduces to a total of 12 (Fig. 3Aiii). The probability that each of the five disulfide bonds are formed in the different scenarios is shown in Fig. 3B. It is apparent that special conditions change the probabilities for disulfide formation from a value of 0.5 for all bonds if there are no dependencies (Fig. 3Ai) to values as low as 0.33 and high as 1.0 if there are dependencies for bond formation (Fig. 3Aaiii).

5 Probability of Protein States Containing Any Number of Disulfide Bonds

The probability of a state containing any number of bonds is the product of the probabilities for each of the bonds:

$$P(a \,\&\, b \,\&\, c\ldots) = P(a)\,.\,P(b)\,.\,P(c)\ldots \tag{3}$$

For the eight FVIII disulfide bonds, the probability that all bonds are formed assuming no special dependencies is 0.004 (Eq. 2). From Eq. 3, the probability that all eight bonds are formed in FVIII is the product of the probabilities for each bond. For the eight experimental values of mean formed state (Table 1), this equates to a probability of 0.098. This result indicates that (1) disulfide bond formation in FVIII is conditional considering the difference in the unconditional value of 0.004 and the experimentally determined value of 0.098 and (2) just ~10% of FVIII is predicted to contain all eight formed disulfide bonds. The conditional nature of disulfide bond formation in FVIII is not known, although as the C172–C198 bond is almost always formed (fraction formed is 0.93), I speculate that formation of one or more other bonds is dependent on the C172–C198 being formed.

Acknowledgments

This research was funded by the National Health and Medical Research Council of Australia (2026631) and a Senior Researcher Grant from the NSW Cardiovascular Research Capacity Program.

References

1. Anfinsen CB, Haber E, Sela M, White FH Jr (1961) The kinetics of formation of native ribonuclease during oxidation of the reduced polypeptide chain. Proc Natl Acad Sci U S A 47:1309–1314. https://doi.org/10.1073/pnas.47.9.1309
2. Schmidt B, Ho L, Hogg PJ (2006) Allosteric disulfide bonds. Biochemistry 45(24):7429–7433
3. Hogg PJ (2003) Disulfide bonds as switches for protein function. Trends Biochem Sci 28(4):210–214
4. Chiu J, Hogg PJ (2019) Allosteric disulfides: sophisticated molecular structures enabling flexible protein regulation. J Biol Chem 294(8):2949–5908. https://doi.org/10.1074/jbc.REV118.005604
5. Flaumenhaft R, Furie B (2016) Vascular thiol isomerases. Blood 128(7):893–901. https://doi.org/10.1182/blood-2016-04-636456
6. Wu Y, Essex DW (2020) Vascular thiol isomerases in thrombosis: the yin and yang. J Thromb Haemost 18(11):2790–2800. https://doi.org/10.1111/jth.15019
7. Pijning AE, Butera D, Hogg PJ (2022) Not one, but many forms of thrombosis proteins. J Thromb Haemost 20(2):285–292. https://doi.org/10.1111/jth.15567
8. Butera D, Hogg PJ (2020) Fibrinogen function achieved through multiple covalent states. Nat Commun 11(1):5468. https://doi.org/10.1038/s41467-020-19295-7
9. Pijning AE, Blyth MT, Coote ML, Passam F, Chiu J, Hogg PJ (2021) An alternate covalent form of platelet αIIbβ3 integrin that resides in focal adhesions and has altered function. Blood 138(15):1359–1372. https://doi.org/10.1182/blood.2021012441
10. Butera D, Wang HJ, Woon HG, Zhao YC, Ju LA, Hogg PJ (2023) Mechano-covalent protection of coagulation factor VIII by von Willebrand factor. Blood Adv 7(10):2117–2128. https://doi.org/10.1182/bloodadvances.2022008650
11. Hogg PJ (2020) Multiple disulfide-bonded states of native proteins: estimate of number using probabilities of disulfide bond formation. Molecules 25(23):5729. https://doi.org/10.3390/molecules25235729
12. Pipe SW, Gonen-Yaacovi G, Segurado OG (2022) Hemophilia A gene therapy: current and next-generation approaches. Expert Opin Biol Ther 22(9):1099–1115. https://doi.org/10.1080/14712598.2022.2002842
13. Pipe SW (2018) Bioengineered molecules for the management of haemophilia: promise and remaining challenges. Haemophilia 24(6):68–75. https://doi.org/10.1111/hae.13507
14. Smith IW, d'Aquino AE, Coyle CW, Fedanov A, Parker ET, Denning G, Spencer HT, Lollar P, Doering CB, Spiegel PC Jr (2020) The 3.2 Å structure of a bioengineered variant of blood coagulation factor VIII indicates two conformations of the C2 domain. J Thromb Haemost 18(1):57–69. https://doi.org/10.1111/jth.14621
15. Fuller JR, Knockenhauer KE, Leksa NC, Peters RT, Batchelor JD (2021) Molecular determinants of the factor VIII/von Willebrand factor complex revealed by BIVV001 cryo-electron microscopy. Blood 137(21):2970–2980. https://doi.org/10.1182/blood.2020009197
16. Ronayne EK, Peters SC, Gish JS, Wilson C, Spencer HT, Doering CB, Lollar P, Spiegel PC Jr, Childers KC (2021) Structure of blood coagulation factor VIII in complex with an anti-C2 domain non-classical, pathogenic antibody inhibitor. Front Immunol 12:697602. https://doi.org/10.3389/fimmu.2021.697602
17. Butera D, Passam F, Ju L, Cook KM, Woon H, Aponte-Santamaria C, Gardiner E, Davis AK, Murphy DA, Bronowska A, Luken BM, Baldauf C, Jackson S, Andrews R, Grater F, Hogg PJ (2018) Autoregulation of von Willebrand factor function by a disulfide bond switch. Sci Adv 4(2):eaaq1477. https://doi.org/10.1126/sciadv.aaq1477
18. Zhou B, Baldus IB, Li W, Edwards SA, Grater F (2014) Identification of allosteric disulfides from prestress analysis. Biophys J 107(3):672–681. https://doi.org/10.1016/j.bpj.2014.06.025
19. Wong JW, Hogg PJ (2010) Analysis of disulfide bonds in protein structures. J Thromb Haemost 8:2345. https://doi.org/10.1111/j.1538-7836.2010.03894.x
20. Gish JS, Jarvis L, Childers KC, Peters SC, Garrels CS, Smith IW, Spencer HT, Doering CB, Lollar P, Spiegel PC (2021) Structure of blood coagulation factor VIII in complex with an anti-C1 domain pathogenic antibody inhibitor. Blood 137(21):2981–2986. https://doi.org/10.1182/blood.2020008940

Chapter 5

Using AlphaFold3 for the Interrogation of Protein Disulfide Bonds

Aster E. Pijning

Abstract

Disulfide bonds play critical roles in stabilizing protein structure and modulating protein function. Traditional approaches to studying these bonds have often relied on generating cysteine mutations to remove disulfide bonds, followed by labor-intensive biochemical characterization of the resulting variants. With the advent of AlphaFold3 (AF3) by Google DeepMind, in silico structural interrogation of disulfide bonds has been made possible. By integrating AF3's predictive capabilities with PyMOL-based structural analysis, researchers can rapidly assess the structural impact of removing disulfide bonds. This workflow provides informed predictions regarding the functional consequences of such modifications and enables investigators to refine hypotheses for other experimental approaches. Here, we describe a practical protocol for employing AF3 to investigate the contribution of disulfide bonds to protein stability and function.

Keywords Disulfide bond, AlphaFold, Integrin, Membrane protein

1 Introduction

The study of disulfide bonds using cysteine mutagenesis has aided the discovery of allosteric disulfide bonds [1–4] and elucidated the function of specific disulfide-bonded states of coagulation proteins [5]. However, the generation of disulfide mutant protein variants can be laborious and expensive, especially when investigating large proteins with many disulfide bonds. Computational methods now offer a more efficient alternative.

In recent years, computational methods have emerged as powerful tools to complement and streamline experimental strategies. Molecular dynamics (MD) simulations, for instance, offer insights into the conformational flexibility and energetics of proteins carrying specific disulfide bond mutations. Most notably, breakthroughs in protein structure prediction by Google DeepMind's AlphaFold have transformed the field's ability to generate highly accurate in silico models directly from primary amino acid sequences. The latest

Philip J. Hogg (ed.), *Functional Disulphide Bonds*, Methods in Molecular Biology, vol. 3016,
https://doi.org/10.1007/978-1-0716-5158-2_5,

iteration, AF3, extends these advances, enabling faster and more reliable predictions of protein structures that can be leveraged to interrogate the conformational consequences of disulfide bond disruption [6].

Schrodinger's PyMOL mutagenesis functionality has long provided a convenient platform for "quick swap" modifications of amino acids in a three-dimensional protein model [7]. However, such an approach typically assumes minimal or localized structural perturbations and does not always account for the broader, dynamic reorganizations a protein may undergo when a disulfide bond is broken or introduced. In contrast, generating a mutated sequence and predicting its structure de novo using AF3 can capture global structural changes, providing a more comprehensive view of the potential impact of specific disulfide bonds on a protein's overall structure.

AF3 outputs, such as the pLDDT confidence score, provide insights into a model's reliability and can hint at structural flexibility [8, 9]. These in silico predictions of disulfide mutant protein structure can also guide molecular modeling studies—for example, evaluating how the removal of a disulfide bond might affect predicted ligand binding. In this chapter, we focus on the blood clotting protein fibrinogen to illustrate how integrating AF3, PyMOL, and downstream computational analysis enables both rapid hypothesis generation and informed experimental design, ultimately revealing new insights into disulfide bond-mediated protein regulation.

2 Materials

The AF3 server is accessible at https://alphafoldserver.com/ and requires a Google account. As of publication, 20 "jobs" (structures) can be submitted to the server per day. AF3 supports the submission of multimeric structures, as well as ions and nucleic acids (*see* **Note 1**). We recommend reading the comprehensive guides provided on the platform prior to using AlphaFold.

A primary amino acid sequence of the protein to be studied is required—AF3 cannot handle unknown residues. Analysis of output structures can be viewed in any molecular viewer, such as PyMOL or ChimeraX. PyMOL requires an active license for full access, available at http://www.pymol.org. Additional molecular modeling of structures can be performed in Maestro (Schrödinger LLC, licensed) or comparable molecular modeling software.

3 Methods

3.1 Preparing Primary Amino Acid Sequences

In this example, we investigated the D domain of the plasma protein fibrinogen, which circulates as a dimer of a trimeric structure comprised of an α-, β-, and γ-strand. The D-domain β-strand contains five disulfide bonds, three intra-strand disulfides, and two inter-strand disulfide bonds, linking the β with the α- and γ-strand (Fig. 1). Using differential cysteine alkylation and mass spectrometry [10], each of these disulfide bonds was found to be unformed in a different fraction of the plasma fibrinogen pool. Considering this, 32 different disulfide states (2^5) of the fibrinogen β-strand may exist in vivo. These 32 states were included in an in silico investigation on the possible structural differences between disulfide-bonded states of the fibrinogen β-strand.

For each disulfide state, the primary amino acid sequence of one or more sets of disulfide-bonded cysteines was replaced with alanine (*see* **Note 2**). The structures of these in silico disulfide mutants and the WT structure were predicted using AF3 (a total of 32 structures).

3.2 Running AF3 Jobs

For each individual AF3 job, the amino acid sequence of the disulfide mutant was copied into the designated field (Fig. 2). Since the fibrinogen D-domain is a trimeric structure that binds two calcium ions, there are three protein entities and two ions (*see* **Note 3**). Each job is seeded for reproducibility.

3.3 Analyzing AF3 Outputs

3.3.1 AF3 Parameters

When AF3 completes a prediction, it provides several output parameters and visual aids that inform both the reliability of the predicted structure and the region(s) most likely to deviate from the submitted sequence's true conformation. Two of the most widely used metrics are pLDDT and PAE.

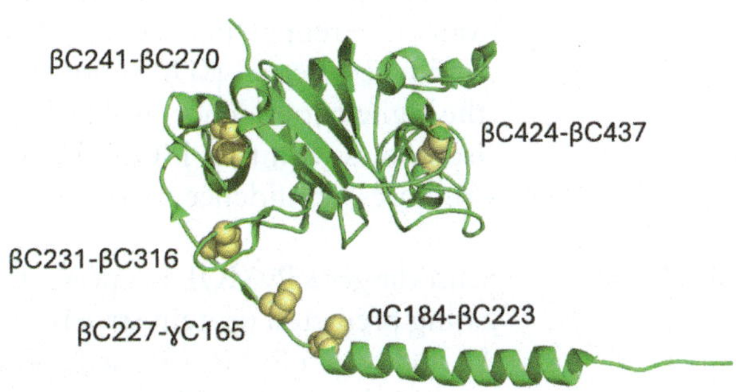

Fig. 1 Crystal structure of the β-nodule (strand) of the D-domain of fibrinogen, with each of the five disulfide bonds indicated as yellow spheres (PDB 2Z4E) [11]

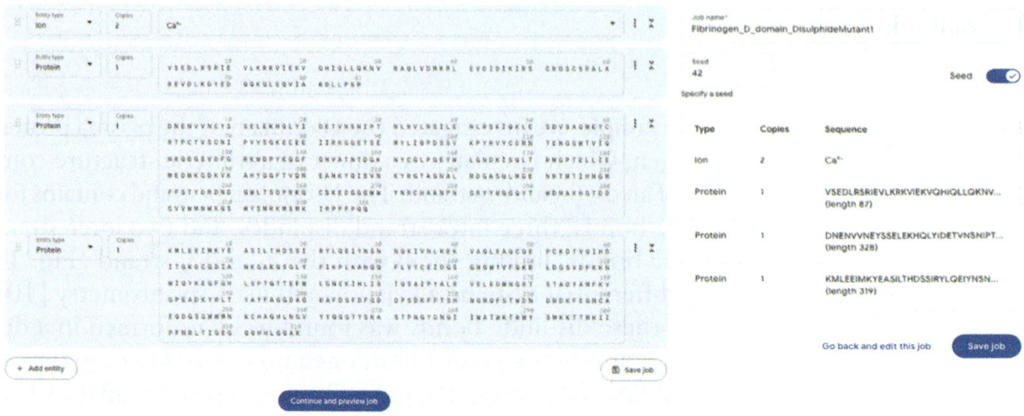

Fig. 2 Inputting sequences and ions on the AF3 server. Clicking "continue and preview job" will result in a pop-up screen to preview, rename, and seed the job prior to saving

Predicted Local Distance Difference Test (pLDDT)

pLDDT is a per-residue confidence metric that ranges from 0 to 100. A higher pLDDT indicates that AF is more confident in the local geometry of that residue and its surrounding environment. Multiple studies have shown that low pLDDT regions tend to correspond to loops, linkers, and disordered or flexible segments of proteins, and can be an indicator of structural flexibility [8, 9, 12, 13]. For example, Wróbleski and Kmiecik showed that the addition of the pLDDT score to the well-established protein flexibility simulator CABS-flex improved flexibility predictions [9].

Predicted Aligned Error (PAE)

The PAE matrix represents AF's predicted positional error between any pair of residues in the protein. It is depicted as a 2D heatmap, where regions with very low PAE indicate high confidence in the relative positioning of those residues. The PAE 2D plot is especially helpful in determining whether domains are well-structured and whether the relative placement of the domains in the predicted structure is correct.

Downloading the AF output files will create a zipped folder containing five different models (usually labeled 0–4), each resulting from a different neural network. The models are ranked by an internal confidence metric (pTM score). In many cases, the model named 0 (the highest confidence) will be used for further analysis. Each model also comes with two JSON files containing numerical data on the structures' confidence scores and predicted errors (Fig. 3).

3.3.2 PyMOL Analysis

Schrödinger's PyMOL is a powerful tool for visualizing and interrogating predicted protein structures, including those exported from AF. Below is a PyMOL-based approach for analyzing the predicted structural impact of 32 different disulfide mutants of the β-chain of the fibrinogen D-domain (Fig. 4).

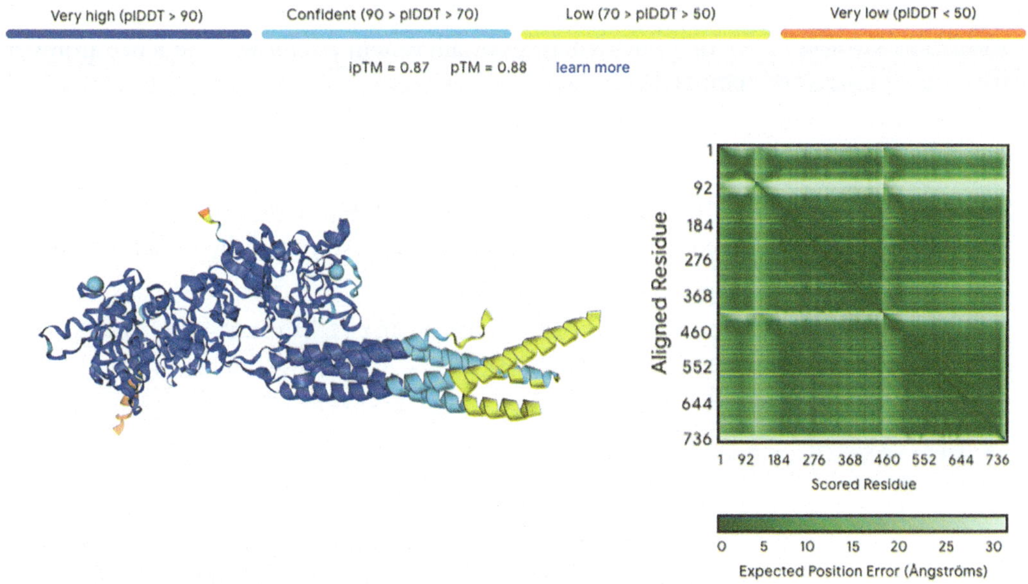

Fig. 3 Output of AF3 job. A 3D model of the heterotrimeric structure with both Ca^{2+} ions in place is portrayed on the left, colored by pLDDT score. A 2D PAE plot representing confidence in the relative placement of residues and domains to each other is shown on the right

Disulfide	Cys 1	Cys 2	Average unformed
1	α184	β223	18%
2	βC227	γC165	37%
3	βC231	βC316	37%
4	βC241	βC270	27%
5	βC424	βC437	34%

All possible β-nodule disulfide states (numbers are bonds missing)	
Fully bonded	1,2,3
1	1,2,4
2	1,2,5
3	1,3,4
4	1,3,5
5	1,4,5
1,2	2,3,4
1,3	2,3,5
1,4	2,4,5
1,5	3,4,5
2,3	1,2,3,4
2,4	1,2,4,5
2,5	1,2,3,5
3,4	1,3,4,5
3,5	2,3,4,5
4,5	1,2,3,4,5

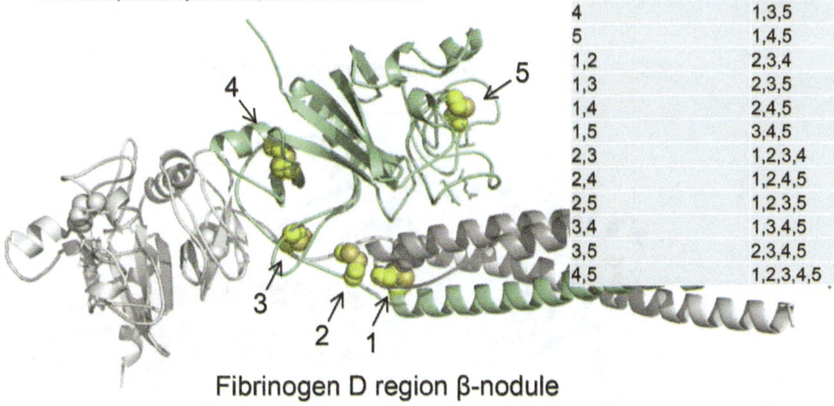

Fig. 4 The trimeric structure of the fibrinogen D-domain is shown with each of the five β-strand disulfide bonds shown as spheres (AF3 structure). The predicted structure of 32 different disulfide states of interest was computed using AF3

Comparing WT and Disulfide Mutant Structures

Predicted AF structures of disulphide mutants are compared to WT AF or X-ray structures by alignment after low-confidence atoms are removed:

```
remove (b < 50)
align WT, mutant
```

The per-atom conformational shift of the mutant structure compared to WT can be calculated and stored in the B-factor column.

```
alter mutant, b=0
iterate mutant, cmd.alter(f"mutant and id {ID}",
f"b = cmd.distance('', 'WT and id {ID}', 'mutant
and id {ID}')")
stored_b_factors = []
iterate mutant, stored_b_factors.append((ID,
resn, resi, name, b))
```

When the B-factor column is now applied as a color spectrum to the mutant structure, any conformational shift as a result of the absence of disulfide bonds is predicted and can be easily assessed (*see* **Note 4**; Fig. 5).

```
spectrum b, blue_white_red, mutant, minimum=0,
maximum=max
ramp_new shift_ramp, mutant, [0, max], [blue,
white, red]
```

Comparing Predicted Changes in Disulfide Mutant Structural Flexibility

While AF-predicted structures do not account for structural flexibility, the pLDDT score has been shown to correlate with areas of intrinsic disorder/flexibility [8, 9, 12, 13]. As such, comparing any changes in pLDDT in disulfide mutant AF structures to WT can

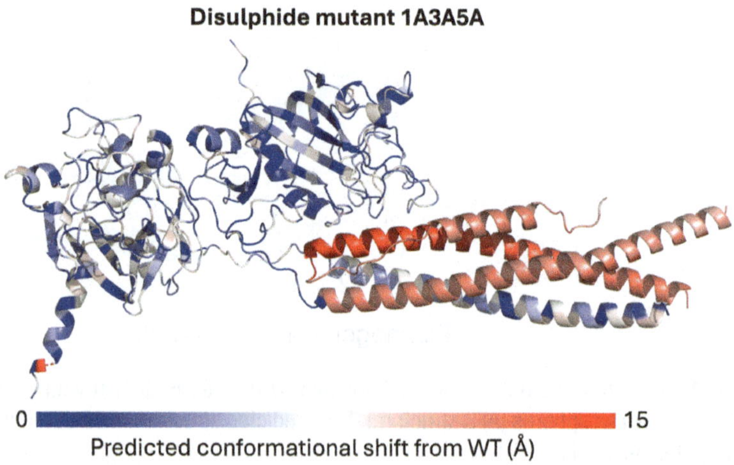

Fig. 5 Predicted conformational shift of the WT fibrinogen D-domain compared to a disulfide mutant (1A3A5A), calculated as the Å shift of each α carbon of the highest confidence structures

inform which disulfide bonds confer structural rigidity. The pLDDT score is stored in the B-factor column in AF output models (unless this column is altered, such as in step 1 above) and can easily be exported from PyMOL.

```
stored_pLDDT = []
iterate  all,  stored_pLDDT.append((ID,  resn,
resi, name, b))
```

Export pLDDT scores to a CSV file for downstream analysis using Python in the PyMOL command line.

```
python
output_file = '/your/file/location/mutant_pLDDT.
csv'
with open(output_file, 'w') as file:
  file.write('Atom    ID,Residue    Name,Residue
Index,Atom Name,pLDDT\n')
  for atom in stored_pLDDT:
    file.write('%s,%s,%s,%s,%.2f\n' % (atom[0],
atom[1], atom[2], atom[3], atom[4]))
python end
```

Using this approach, we found that the absence of a single disulfide bond (C424–C437) in the fibrinogen D-domain β-chain is predicted to result in increased flexibility of the β-chain GHRP binding pocket and Ca^{2+} ion binding pockets (Fig. 6). We also predict that the absence of the other four disulfide bonds in the D-domain β-chain generally do not cause an additive increase in flexibility.

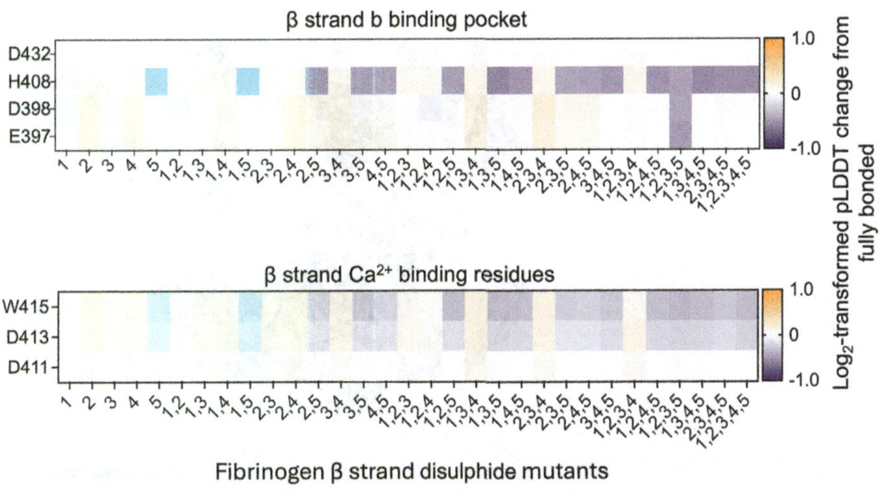

Fig. 6 Relative pLDDT scores of four β-chain residues forming a crucial peptide (GHRP) binding pocket in the fibrinogen D-domain and three residues forming the Ca^{2+} binding pocket in the β-chain, compared to WT. Note that a lower score indicates higher intrinsic disorder and possible flexibility in the structure

Analyzing Unmapped Disulfide Bonds

In the absence of a high-resolution crystal or cryo-EM structure, an AF3 predicted model of von Willebrand factor (vWF) provides an opportunity to examine disulfide bonds that remain experimentally unresolved. UniProt entry P04275 reports 31 mapped disulfide bonds in vWF. In contrast, the full-length AF3 structure predicts 114 disulfide bonds using the below PyMOL command line inputs (*see* **Notes 5 and 6**; Fig. 7).

```
python
pairs = cmd.find_pairs("resn cys and name SG",
"resn cys and name SG", cutoff=2.3)

python seen = set()
for p1, p2 in pairs:
     if p2 < p1:
        p1, p2 = p2, p1
     if (p1, p2) in seen:
        continue
     seen.add((p1, p2))
     atom1 = cmd.get_model(f"{p1[0]} and index
{p1[1]}").atom[0]
     atom2 = cmd.get_model(f"{p2[0]} and index
{p2[1]}").atom[0]
     print(f"Disulfide bond between {atom1.resn}
{atom1.chain}{atom1.resi}   and   {atom2.resn}
{atom2.chain}{atom2.resi}")
python end
```

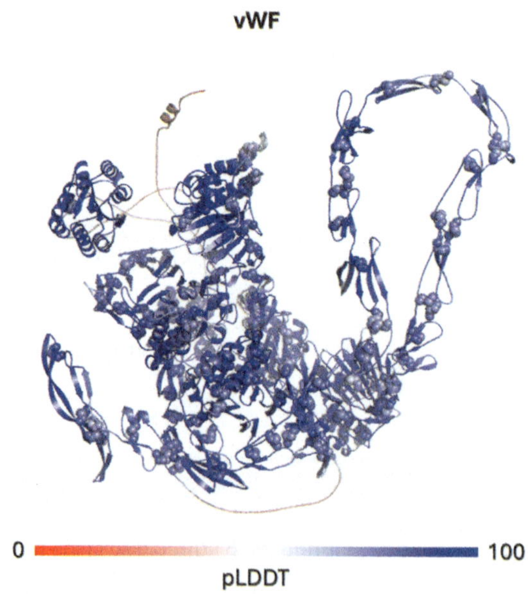

Fig. 7 AF3 predicted structure of full-length vWF, disulfide bonds are shown as spheres and the structure is colored by pLDDT score

This highlights the ability of AF3 to reveal additional linkages that may have been missed or are difficult to characterize experimentally. Each bond can be further investigated, for instance, by calculating the solvent-accessible surface area (SASA, Å2) and relative solvent exposure (RSA) of each cysteine (Table 1).

```
python
from pymol import cmd
# Important SASA settings
cmd.set("dot_solvent", 1)      # 1 = SASA, 0 =
molecular/solvent-excluded surface
cmd.set("dot_density", 3)      # higher = more
accurate, slower
cmd.set("solvent_radius", 1.4) # water probe
radius in Å

cys_residues = []
cmd.iterate("name CA and resn CYS", "cys_
residues.append((chain, resi, resn))")
cys_residues = list(set(cys_residues))
cys_residues.sort(key=lambda x: (x[0], int(x[1])
if x[1].isdigit() else 9999))

print("chain resi resn SASA_A2 RSA_approx
exposure")
for (ch, resi, resn) in cys_residues:
  sel_string = f"chain {ch} and resi {resi}
and resn CYS"
  sasa = cmd.get_area(sel_string, state=1)

  # Approximate cysteine maximum ASA ≈ 150 Å².
  rsa = sasa / 150

  if rsa < 0.05:
     exposure = "buried"
  elif rsa < 0.20:
     exposure = "mostly_buried"
  elif rsa < 0.25:
     exposure = "borderline"
  elif rsa < 0.50:
     exposure = "exposed"
  else:
     exposure = "highly_exposed"

  print(ch,resi,resn,round(sasa,2),round(rsa,3),
  exposure)
python end
```

By mapping predicted disulfide bonds and their exposure levels, researchers can gain deeper insights into how these linkages

Table 1

All predicted disulfide bonds in full-length monomeric vWF and the relative solvent exposure (RSA) of disulfide-bonded cysteines (0 is buried, 1 is fully exposed). In bold are the disulfide bonds that are described in UniProt entry P04275, underlined are disulfide bonds that are paired differently by AF compared to UniProt

Disulfide bond		RSA		Disulfide bond		RSA	
Cys1	Cys2	Cys1	Cys2	Cys1	Cys2	Cys1	Cys2
35	**162**	0	0.003	**1272**	**1458**	0.011	0.161
57	**200**	0.118	0.258	**1669**	**1670**	0.001	0.392
65	159	0.355	0.174	**1686**	**1872**	0.082	0.266
210	255	0.416	0.404	**1879**	**1904**	0.041	0
225	250	0	0.003	**1899**	**1940**	0.055	0.095
237	275	0	0	1919	1942	0.023	0.129
257	263	0.384	0.067	1927	1946	0.1	0.064
265	291	0.194	0.033	1948	2088	0.096	0.082
295	329	0.211	0.105	1950	2085	0	0.004
304	325	0.297	0.356	1972	2123	0.05	0.502
308	321	0.302	0.305	**1993**	**2001**	0.305	0.15
312	348	0.044	0	2131	2168	0.36	0.377
331	342	0.164	0.193	2139	2163	0	0.011
350	372	0	0.002	2150	2184	0.003	0
367	384	0.28	0.27	2174	2199	0.144	0.105
370	379	0.016	0.298	2190	2212	0.267	0.169
388	**524**	0.009	0.002	2203	2235	0.268	0.05
410	**559**	0.053	0.219	2216	2227	0.296	0.369
418	521	0.315	0.21	2220	2254	0.127	0.116
432	**440**	0.139	0.296	2237	2248	0.215	0.118
570	613	0.07	0.03	2257	2283	0.003	0
584	608	0.003	0	2278	2319	0.259	0.22
595	633	0	0.001	2281	2291	0.029	0.355
615	621	0.135	0.202	2296	2320	0.347	0.329
623	648	0.191	0.113	2304	2325	0.264	0.337
652	687	0.255	0.089	2307	2332	0.213	0.111
661	683	0.343	0.319	2327	2354	0.487	0.277
665	679	0.166	0.194	2340	2360	0.153	0.393
669	707	0	0.002	2362	2388	0.352	0.161
689	701	0.107	0.17	2367	2389	0.353	0.313

Disulfide bond		RSA		Disulfide bond		RSA	
Cys1	Cys2	Cys1	Cys2	Cys1	Cys2	Cys1	Cys2
709	731	0	0.041	2375	2394	0.263	0.292
729	738	0.005	0.37	2398	2420	0.434	0.2
767	**808**	0.228	0.267	2405	2425	0.268	0.385
776	**804**	0.215	0.244	2431	2453	0.006	0.033
788	799	0.118	0.125	2448	2490	0.419	0.344
792	**827**	0.057	0.013	2451	2468	0.008	0.473
810	**821**	0.186	0.218	2473	2491	0.4	0.283
829	851	0	0.022	2477	2494	0.318	0.365
846	863	0.414	0.262	2499	2533	0.036	0.013
849	858	0.01	0.333	2528	2570	0.226	0.251
867	**996**	0.009	0	2549	2571	0.441	0.193
889	**1031**	0.053	0.138	2557	2576	0.249	0.31
898	**993**	0.199	0.134	2565	2574	0.508	0.158
914	**921**	0.027	0.163	2582	2605	0.051	0.009
1046	1089	0.044	0.013	2600	2640	0.229	0.261
1060	**1084**	0	0.001	2603	2619	0.002	0.507
1071	**1111**	0.009	0.001	2624	2641	0.424	0.306
1091	1099	0.194	0.095	2627	2644	0.234	0.352
1101	1126	0.097	0.027	2649	2676	0.042	0.016
1130	1173	0.066	0.096	2671	2715	0.175	0.255
1149	**1169**	0.225	0.057	2693	2716	0.395	0.048
1153	**1165**	0.189	0.283	2701	2719	0.042	0.298
1157	1196	0.014	0.065	**2724**	**2774**	0.203	0.16
1177	1190	0.191	0.049	**2739**	**2788**	0.129	0.111
1199	1227	0.029	0.017	**2750**	**2804**	0.147	0.068
1222	1237	0.091	0.26	**2754**	**2806**	0.175	0.343
1225	1234	0.011	0.421	2771	2811	0.261	0.338

might influence vWF's stability, multimer assembly, and overall function—even in the absence of a classic high-resolution experimental structure.

3.4 Other Molecular Modeling Approaches

AF3 structures can be used for docking approaches in molecular modeling software such as Maestro's Schrodinger. In fact, AF structures were shown to be comparable and complementary to apo structures for docking purposes when prepared correctly, i.e., by removing low confidence regions and enabling flexibility of side chains [14].

In this example, we were interested in predicted functional differences between the possible fibrinogen D-domain disulfide states. Fibrinogen forms fibrin after thrombin cleavage of fibrinopeptides A and B from the N-termini of the Aα and Bβ chains of fibrinogen, exposing "knobs" that bind to "holes" in the γ- and β-chains to initiate fibrin fiber formation. The β-chain "hole" that binds GHRP peptide (as well as the Ca^{2+} binding pocket) is predicted to be more flexible in disulfide states where disulfide bond 5 is unformed (Fig. 6). The following process in Maestro aims to explore whether this could have consequences for GHRP peptide binding affinity.

3.4.1 Protein Preparation

The WT and mutant AF3 predicted protein structures were prepared using the Protein Preparation Wizard in Schrödinger's Maestro. Hydrogens were added, and the protonation states of residues were

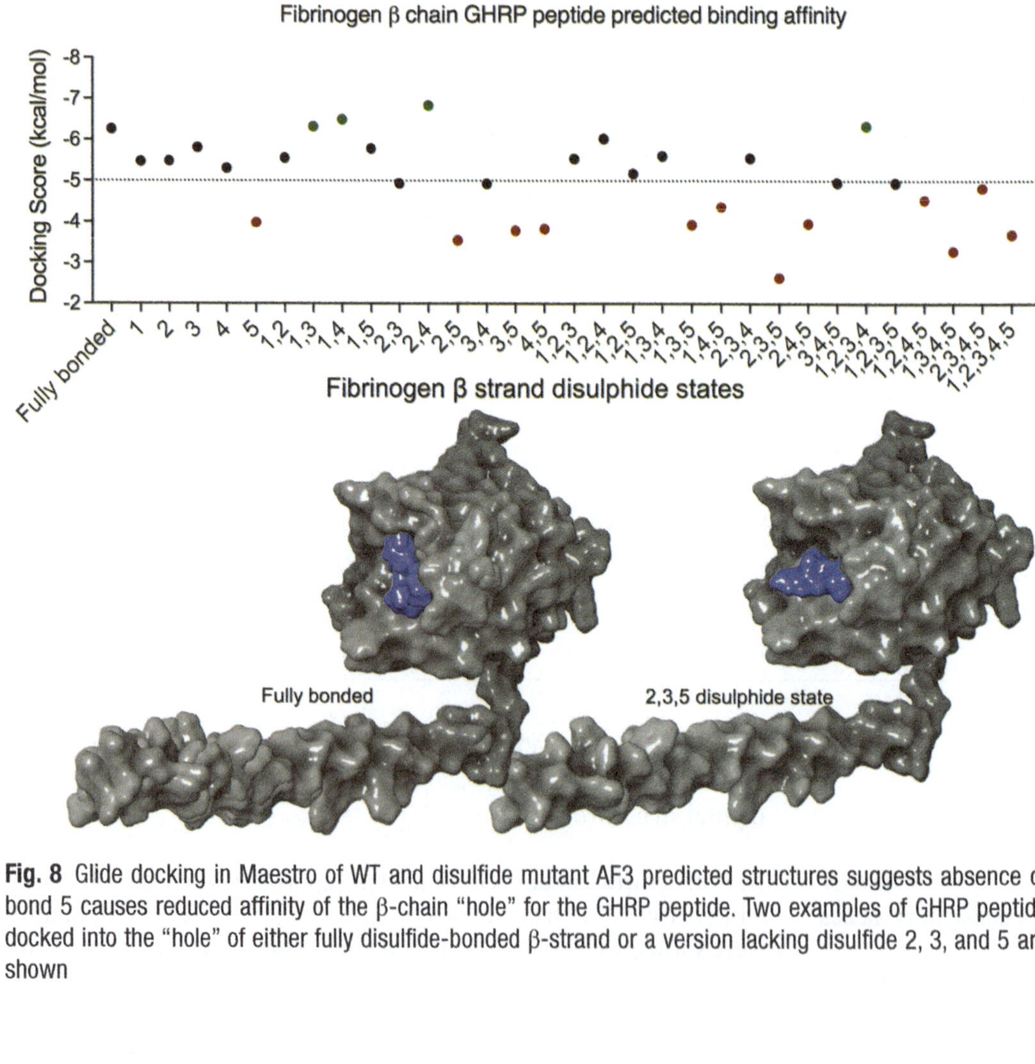

Fig. 8 Glide docking in Maestro of WT and disulfide mutant AF3 predicted structures suggests absence of bond 5 causes reduced affinity of the β-chain "hole" for the GHRP peptide. Two examples of GHRP peptide docked into the "hole" of either fully disulfide-bonded β-strand or a version lacking disulfide 2, 3, and 5 are shown

adjusted to pH 7.4 using Epik. The peptide ligand was prepared using a combination of LigPrep and ConfGen. LigPrep was used to generate a set of possible tautomeric and protonation states at pH 7.4, ensuring chemical diversity in the ligand conformations. The ligand was then subjected to ConfGen to generate an ensemble of low-energy conformers, capturing potential structural flexibility. Protein and peptide conformations were minimized using the OPLS3e force field.

3.4.2 Docking Grid Generation

A docking grid was generated to define the binding site on the protein where the peptide is predicted to bind. The grid was centered around four key binding residues in the beta chain (E427, D428, H438, and D462) [15]. A cubic grid box was set to encompass the entire binding region, with flexibility to accommodate the peptides maximum size of 14 Å. Default settings were used to allow optimal sampling within the defined site.

3.4.3 Docking

Peptide–protein docking was performed using Glide's peptide docking protocol. The Docking Score scoring function was employed to rank binding poses based on predicted binding affinity. Ligand flexibility was allowed during docking, with the sampling options set to consider various peptide conformations and orientations in the binding pocket. The most favorable peptide binding pose was selected based on the lowest Docking Score and used for comparison (Fig. 8).

4 Notes

1. Note that the ability to add ions to structures may be important when using the predicted structures for downstream docking simulations.
2. When deleting a disulfide bond, we prefer investigating structures in which both contributing cysteines of the bond have been mutated, in order to prevent unintended effects of having a free thiol in the system.
3. Post-translational modifications can be added to the primary sequence here; click on the three dots.
4. Change "max" to a maximum shift that makes sense in your analysis.
5. In AF3, disulfide bonds are inferred through sulfur–sulfur distance (average 2.2 Å), rather than encoded in the structure file.
6. The most stable VWF conformation is a homodimer. However, a dimerized structure of full-length VWF currently exceeds the maximum capacity of AF3.

Acknowledgments

This study was funded by the National Health and Medical Research Council of Australia (2026631) and a Senior Researcher Grant from the NSW Cardiovascular Research Capacity Program to Philip Hogg.

References

1. Passam F, Chiu J, Ju L, Pijning A, Jahan Z, Mor-Cohen R, Yeheskel A, Kolsek K, Tharichen L, Aponte-Santamaria C, Grater F, Hogg PJ (2018) Mechano-redox control of integrin de-adhesion. eLife 7:e34843. https://doi.org/10.7554/eLife.34843
2. Chiu J, Hogg PJ (2019) Allosteric disulfides: sophisticated molecular structures enabling flexible protein regulation. J Biol Chem 294(8):2949–5908. https://doi.org/10.1074/jbc.REV118.005604
3. Cook KM, Hogg PJ (2013) Post-translational control of protein function by disulfide bond cleavage. Antioxid Redox Signal 18(15):1987–2015. https://doi.org/10.1089/ars.2012.4807
4. Schmidt B, Ho L, Hogg PJ (2006) Allosteric disulfide bonds. Biochemistry 45(24):7429–7433
5. Pijning AE, Blyth MT, Coote ML, Passam F, Chiu J, Hogg PJ (2021) An alternate covalent form of platelet αIIbβ3 integrin that resides in focal adhesions and has altered function. Blood 138(15):1359–1372. https://doi.org/10.1182/blood.2021012441
6. Abramson J, Adler J, Dunger J, Evans R, Green T, Pritzel A, Ronneberger O, Willmore L, Ballard AJ, Bambrick J, Bodenstein SW, Evans DA, Hung CC, O'Neill M, Reiman D, Tunyasuvunakool K, Wu Z, Zemgulyte A, Arvaniti E, Beattie C, Bertolli O, Bridgland A, Cherepanov A, Congreve M, Cowen-Rivers AI, Cowie A, Figurnov M, Fuchs FB, Gladman H, Jain R, Khan YA, Low CMR, Perlin K, Potapenko A, Savy P, Singh S, Stecula A, Thillaisundaram A, Tong C, Yakneen S, Zhong ED, Zielinski M, Zidek A, Bapst V, Kohli P, Jaderberg M, Hassabis D, Jumper JM (2024) Accurate structure prediction of biomolecular interactions with AlphaFold 3. Nature 630(8016):493–500. https://doi.org/10.1038/s41586-024-07487-w
7. Schrodinger LLC (2015) The PyMOL molecular graphics system, version 1.8. DeLano Scientific, San Carlos, CA, USA
8. Ma P, Li DW, Bruschweiler R (2023) Predicting protein flexibility with AlphaFold. Proteins 91(6):847–855. https://doi.org/10.1002/prot.26471
9. Wroblewski K, Kmiecik S (2024) Integrating AlphaFold pLDDT Scores into CABS-flex for enhanced protein flexibility simulations. Comput Struct Biotechnol J 23:4350–4356. https://doi.org/10.1016/j.csbj.2024.11.047
10. Butera D, Hogg PJ (2020) Fibrinogen function achieved through multiple covalent states. Nat Commun 11(1):5468. https://doi.org/10.1038/s41467-020-19295-7
11. Doolittle RF, Pandi L (2007) Probing the beta-chain hole of fibrinogen with synthetic peptides that differ at their amino termini. Biochemistry 46(35):10033–10038. https://doi.org/10.1021/bi7010916
12. Guo HB, Perminov A, Bekele S, Kedziora G, Farajollahi S, Varaljay V, Hinkle K, Molinero V, Meister K, Hung C, Dennis P, Kelley-Loughnane N, Berry R (2022) AlphaFold2 models indicate that protein sequence determines both structure and dynamics. Sci Rep 12(1):10696. https://doi.org/10.1038/s41598-022-14382-9
13. Saldano T, Escobedo N, Marchetti J, Zea DJ, Mac Donagh J, Velez Rueda AJ, Gonik E, Garcia Melani A, Novomisky Nechcoff J, Salas MN, Peters T, Demitroff N, Fernandez Alberti S, Palopoli N, Fornasari MS, Parisi G (2022) Impact of protein conformational diversity on AlphaFold predictions. Bioinformatics 38(10):2742–2748. https://doi.org/10.1093/bioinformatics/btac202
14. Holcomb M, Chang YT, Goodsell DS, Forli S (2023) Evaluation of AlphaFold2 structures as docking targets. Protein Sci 32(1):e4530. https://doi.org/10.1002/pro.4530
15. Everse SJ, Spraggon G, Veerapandian L, Riley M, Doolittle RF (1998) Crystal structure of fragment double-D from human fibrin with two different bound ligands. Biochemistry 37(24):8637–8642. https://doi.org/10.1021/bi9804129

Check for updates

Chapter 6

Replacing of Cysteine with Selenocysteine: Biochemical Considerations, Computational Modeling, and Protein Engineering Applications

Wick Lakshantha

Abstract

Selenocysteine (Sec), the 21st amino acid, combines the structural compatibility of cysteine with the unique chemical properties of selenium, enabling exceptional redox and catalytic performance in both natural and engineered systems. Its low pK_a, high nucleophilicity, and increased polarizability underpin the activity of selenoproteins involved in antioxidant defence, redox regulation, and metabolic processes. This work integrates biochemical, structural, and evolutionary perspectives on Sec with advances in protein engineering and synthetic biology, where strategic Sec incorporation in place of Cys accelerates folding, enhances stability, and expands catalytic capabilities. Emphasis is placed on computational and quantum chemical studies that elucidate Sec's acid–base behavior, bond dissociation energies, and metal-binding preferences, revealing how subtle atomic substitution drives substantial functional gains. Comparative analysis of diselenide and disulfide bonds highlights Sec's potential to improve protein performance under both oxidative and reductive stress. Finally, this chapter demonstrates the application of computational chemistry tools to calculate key physicochemical parameters, illustrating how such methods can guide rational selenoprotein engineering for therapeutic, industrial, and bioengineering applications.

Keywords Disulfide bond, Selenocysteine, Cysteine substitution, Diselenide bonds, Redox biology, Density functional theory, Bond dissociation energy, pK_a calculation, Computational hydrogen electrode, Protein folding, Rational protein design

1 Introduction

Selenocysteine (Sec) is structurally analogous to cysteine (Cys) but is distinguished by the substitution of selenium for sulfur in its side chain. This seemingly minor atomic change results in profound biochemical consequences: a much lower pK_a, greater nucleophilicity, and enhanced polarizability of the selenol group. Collectively, these attributes translate into faster redox turnover and more versatile catalytic chemistry in enzymes [1, 2].

The discovery of Sec challenged the traditional view of the genetic code as immutable. Instead of being restricted to the canonical twenty

Philip J. Hogg (ed.), *Functional Disulphide Bonds*, Methods in Molecular Biology, vol. 3016,
https://doi.org/10.1007/978-1-0716-5158-2_6,

amino acids, cells have evolved a sophisticated recoding strategy that reassigns the UGA stop codon to encode Sec. This recoding depends on a unique translational apparatus, including a selenocysteine insertion sequence (SECIS) element in mRNA, a specialized $tRNA^{Sec}$, and elongation factors such as SelB in bacteria and eEFSec in eukaryotes [1–5]. Such dedicated machinery is metabolically expensive, yet its persistence across all three domains of life demonstrates the strong selective advantage conferred by Sec chemistry.

Biologically, Sec is incorporated into a specialized class of proteins known as selenoproteins, which are critical for maintaining redox balance, regulating thyroid hormone metabolism, and protecting against oxidative stress [6–8]. Enzymes such as glutathione peroxidases and thioredoxin reductases exploit Sec's unique redox profile to catalyze otherwise energetically challenging electron transfer reactions [7, 8]. Evolutionarily, this suggests that the presence of selenium-based catalysis offered a survival benefit during transitions between anaerobic and aerobic environments, where oxidative stress was an existential challenge.

From the perspective of protein engineering and synthetic biology, Sec represents a highly attractive molecular design element. Recent advances in genetic code expansion and site-directed mutagenesis now enable the deliberate incorporation of Sec into recombinant proteins, opening opportunities to design enzymes with enhanced activity, greater stability under stress, and novel catalytic functions [9]. This Chapter integrates biochemical, structural, and computational perspectives on Sec, culminating in a simple quantum chemical case study that illustrates how theoretical tools can guide rational strategies for selenoprotein engineering.

2 Biochemical and Structural Properties of Selenocysteine

2.1 Acid–Base Properties and Redox Activity

The most defining feature of Sec is its selenol (–SeH) group. Compared to the thiol (–SH) of Cys, the selenol exhibits a strikingly lower pK_a (~5.2 versus ~8–10 for Cys) [10, 11]. This difference means that at physiological pH (7.4), Sec exists predominantly as the deprotonated selenolate (Se^-), while Cys remains largely protonated and less nucleophilic. This single physicochemical distinction accounts for much of the catalytic advantage conferred by Sec: it ensures a much higher proportion of the side chains are in their reactive, anionic state under normal cellular conditions.

Electrochemical measurements corroborate these trends. The half-wave potential for Sec is −0.212 V compared to +0.021 V for Cys [2], highlighting its increased susceptibility to oxidation. In enzymatic active sites, this translates into faster electron transfer rates and a greater tolerance for redox cycling. For example, in glutathione peroxidases, Sec enables rapid detoxification of peroxides by efficiently shuttling

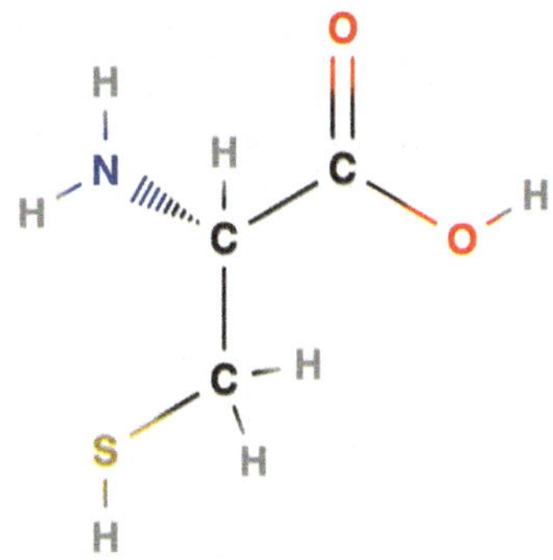

Fig. 1 Structure of cysteine

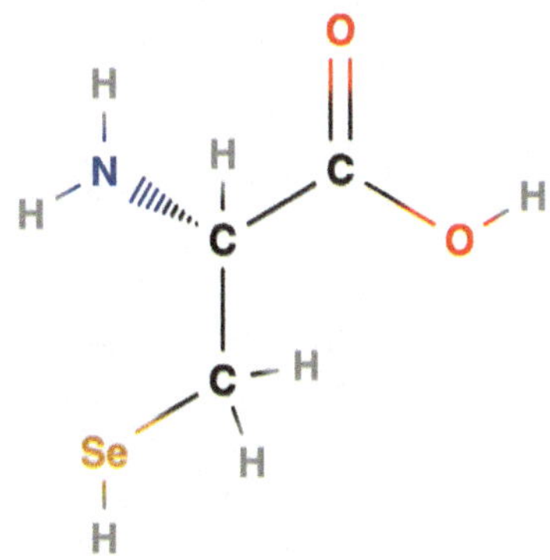

Fig. 2 Structure of selenocysteine

electrons from glutathione. The molecular structures of cysteine and selenocysteine are shown for comparison (Figs. 1 and 2).

2.2 Bonding and Folding Behavior

The differences extend to bond energetics. The Se–H bond is weaker than the S–H bond, making Sec easier to oxidize but also more reactive in forming intermediates during catalysis. When oxidized, Sec can form diselenide (Se–Se) bonds, which differ significantly from disulfides (S–S). Diselenides form readily under mild oxidative conditions but are thermodynamically harder to reduce, creating a paradoxical but advantageous combination: they fold quickly and efficiently into oxidized states but resist reductive cleavage in environments rich in reducing agents such as glutathione [12–15].

This behavior has been exploited experimentally. In Sec-substituted insulin, epidermal growth factor, and several cystine-rich toxins and inhibitors, substitution accelerates oxidative folding, increases thermodynamic stability, and enhances resistance to reductive unfolding [15–19]. For protein engineers, these properties suggest that targeted Sec incorporation can be a powerful strategy to stabilize proteins in otherwise destabilizing environments.

2.3 Structural Compatibility with Cysteine

Despite profound differences in electronic behavior, Sec is structurally very similar to Cys. Selenium and sulfur have comparable atomic radii, and crystallographic studies show that Sec fits seamlessly into protein backbones and side-chain orientations originally designed for Cys. Density functional theory (DFT) models confirm that the preferred conformations of Sec, both in neutral and anionic states, overlap almost perfectly with those of Cys [20].

This near-isosterism extends to metal-binding sites. Sec can coordinate with metals in geometries nearly identical to Cys, which is particularly important for enzymes such as thioredoxin reductase, where metal coordination plays a key catalytic role. Consequently, the substitution of Sec for Cys typically alters electronic and redox properties without introducing significant steric strain, making Sec an "invisible" yet functionally transformative substitution.

3 Computational and Theoretical Insights

3.1 Unified Physicochemical Trends

Computational chemistry offers a complementary lens to experiment, providing energetic and electronic detail at atomic resolution. Quantum chemical calculations consistently reproduce the trends observed in the laboratory: Sec's lower pK_a reflects the stabilization of the selenolate anion in aqueous solution; its increased polarizability accounts for both its higher nucleophilicity and its weaker bond dissociation energies; and its redox behavior aligns with its experimentally determined half-wave potential [11, 20, 21].

Importantly, calculations allow separation of these effects into their underlying causes. For example, comparative DFT studies reveal that the dominant driver of Sec's enhanced reactivity is its lower pK_a, while increased nucleophilicity and softer polarizability contribute additional, but secondary, effects. This insight is not always accessible to experiment and underscores the predictive power of computational approaches in rational protein design.

3.2 Enzyme-Like Environments

More advanced computational techniques, such as hybrid quantum mechanics/molecular mechanics (QM/MM) simulations, extend this analysis into enzyme active sites. In such contexts, Sec behaves as predicted by small-molecule models, but in ways that are shaped by the local environment. For example, simulations comparing methaneselenolate and methanethiolate reacting with peroxynitrite show that the Se analog reacts more rapidly, not simply because of greater nucleophilicity but because its deprotonation equilibrium at physiological pH ensures a much higher proportion of reactive species [21].

Docking and QM/MM studies of thioredoxin reductase reveal that Sec strongly coordinates to gold(I) and gold(III) complexes, rationalizing the high sensitivity of selenoenzymes to gold-based therapeutic inhibitors such as auranofin [22, 23]. Such a mechanistic understanding is invaluable when considering how Sec substitution might alter drug susceptibility in engineered or therapeutic proteins.

3.3 Structural Isosterism in Computation

Computational models also reinforce the steric compatibility of Sec with Cys. Bond lengths, bond angles, and side-chain orientations show only minimal differences between sulfur- and selenium-containing residues [20]. Simulations of metal–ligand interactions confirm that S → Se substitution alters electronic interactions without significantly disturbing geometry. This reinforces the principle that Sec substitution primarily modifies reactivity, not structure, a key reason why it is tolerated in diverse protein folds.

4 Computational Case Study: Dimethyl Disulfide and Dimethyl Diselenide as Models for Disulfide and Diselenide Bonds

To connect the biochemical and theoretical concepts discussed above with a practical demonstration, a model systems approach is employed to quantify the physicochemical differences between sulfur–sulfur and selenium–selenium bonds. In this case study, dimethyl disulfide (DMDS, CH_3–S–S–CH_3) and dimethyl diselenide (DMDSe, CH_3–Se–Se–CH_3) are selected as representative analogs for cystine (Cys–Cys) and selenocystine (Sec–Sec), respectively. The structures of dimethyl disulfide and dimethyl diselenide are illustrated in both 2D and 3D forms (Figs. 3–6).

These small, well-defined molecules preserve the essential bonding and electronic features of their amino acid counterparts, while being simple enough for high-level computational treatment.

By modeling DMDS and DMDSe under identical quantum chemical conditions, direct comparison of key parameters, including bond dissociation energies, acid–base equilibria, and nuclear magnetic resonance (NMR) shielding constants is possible. Such calculations not only reproduce known experimental trends—such as the lower BDE and greater polarizability of Se–Se bonds—but also provide detailed energetic and electronic explanations for these observations.

This approach highlights how computational chemistry can act as a predictive design tool for selenoprotein engineering. By linking fundamental electronic structure properties of model compounds to the folding, stability, and reactivity of real proteins, researchers can make informed decisions on when and where Sec incorporation

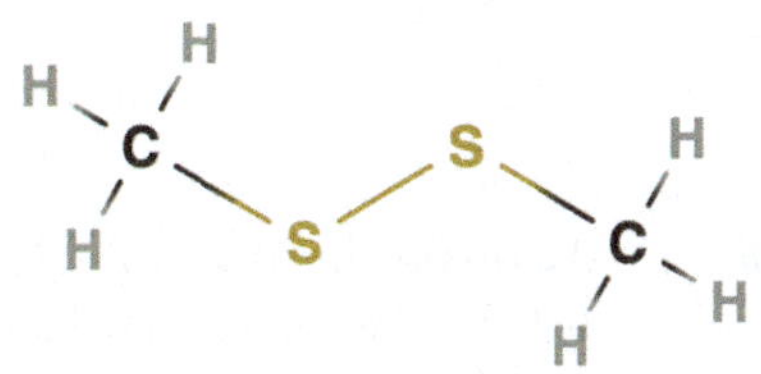

Fig. 3 Dimethyl disulfide—structural formula

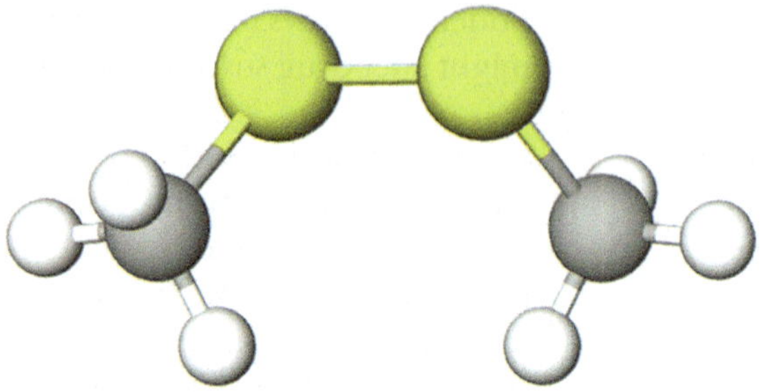

Fig. 4 Dimethyl disulfide—3D model

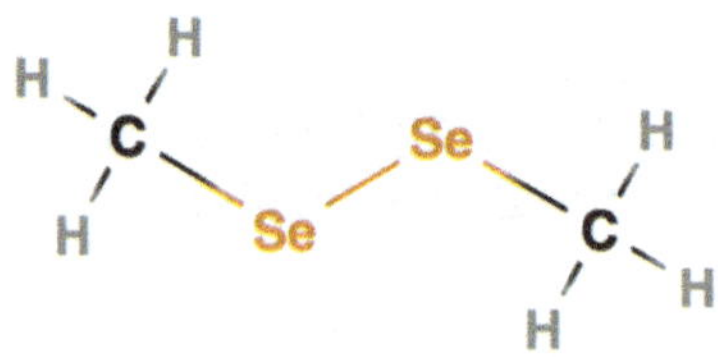

Fig. 5 Dimethyl diselenide—structural formula

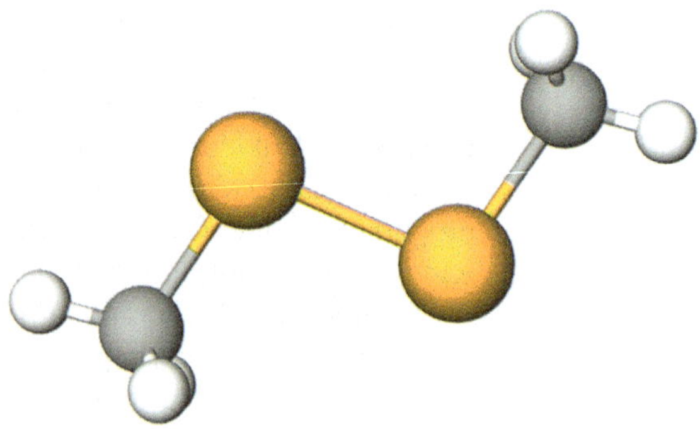

Fig. 6 Dimethyl diselenide—3D model

is likely to confer a functional advantage. The following section describes the computational methods used to carry out this analysis.

5 Computational Methods

5.1 Summary of Computational Parameters

5.2 Computational Setup

All quantum chemical calculations were performed using the ORCA 6.1.0 software package [24]. Geometry optimizations and vibrational frequency analyses were conducted at the B3LYP-D3(BJ)/def2-SVP level of theory, employing the RIJCOSX approximation [25] and the def2/J auxiliary basis set for Coulomb fitting. Solvent effects were included via the conductor-like polarizable continuum model (CPCM) with water as the solvent [26]. The full computational setup and parameters are summarized in Table 1.

For selenium-containing species, scalar-relativistic effects were incorporated using the zero-order regular approximation (ZORA) [27] with the segmented all-electron relativistically contracted (SARC) basis set for selenium [28]. Frequency calculations verified that all optimized structures corresponded to minima on the potential energy surface (no imaginary frequencies).

5.3 Bond Dissociation Energy Calculations

Bond dissociation energies were computed using:

$$\mathrm{BDE(X-H)=E(X)+E(H)-E(X-H)},$$

where EE denotes the single-point energy from the optimized geometry, with $E(H) = -0.5000$Eh representing the exact non-relativistic one-electron energy. Thermal and entropic corrections were included via the thermochemistry summaries from the frequency calculations.

5.4 pK_a Estimations

A thermodynamic cycle approach [30, 31] was used to estimate pK_a values. Gas-phase deprotonation free energies were corrected with solvation free energies obtained from CPCM single-point calculations. The aqueous proton free energy was taken as 1112.5 kJ mol^{-1}, following Kelly et al. [32], and converted to pK_a via:

Table 1
Computational parameters

Parameter	Details
Software	ORCA 6.1.0 [24]
Level of theory	B3LYP-D3(BJ)/def2-SVP; RIJCOSX approximation with def2/J auxiliary basis set
Relativistic treatment	ZORA scalar-relativistic Hamiltonian for Se [27] with SARC basis set [28]
Solvent model	CPCM (water) [26]
Integration/grid	TightSCF convergence; Grid5
Geometry optimization	Tight convergence; frequency calculations to confirm true minima (no imaginary frequencies)
Thermochemistry	Quasi-rigid rotor–harmonic oscillator (QRRHO) corrections at 298.15 K
NMR calculations	GIAO method [29]; ZORA–GIAO

$$\mathrm{pKa} = \frac{\Delta G^{\mathrm{o}}_{\mathrm{deprot}}}{2.303RT},$$

where $2.303RT = 5.7082.303RT = 5.708$ kJ mol-1 at 298.15 K. Standard-state corrections (1 atm → 1 M, +0.00303Eh) were applied to all deprotonation reactions.

5.5 NMR Shielding Calculations

Isotropic nuclear magnetic shielding constants (σ) were obtained from single-point GIAO calculations on the optimized geometries [29]. For selenium nuclei, ZORA–GIAO was employed to capture scalar-relativistic effects. Chemical shifts (δ) were computed relative to internal references (mean shielding within the same molecule), allowing qualitative comparisons between sulfur- and selenium-containing systems in the absence of absolute reference compound calculations.

5.6 Theoretical Redox Potential Calculation

Redox potential is calculated using the computational hydrogen electrode (CHE) model, which substitutes the proton + electron pair ($H^+ + e^-$) with half a hydrogen molecule ($\frac{1}{2}H_2$ (g,1 bar)) as the reference. This method effectively anchors calculated redox potentials to the standard hydrogen electrode (SHE) while avoiding explicit calculations of solvated protons, ensuring thermodynamic consistency and experimental comparability [33].

For the half-reaction:

$$[R-X-X-R]+2H^{+}+2e^{-}\rightarrow 2[R-XH].$$

(Corresponding chemical reaction:)

$$[R-X-X-R]+H_2(g,1\ \mathrm{bar})\rightleftharpoons 2[R-XH].$$

The standard reaction free energy is:

$$\Delta G^{\mathrm{o}}_{\mathrm{r}} = 2G_{\mathrm{aq}}(R-XH) - G_{\mathrm{aq}}(R-X-X-R) - G^{\mathrm{o}}\left(H_2,\ 1\ \mathrm{bar}\right).$$

And the standard potential relative to SHE at pH 0 is:

$$E^{\mathrm{o}} = -\frac{\Delta G^{\mathrm{o}}_{\mathrm{r}}}{nF}\left(n=2,\ F=96485\ \mathrm{C\ mol^{-1}}\right).$$

To align with physiological conditions (pH 7, biochemical standard, noted $E^{\mathrm{o}\prime}$), the Nernst-based Alberty transformation has been applied [34]:

$$E^{o'}(\text{pH}) = E^{o} - \frac{RT}{nF}\ln(10) \times m\text{pH}.$$

Using $m = 2$ (two protons transferred) and evaluating at 298.15 K yields:

$$E^{o'}(7) = E^{o} - 0.414 \text{ V}.$$

6 Computational Results

6.1 Bond Dissociation Energies (BDEs)

The computed homolytic BDEs clearly show that the Se–Se bond in DMDSe is weaker than the S–S bond in DMDS under identical computational conditions. The calculated bond dissociation energies are provided in Table 2.

The reduced BDE for Se–Se reflects the greater diffuseness and polarizability of selenium valence orbitals, which allow bond cleavage at lower energy cost. This property directly correlates with the faster redox cycling observed in selenoproteins containing diselenide bonds.

6.2 pK_a Values

The calculated pK_a values for the thiol and selenol moieties in the corresponding monomers (methanethiol and methaneselenol) confirm the experimental trend. The estimated pK_a values are summarized in Table 3.

Table 2
Calculated electronic BDEs (no thermal corrections) for S–H and Se–Se bonds

Bond	ΔE (Eh)	BDE (kJ mol^{-1})
S–H (CH_3SH)	0.140113	367.868
Se–H (CH_3SeH)	0.127932	335.885
Se–Se (DMSeSe)	0.077057	202.313

Table 3
Calculated pK_a values

Molecule	Functional group	pK_a (calc.)
Methanethiol	–SH	18.60
Methaneselenol	–SeH	9.84

Although the absolute pK_a values are overestimated compared to experiment ($CH_3SH \approx 10$–11; $CH_3SeH \approx 5$–6), the relative acidity trend is reproduced, with the selenol being substantially more acidic than the thiol. The computed ΔpK_a (Se–S) from these data is −8.76, corresponding to a free energy difference of −50.0 kJ mol^{-1} in deprotonation, consistent with the enhanced stabilization of selenolate anions in aqueous solution.

The absolute deviations from experiment are attributed primarily to limitations of the CPCM continuum model in describing anion solvation, as well as the modest basis set used for the final energies. However, relative acidities are more robust to systematic errors, and the observed ΔpK_a captures the expected qualitative behavior. Improvements in quantitative agreement would likely result from larger-basis single-point calculations in conjunction with more advanced solvation models such as SMD(Water) [35] or LR-CPCM, and/or the inclusion of explicit hydrogen-bonded water molecules in a cluster–continuum framework [36].

The selenol group is significantly more acidic, leading to higher concentrations of the deprotonated, nucleophilic species at physiological pH. This underlies the greater reactivity of Sec relative to Cys in enzymatic active sites.

6.3 NMR Shielding Constants

NMR calculations show distinct isotropic shielding values for sulfur and selenium atoms in the disulfide and diselenide forms. The computed isotropic shielding values are presented in Table 4.

The higher shielding observed for selenium is consistent with its greater electron density and relativistic effects. These differences can aid in experimental identification of Se-containing analogs in NMR spectra.

Table 4
Selected absolute isotropic shielding (σ)

Molecule	Atom	σ (ppm)
DMS	S(4)	1056.751
DMS	S(5)	1052.889
DMS	C(0)	264.937
DMS	C(6)	244.809
DMSeSe	Se(4)	2995.163
DMSeSe	Se(5)	2957.891
DMSeSe	C(0)	275.135
DMSeSe	C(6)	236.869

6.4 Redox Potential

Geometry optimizations and frequency calculations for:

- Oxidized species: DMDS (CH_3–S–S–CH_3) and DMDSe (CH_3–Se–Se–CH_3),
- Reduced species: 2 CH_3SH and 2 CH_3SeH,
- Hydrogen molecule: H_2 (g, 1 bar).

All calculations used B3LYP-D3(BJ)/def2-SVP (with ZORA for Se systems) and the CPCM(Water) solvent model [27]. Thermal corrections and Gibbs free energies (G^o) at 298.15 K were extracted from each output.

The computed potentials confirm that both redox reactions are favorable at physiological pH, but critically, the diselenide couple (Se–Se) exhibits a significantly lower $E^{o\prime}$ compared to disulfide (S–S), indicating it is thermodynamically harder to reduce, despite having a weaker homolytic bond dissociation energy. The calculated redox potentials for sulfur and selenium systems are reported in Table 5.

6.5 Interpretation in the Context of Protein Engineering

The combined computational findings—lower BDE of Se–Se versus S–S, reduced pK_a of selenols, sequestered NMR shifts, and now the calculated redox potentials—paint a compelling picture of why selenocysteine (Sec) substitution accelerates oxidative folding, stabilizes protein structures under reductive environments, and enhances catalytic turnover. Our calculations show that the diselenide couple (Se–Se → 2 R–SeH) has a biochemical standard potential E′°(pH 7) of approximately +0.234 V, significantly lower than the corresponding disulfide couple (S–S), which is at +0.735 V. This highlights that although the Se–Se bond is mechanically weaker (lower BDE), it is thermodynamically harder to reduce in the biological setting. This aligns with experimental observations suggesting that Sec-incorporated diselenide bonds resist reductive cleavage better than their cysteine–counterpart disulfides, enabling proteins to maintain folded states under reductive stress [19]. Altogether, weaker bond strength and higher acidity combine with a more negative redox potential to support rapid oxidative bond formation and robust redox resilience ideal for engineered redox-active proteins.

Table 5
Computed potentials

Couple	ΔG_{or} (Eh)	ΔG_{or} (kJ mol^{-1})	E^o (V, pH 0)	E'^o (V, pH 7)
S–S (DMDS → 2 CH_3SH)	−0.084485	−221.8	+1.149	+0.735
Se–Se (DMDSe → 2 CH_3SeH)	−0.047630	−125.1	+0.648	+0.234

7 Further Developments

While the present case study demonstrates the power of small-molecule modeling, further refinement could be achieved by:

- Incorporating explicit solvent molecules in molecular dynamics simulations to better capture hydrogen-bonding and hydration effects.
- Using larger peptide-based models to assess the influence of backbone constraints on S–S versus Se–Se bond energetics.
- Applying QM/MM methods to model Sec substitution in full protein structures, linking quantum-level reactivity to macromolecular folding and dynamics.

Such approaches would bridge the gap between gas-phase or continuum-solvent calculations and the complex environments of real proteins, enabling more predictive design of Sec-containing biocatalysts, therapeutics, and biomaterials.

8 Conclusion

Selenocysteine (Sec) exemplifies how a single-atom substitution can yield profound changes in biochemical reactivity, redox behavior, and catalytic potential. Its lower pK_a, greater nucleophilicity, and enhanced polarizability compared to cysteine underpin its central role in diverse selenoproteins, from antioxidant enzymes to metabolic regulators. These properties not only confer advantages in natural evolution but also present opportunities for rational protein engineering.

By integrating experimental evidence, theoretical models, and computational chemistry, this review has shown how Sec incorporation can accelerate folding, enhance structural stability under both oxidative and reductive stress, and expand catalytic capabilities. The computational case study using dimethyl disulfide (DMDS) and dimethyl diselenide (DMDSe) illustrates how model compounds can quantitatively capture the key physicochemical differences between S–S and Se–Se bonds. These differences—lower bond dissociation energies, reduced acid–base thresholds, and distinct NMR signatures—map directly onto experimentally observed protein behaviors.

Looking forward, advances in quantum chemical methods, QM/MM simulations, and explicit solvent modeling promise increasingly accurate predictions of Sec's effects in complex protein environments. Coupling these tools with genetic code expansion and site-specific incorporation strategies will enable precise control over Sec placement and function, accelerating the development of novel selenoproteins for therapeutic, industrial, and bioengineering applications.

In sum, Sec stands as a powerful, versatile design element—one whose unique chemistry continues to inspire both deeper fundamental understanding and innovative applications at the interface of biology and chemistry.

References

1. Chung CZ, Krahn N (2022) The selenocysteine toolbox: a guide to studying the 21st amino acid. Arch Biochem Biophys 726:109421. https://doi.org/10.1016/j.abb.2022.109421
2. Chaudiere J (2023) Biological and catalytic properties of selenoproteins. Int J Mol Sci 24(12):10109. https://doi.org/10.3390/ijms241210109
3. Kotini SB, Peske F, Rodnina MV (2015) Partitioning between recoding and termination at a stop codon–selenocysteine insertion sequence. Nucleic Acids Res 43(14):7136–7148. https://doi.org/10.1093/nar/gkv558
4. Aldag C, Bröcker MJ, Hohn MJ, Prat L, Hammond G, Plummer A, Söll D (2013) Rewiring translation for elongation factor Tu-dependent selenocysteine incorporation. Angew Chem Int Ed Engl 52(49):14441–14444. https://doi.org/10.1002/anie.201207567
5. Busch A, Will S, Backofen R (2005) SECISDesign: a server to design SECIS elements within the coding sequence. Bioinformatics 21(15):3312–3313. https://doi.org/10.1093/bioinformatics/bti507
6. Gladyshev VN, Hatfield DL (2009) Selenocysteine biosynthesis, selenoproteins, and selenoproteomes. In: Atkins JF, Gesteland RF (eds) Recoding: expansion of decoding rules enriches gene expression. Springer, New York, pp 3–27. https://doi.org/10.1007/978-0-387-89382-2_1editors
7. Mannes AM (2019) Dissecting the role of selenothiol- versus thiol-based catalysis using the model enzyme glutathione peroxidase 4 (GPx4). Dissertation, Ludwig Maximilian University of Munich
8. Marino SM, Roos G, Gladyshev VN (2013) Computational redox biology: methods and applications. In: Jacob C, Winyard PG (eds) Oxidative stress and redox regulation. Springer, Dordrecht, pp 187–211. https://doi.org/10.1007/978-94-007-5787-5_7editors.
9. Donovan J, Copeland PR (2010) Threading the needle: getting selenocysteine into proteins. Antioxid Redox Signal 12(7):881–892. https://doi.org/10.1089/ars.2009.2878
10. Phadnis PP, Mugesh G (2005) Internally stabilised selenocysteine derivatives: syntheses, 77Se NMR and biomimetic studies. Org Biomol Chem 3(17):3096–3102. https://doi.org/10.1039/b505299h
11. Huber RE, Criddle RS (1967) Comparison of the chemical properties of selenocysteine and selenocystine with their sulfur analogs. Arch Biochem Biophys 122(1):164–173. https://doi.org/10.1016/0003-9861(67)90222-2
12. Itoh Y, Sekine S, Yokoyama S (2015) Crystal structure of the full-length bacterial selenocysteine-specific elongation factor SelB. Nucleic Acids Res 43(18):9028–9038. https://doi.org/10.1093/nar/gkv833
13. Itoh Y, Sekine SI, Matsumoto E, Akasaka R, Takemoto C, Shirouzu M, Yokoyama S (2009) Structure of selenophosphate synthetase essential for selenium incorporation into proteins and RNAs. J Mol Biol 385(5):1456–1469. https://doi.org/10.1016/j.jmb.2008.08.042
14. Itoh Y, Chiba S, Sekine SI, Yokoyama S (2009) Crystal structure of human selenocysteine tRNA. Nucleic Acids Res 37(19):6259–6268. https://doi.org/10.1093/nar/gkp648
15. Weil-Ktorza O, Rege N, Lansky S, Shalev D, Shoham G, Weiss M, Metanis N (2019) Substitution of an internal disulfide bridge with a diselenide enhances both foldability and stability of human insulin. Chem Eur J 25(35):8513–8521. https://doi.org/10.1002/chem.201900892
16. Mousa R, Hidmi T, Pomyalov S, Lansky S, Khouri L, Shalev DE, Shoham G, Metanis N (2021) Diselenide crosslinks for enhanced and simplified oxidative protein folding. Commun Chem 4:30. https://doi.org/10.1038/s42004-021-00463-9
17. Metanis N, Hilvert D (2015) Harnessing selenocysteine reactivity for oxidative protein folding. Chem Sci 6(1):322–325. https://doi.org/10.1039/C4SC02379J
18. Arai K, Okumura M, Lee YH, Katayama H, Mizutani K, Lin Y, Park SY, Sawada K, Toyoda M, Hojo H, Inaba K (2023) Diselenide-bond replacement of the external disulfide bond of insulin increases its oligomerisation, leading to sustained activity. Commun Chem 6:258. https://doi.org/10.1038/s42004-023-01056-4
19. He Y, Hojo H (2024) Chemical synthesis and diselenide metathesis in selenocysteine-containing peptides. Met Res 4(3):14–28. https://

doi.org/10.11299/metallomicsresearch.MR202407 rev
20. Spezia R, Tournois G, Cartailler T, Tortajada J, Jeanvoine Y (2006) Co^{2+} binding cysteine and selenocysteine: a DFT study. J Phys Chem A 110(29):9727–9735. https://doi.org/10.1021/jp0614998
21. Messias A, Rauber A, Vuletich S, Zeida A, Semelak JA, Estrin DA (2024) Comparing thiol and selenol reactivity towards peroxynitrite by computer simulation. Redox Biochem Chem 9:100035. https://doi.org/10.1016/j.rbc.2024.100035
22. Dardashti RN, Dery L, Mousa R, Dery S, Reddy PS, Metanis N (2016) The chemistry of selenocysteine in proteins. In: Hatfield DL, Schweizer U, Tsuji PA, Gladyshev VN (eds) Selenium: its molecular biology and role in human health, 3rd ed. Springer, Cham, pp 73–83. https://doi.org/10.1007/978-3-319-41283-2_7
23. Saccoccia F, Angelucci F, Boumis G, Brunori M, Miele AE, Williams DL, Bellelli A (2012) On the mechanism and rate of gold incorporation into thiol-dependent flavoreductases. J Inorg Biochem 108:105–111. https://doi.org/10.1016/j.jinorgbio.2011.11.005
24. Neese F (2012) The ORCA program system. Wiley Interdiscip Rev Comput Mol Sci 2(1):73–78. https://doi.org/10.1002/wcms.81
25. Weigend F, Ahlrichs R (2005) Balanced basis sets of split valence, triple zeta valence and quadruple zeta valence quality for H to Rn: design and assessment of accuracy. Phys Chem Chem Phys 7(18):3297–3305. https://doi.org/10.1039/b508541a
26. Klamt A, Schüürmann G (1993) COSMO: a new approach to dielectric screening in solvents with explicit expressions for the screening energy and its gradient. J Chem Soc Perkin Trans 2(5):799–805. https://doi.org/10.1039/p29930000799
27. van Lenthe E, Baerends EJ, Snijders JG (1993) Relativistic regular two-component Hamiltonians. J Chem Phys 99(6):4597–4610. https://doi.org/10.1063/1.466059
28. Pantazis DA, Chen XY, Landis CR, Neese F (2008) All-electron scalar relativistic basis sets for third-row transition metal atoms. J Chem Theory Comput 4(6):908–919. https://doi.org/10.1021/ct800047t
29. Wolinski K, Hinton JF, Pulay P (1990) Efficient implementation of the gauge-independent atomic orbital method for NMR chemical shift calculations. J Am Chem Soc 112(23):8251–8260. https://doi.org/10.1021/ja00179a005
30. Liptak MD, Shields GC (2001) Accurate pKa calculations for carboxylic acids using complete basis set and Gaussian-n models combined with CPCM continuum solvation methods. J Am Chem Soc 123(31):7314–7319. https://doi.org/10.1021/ja010509j
31. Ho J, Coote ML (2010) pKa calculation of small molecules: a review of recent approaches. Aust J Chem 63(1):1–13. https://doi.org/10.1071/CH09320
32. Kelly CP, Cramer CJ, Truhlar DG (2006) Aqueous solvation free energies of ions and ion–water clusters based on an accurate value for the absolute aqueous solvation free energy of the proton. J Phys Chem B 110(32):16066–16081. https://doi.org/10.1021/jp063552y
33. Busch M, Ahlberg E, Laasonen K (2021) From absolute potentials to a generalised computational standard hydrogen electrode for aqueous and non-aqueous solvents. Phys Chem Chem Phys 23(20):11727–11737. https://doi.org/10.1039/d1cp01126a
34. Alberty RA (1992) Calculation of transformed thermodynamic properties of biochemical reactants at specified pH and pMg. Biophys Chem 43(3):239–254. https://doi.org/10.1016/0301-4622(92)87031-8
35. Marenich AV, Cramer CJ, Truhlar DG (2009) Universal solvation model based on solute electron density and on a continuum model of the solvent defined by the bulk dielectric constant and atomic surface tensions. J Phys Chem B 113(18):6378–6396. https://doi.org/10.1021/jp810292n
36. Cramer CJ, Truhlar DG (1999) Implicit solvation models: equilibria, structure, spectra, and dynamics. Chem Rev 99(8):2161–2200. https://doi.org/10.1021/cr960149m

Check for updates

Chapter 7

Quantification of the Redox State of Protein Disulfide Bonds

Diego Butera and Philip J. Hogg

Abstract

To elucidate how a functional disulfide bond controls protein activity, it is essential that the redox state of the bond in the population of protein molecules is known. A differential cysteine alkylation and mass spectrometry technique is described that affords precise quantification of protein disulfide bond redox state. The utility of the technique is demonstrated by quantifying the redox states of the eight disulfide bonds in human coagulation factor VIII.

Keywords Disulfide bond, Cysteine alkylation, Mass spectrometry, Redox state, FVIII

1 Introduction

Our understanding of how disulfides regulate protein function is limited by the molecular techniques available to detect changes in cysteines and disulfide bonds. Measurement of changes in disulfide redox state has mainly relied on the use of thiol probes such as biotin-conjugated maleimide (3-(*N*-maleimidylpropionyl)biocytin), followed by detection using streptavidin conjugated-peroxidase [1]. This provides a qualitative indication of a change in redox state and does not identify which cysteines are involved. Differential cysteine alkylation and mass spectrometry allows for quantitation of the fraction of a protein disulfide bond that is reduced or unformed in the protein preparation [2].

In the method described herein, reduced disulfide bond cysteines are alkylated with 2-iodo-N-phenylacetamide (^{12}C-IPA) and the oxidized disulfide bond cysteines with a stable carbon-13 isotope of IPA (^{13}C-IPA) following reduction with dithiothreitol (Fig. 1). The ratio of alkylation of peptides containing the disulfide bond cysteines labeled with ^{12}C-IPA compared to ^{13}C-IPA represents the fraction of the disulfide in the population that is in the unformed state. The advantage of this pair of cysteine alkylators is that they have the same chemical reactivity and the same structure, which enhances the reliability of alkylation, resolution of the

Philip J. Hogg (ed.), *Functional Disulphide Bonds*, Methods in Molecular Biology, vol. 3016,
https://doi.org/10.1007/978-1-0716-5158-2_7,

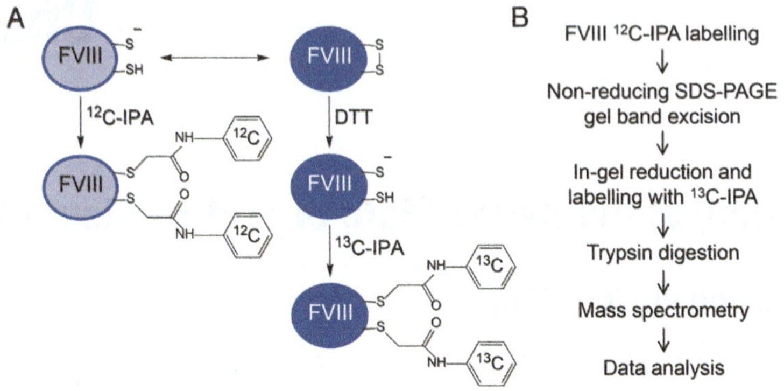

Fig. 1 Overview of the method for measuring the redox state of disulfide bonds in FVIII. (**A**) Differential cysteine alkylation for measuring the redox state of cysteines in FVIII by mass spectrometry. Unpaired cysteine thiols in purified FVIII are alkylated with ^{12}C-IPA and the disulfide-bonded cysteine thiols with ^{13}C-IPA following reduction with DTT. (**B**) The workflow described in this chapter, which consists of ^{12}C-IPA alkylation of FVIII thiols, in-gel reduction and alkylation with ^{13}C-IPA, trypsin digest, to mass spectrometry and data analysis

alkylated peptides by liquid chromatography and their detection by mass spectrometry [3–10]. A mass difference of 6 Da is the only change in a cysteine labeled with ^{12}C-IPA or ^{13}C-IPA.

To exemplify this technique, the redox state of the eight human coagulation factor VIII (FVIII) disulfide bonds is calculated [7]. Seventeen IPA-labeled peptides encompassing 11 of the 16 cysteines in FVIII were resolved, which covers all eight FVIII disulfide bonds.

2 Materials

As mass spectrometry analysis is highly sensitive, uttermost care should be taken to ensure clean working reagents and environment to avoid keratin contaminations that originate from skin, hair, and dust. All chemicals and reagents should be of LC-grade.

2.1 Processing of FVIII for Mass Spectrometry Analysis

1. Recombinant FVIII Advate (Takeda)
2. Phosphate buffered saline (PBS): 137 mM NaCl, 2.7 mM KCl, 10 mM Na_2HPO_4, 1.8 mM KH_2PO_4, pH 7.4
3. Freshly prepared 40 mM ^{12}C-IPA (Cambridge isotopes) in DMSO (*see* **Note 1**)
4. NuPAGE™ 4× LDS sample loading buffer (Thermo Scientific)
5. NuPAGE™ 4%–12% Bis-Tris gel (Thermo Scientific)
6. 1× NuPAGE™ MOPS running buffer (Thermo Scientific)
7. Coomassie Blue G solution (Sigma)
8. Gel destaining solution: 10% (v/v) methanol, 5% (v/v) acetic acid

9. 25 mM ammonium bicarbonate
10. 25 mM ammonium bicarbonate, 50% (v/v) acetonitrile
11. Freshly prepared 40 mM dithiothreitol in 25 mM ammonium bicarbonate
12. Freshly prepared 40 mM ^{13}C-IPA (Cambridge isotopes) in DMSO (*see* **Note 1**)
13. Protease stop buffer: 5% (v/v) formic acid
14. Peptide elution solution: 5% formic acid, 50% acetonitrile
15. Peptide reconstitution solution: 0.1% formic acid
16. Mass spectrometry grade Trypsin Gold (Promega) reconstituted to 1 mg/mL in 50 mM acetic acid and diluted 80-fold to 12.5 μg/mL in 25 mM ammonium bicarbonate
17. Speedvac SC210P1-15 (Thermo Scientific)

2.2 Mass Spectrometry Analysis

1. HPLC Buffer A: 95% LC-grade water, 5% acetonitrile, 0.1% formic acid
2. HPLC Buffer B: 20% LC-grade water, 80% acetonitrile, 0.1% formic acid
3. Analysis software of mass spectrometry data
4. Mascot Daemon (*see* **Note 2**)
5. Qual Browser from XCalibur (Thermo Scientific)

3 Methods

3.1 Alkylation of FVIII (^{12}C-IPA)

The following protocol describes ^{12}C-IPA alkylation of unpaired protein cysteines followed by gel fractionation of the protein.

1. Alkylate unpaired thiols in FVIII (5 μg) with ^{12}C-IPA at final concentration of 4 mM in 10% DMSO in PBS. Incubate at 22°C for 1 h in the dark.
2. Add 4× LDS sample loading buffer and heat samples at 60°C for 30 min.
3. Resolve samples on nonreducing 4%–12% Bis-Tris NuPAGE gel in 1× MOPS running buffer at 150 volts for 75 min.
4. Stain protein gel with 40 mL of Coomassie Blue G solution for 30 min. Remove Coomassie Blue G solution.
5. Destain gel in destaining solution overnight.
6. Excise the protein gel bands corresponding to FVIII (Fig. 2) with a clean scalpel.

3.2 In-Gel Reduction and Alkylation (DTT and ^{13}C-IPA)

The following protocol describes reduction of protein disulfide bonds followed by alkylation of the new unpaired cysteines with ^{13}C-IPA.

1. Dice gel band into small pieces of 1–2 mm^3. Place gel pieces into a clean tube.
2. Add 150 µL of 25 mM ammonium bicarbonate/50% acetonitrile to gel pieces.
3. Vortex gel pieces for 10 min at 22°C using Thermomixer (Eppendorf). Remove supernatant by pipetting.
4. Repeat Steps 3 and 4 for another two washes of the gel pieces.
5. Dry gel pieces by adding 100 µL of 100% acetonitrile. Remove acetonitrile by pipetting (*see* **Note 3**).
6. Cover gel pieces with 100 µL of 40 mM DTT in 25 mM ammonium bicarbonate. Incubate at 56°C for 30 min. Remove supernatant.
7. Wash three times with 150 µL of 25 mM ammonium bicarbonate/50% acetonitrile. Vortex each time for 10 min at 22°C.
8. Dry gel slices by adding 100 µL of 100% acetonitrile. Remove acetonitrile by pipetting (*see* **Note 4**).

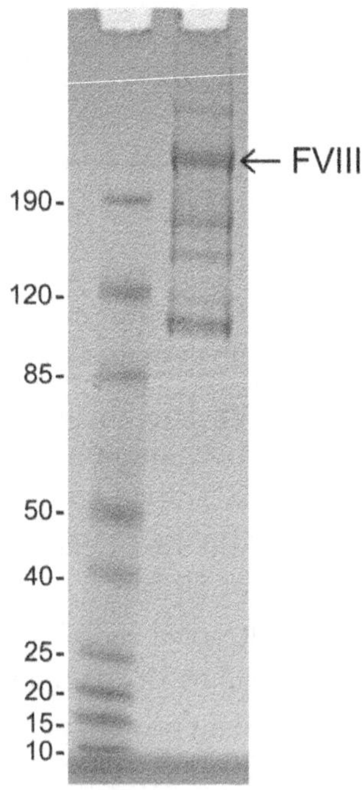

Fig. 2 Resolution of FVIII by nonreducing SDS-PAGE. Example of ^{12}C-IPA-labeled FVIII resolved on SDS-PAGE and stained with Coomassie blue. For reference, FVIII has a molecular mass of 340 kDa. Molecular mass standards are shown in the left-hand lane. The band corresponding to FVIII, indicated with the arrow, is excised for subsequent in-gel processing

9. Cover gel pieces with 100 μL of 4 mM ^{13}C-IPA in 25 mM ammonium bicarbonate and incubate in the dark for 1 h at 22°C. Remove supernatant.
10. Wash three times with 150 μL of 25 mM ammonium bicarbonate/50% acetonitrile. Vortex each time for 10 min at 22°C.
11. Dry gel slices by adding 100 μL of 100% acetonitrile. Remove acetonitrile by pipetting.
12. Dry gel pieces in Speedvac for 15–20 min to remove traces of acetonitrile.

3.3 In-Gel Digestion of FVIII

1. Add 40 μL of freshly diluted trypsin to gel pieces and incubate at 30°C overnight.
2. Stop digestion of protein by addition of 10 μL of 5% formic acid.
3. Transfer digest solution which contains peptides to fresh tubes.
4. Elute remaining peptides from gel slices by adding 30 μL of 5% formic acid/50% acetonitrile. Vortex for 20 min at 25°C.
5. Transfer supernatant to tubes containing peptides.
6. Dry peptides with Speedvac.

3.4 LC-MS/MS analysis

1. For LC MS/MS analysis, reconstitute peptides in 12 μL of 0.1% formic acid (*see* **Note 5**).
2. This protocol is based on the use of a Thermo Scientific Dionex Ultimate 3000. Peptides (3 μL) are injected and resolved on a 35 cm × 75 μm C18 reverse phase analytical column with integrated emitter using a gradient of 2%–35% acetonitrile over 22 min (Table 1) with a flow rate of 300 nL/min (*see* **Note 6**).
3. Peptides are ionized by electrospray ionization at +2.0 kV. Tandem mass spectrometry analysis is performed on a Q-Exactive Plus mass spectrometer (Thermo Scientific) using CID fragmentation. The scan range for full MS is set at 350–1750 m/z. The spectrum data type for full MS is profile mode, whereas for tandem MS is centroid mode.
4. Data-dependent acquisition method is used to obtain MS/MS spectra of the top five most abundant ions at any one point during the gradient.

3.5 Database Search and Quantification of Cys-Containing Peptides

1. Data files generated from LC-MS/MS are converted to mascot generic format (.mgf) using RawConverter available for free download at http://fields.scripps.edu/rawconv/ [11].
2. Data are analyzed using database searching tool Mascot Daemon (Matrix Science) against Swissprot database. Search parameters are as follows: precursor tolerance of 6 ppm and product ion tolerances of ± 0.4 Da, iodoacetanilide derivative

Table 1
Solvent gradient used in LC

Time (min)	% Buffer A	% Buffer B
0	97	3
20	97	3
22	**90**	**10**
44	**60**	**40**
47	5	95
50	5	95
50.1	95	5
60	95	5

Acetonitrile gradient for peptide elution is emphasized in bold

(Cys), iodoacetanilide ^{13}C derivative (Cys), and oxidation (Met) selected as variable modifications with full tryptic cleavage of up to three missed cleavages.

3. With the list of cysteine-containing peptides identified using Mascot, their monoisotopic mass/charge ratio (m/z) of +1, +2, and +3 are calculated using Peptide/Protein MS Product Utility Program MS-Product available via Protein Prospector Tools (http://prospector.ucsf.edu/prospector/cgi-bin/msform.cgi?form=msproduct). Cys-iodoacetanilide has a mass adduct of 133.05276, and Cys-^{13}C-iodoacetanilide has a mass adduct of 139.07289, which differ by 6 Da due to the six ^{13}C-carbon ring of IPA (Table 2). For example, to calculate the m/z ratios of the peptide SLPGLIGCHR from FVIII, the sequence should be SLPGLIGC(133.05276)HR and SLPGLIGC(139.07289)HR for ^{12}C-IPA- and ^{13}C-IPA-labeled cysteine, respectively. Their monoisotopic m/z ratios calculated using MS-Product from Protein Prospector is shown in Table 3. Values for m/z ratios should be at least four decimal places.

4. To calculate ion abundance of peptides, extracted ion chromatograms are generated using XCalibur Qual Browser software (v2.1.0; Thermo Scientific). Set autoprocessing parameters and display options for chromatogram and spectra as follows:

   (i) Make the chromatogram cell (top panel) active by selecting the pin  at the upper right corner. To define parameters for auto processing for chromatograms, choose **Display** and then **Ranges** from the drop-down menu. When the Ranges dialog box opens, select

Table 2
Monoisotopic mass related to amino acid modification

Amino acid	Modification	Monoisotopic mass*	Composition*
Cys	Iodoacetanilide	133.052764	H(7) C(8) N O
Cys	^{13}C-Iodoacetanilide	139.072893	H(7) C(2) ^{13}C(6) N O
Met	Oxidation	15.994915	O

*Values and composition are obtained from Unimod (www.unimod.org/)

Table 3
Monoisotopic m/z ratios for SLPGLIGCHR peptide from FVIII

		m/z			
Cys-label	Sequence	+1	+2	+3	+4
^{12}C-IPA	SLPGLIGC(133.05276)HR	1185.62	593.314	395.878	297.16
^{13}C-IPA	SLPGLIGC(139.072893)HR	1191.64	596.324	397.885	298.665

m/z values are calculated using MS-Product tool from Protein Prospector

Autoprocessing and click **Use user defined** for Mass Tolerance and set it as 6 ppm. Mass precision is set at 5 Decimals. For display options, choose **Display** and then **Display Options**. Select **Labels** and choose **Retention time** (2 Decimals) and **Area**.

(ii) Make the spectrum cell (bottom panel) active by selecting the pin at the upper right corner. To define parameters for auto processing for spectra, choose **Display** and then **Ranges** from the drop-down menu. When the Ranges dialog box opens, select **Autoprocessing** and click **Use user defined** for Mass Tolerance and set it as 6 ppm. Mass precision is set at 5 Decimals. For display options, choose **Display** and then **Display Options**. Select **Labels** and choose **Mass** (5 Decimals).

(iii) To save the settings, select **File** and then **Layout** from the drop-down menu, choose **Save as Default**. The above settings for autoprocessing and display will be used for subsequent quantification of MS peaks.

5. The m/z ratios (+2 and +3) calculated from MS-Product are used to search ions containing the corresponding m/z values. Select the chromatogram cell and choose **Display** then **Ranges** from the drop-down menu. When the Ranges dialog box opens, in the Plot properties, choose Scan filter

FTMS+p NSI Full mass [350.00–1750.00] from the drop-down menu. Enter the values for +1, +2, and +3 m/z in Ranges.

6. Click the automated peak detection function built into the program to calculate peak area (Fig. 3).
7. Automated peaks from Qual Browser are verified by the followings:
   (i) The retention time of Cys-iodoacetanilide or Cys-^{13}C-iodoacetanilide labeled peptides should be the same. ^{12}C-IPA and ^{13}C-IPA are identical in their chemical properties, and hence, the two species of peptides of the same sequence should resolve on the column in identical fashion (Fig. 3).

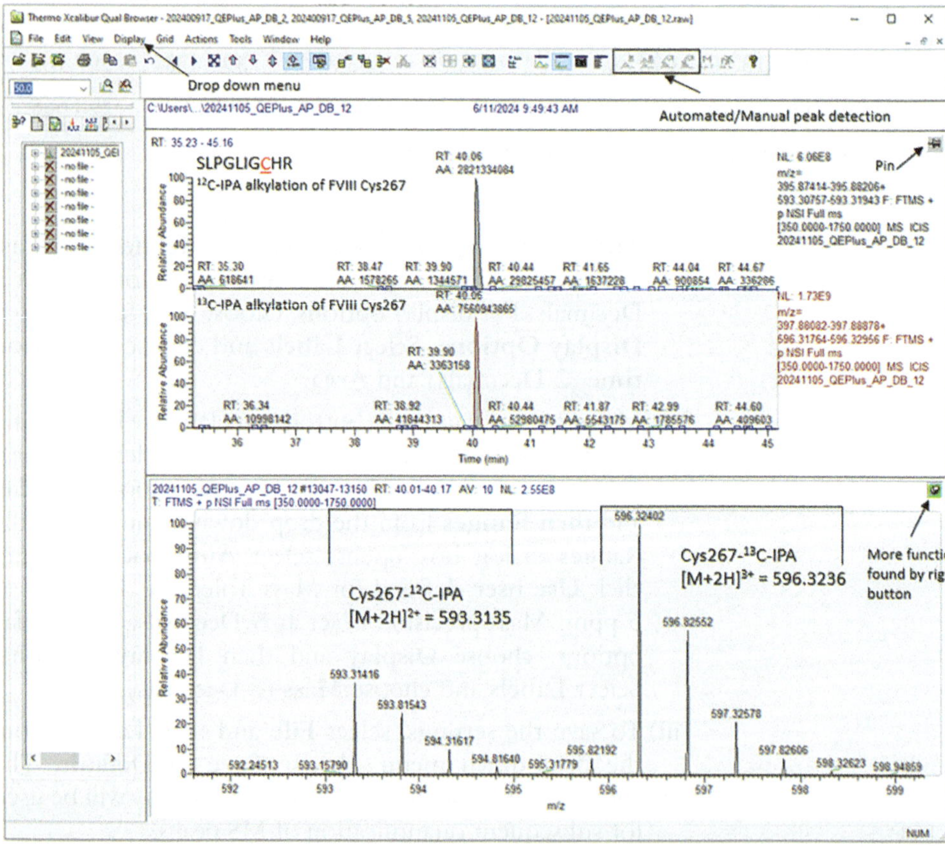

Fig. 3 Screenshot of the XCalibur Qual Browser Window showing automated peaks detected for the peptide SLPGLIGCHR labelled with ^{12}C-IPA or ^{13}C-IPA. The two peptide species have identical retention time and the automated area is used for quantification of their abundance. The automated peaks correspond to m/z of 593.3135 and 596.3236 for SLPGLIGC(133.05276)HR and SLPGLIGC(139.07289)HR, respectively, with an isotopic mass distribution of 0.5 confirming their m/z ratios as +2. Utilities such as drop-down menu, pin button and automated peak detection are shown by black arrows

(ii) The retention time of a peptide identified by Mascot database search and Qual Browser should be of similar value. The retention time from Mascot search is often used as a guide to identify peaks in the Qual Browser.

(iii) Check the isotopic distribution of the m/z ratios (+2 or +3 or both). For m/z ratio of +2, the isotopic peaks should differ by 0.5 (Fig. 3) and for m/z ratio of +3, the isotopic peaks should differ by 0.33.

8. The fraction of unformed disulfide bond is expressed by the area for Cys-iodoacetanilide labeled peptide over the sum of areas for Cys-iodoacetanilide and Cys-^{13}C-iodoacetanilide labeled peptides.

$$\text{Relative abundance of Reduced Cys} = \frac{\text{Area of Cys} - \text{iodoacetanilide}}{\begin{array}{c}\text{Sum of areas of Cys} - \text{iodoacetanilide}\\ \text{and}\\ \text{Cys} - 13\text{C} - \text{iodoacetanilide}\end{array}}$$

Table 4
List of cysteine containing peptides detected by mass spectrometry

Cys measured	Disulfide bond	*Peptide
198	172–198	DLNSGLIGALLVCR
267	267–348	SLPGLIGCHR SLPGLIGCHRK
348	267–348	VDSCPEEPQLR
547	547–573	CLTR
573	547–573	DLASGLIGPLLICYK
730	649–730	VSSCDK
1851	1851-1877	DEFDCK VQHHMAPTKDEFDCK †VQHHMAPTKDEFDCK
1877	1851–1877	DVHSGLIGPLLVCHTNTLNPAHGR
1922	1918–1922	APCNIQMEDPTFK †APCNIQMEDPTFK
2040	2040–2188	CQTPLGMASGHIR †CQTPLGMASGHIR
2345	2193–2345	MEVLGCEAQDLY †MEVLGCEAQDLY

* Cys was labeled with ^{12}C-iodoacetanilide (133.05276) or ^{13}C-iodoacetanilide (139.07289). Labeled Cys are underlined

† Met is oxidized

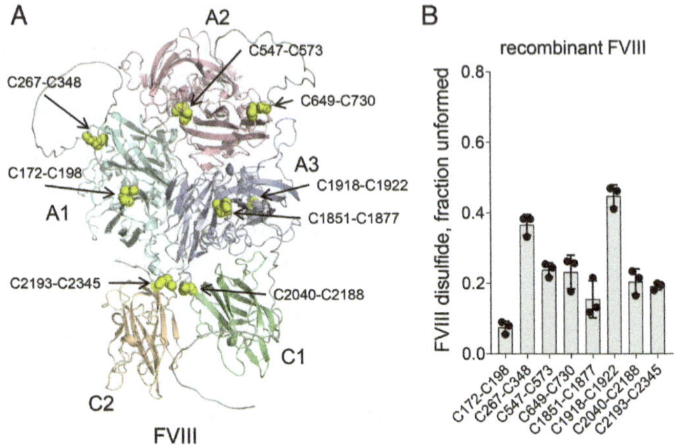

Fig. 4 Incidence of unformed disulfide bonds in human plasma FVIII. (**A**) Ribbon representation of the AlphaFold structure of human VIII lacking the unstructured B domain that does not contain disulfide bonds. The domains (A1, A2, A3, C1, C2) are color coded, and the eight disulfide bonds are shown as yellow spheres and residue numbers indicated (UniProt P00451 numbering). (**B**) Incidence of unformed disulfide bonds in recombinant FVIII ($n = 3$ preparations). For example, the A3 domain C1918-C1922 bond is unformed in ~50% of molecules of the FVIII samples. The bars and the errors are mean ± SD

9. Cysteine-containing peptides identified by differential cysteine alkylation and mass spectrometry are listed in Table 4. A total of 17 cysteine-containing peptides were identified by this methodology with 11 cysteines representing eight of the eight disulfide bonds in FVIII. (Fig. 4A).
10. The redox state of cysteines in FVIII is shown in Fig. 4B.

4 Notes

1. Thiol alkylators are light sensitive.
2. Mascot Server is not a freely available database search engine. MS-Fit from Protein Prospector (http://prospector.ucsf.edu/prospector/cgi-bin/msform.cgi?form=msfitupload) can be used as alternative.
3. Gel pieces should be completely dried out within 1–2 min. Can store dried gel slices after this step at –30°C.
4. It is not recommended to store gel pieces with proteins that contain free thiols generated from DTT reduction prior to alkylation with ^{13}C-IPA. Gel pieces can be stored at –30°C after alkylation with ^{13}C-IPA, washed and dehydrated with 100% acetonitrile.

5. Reconstitute in a larger volume if peak values from the intensity of the total ion chromatogram (TIC) is too high. Alternatively, dilute the reconstituted peptide 3-5 fold with 0.1% formic acid prior to injection to the nanoLC. The intensity of the TIC peaks should be in the range of 1E9–1E10.
6. Thermo LTQ Orbitrap Velos has also been used successfully to quantify peptides. If Velos is used, MS/MS spectra of the top 15 most abundant ions at any one point during the gradient are acquired.

References

1. Jiang XM, Fitzgerald M, Grant CM, Hogg PJ (1999) Redox control of exofacial protein thiols/disulfides by protein disulfide isomerase. J Biol Chem 274(4):2416–2423
2. Chiu J, Hogg PJ (2019) Allosteric disulfides: sophisticated molecular structures enabling flexible protein regulation. J Biol Chem 294(8): 2949–5908. https://doi.org/10.1074/jbc.REV118.005604
3. Pasquarello C, Sanchez JC, Hochstrasser DF, Corthals GL (2004) N-t-butyliodoacetamide and iodoacetanilide: two new cysteine alkylating reagents for relative quantitation of proteins. Rapid Commun Mass Spectrom 18(1):117–127. https://doi.org/10.1002/rcm.1286
4. Pijning AE, Butera D, Hogg PJ (2022) Not one, but many forms of thrombosis proteins. J Thromb Haemost 20(2):285–292. https://doi.org/10.1111/jth.15567
5. Butera D, Hogg PJ (2020) Fibrinogen function achieved through multiple covalent states. Nat Commun 11(1):5468. https://doi.org/10.1038/s41467-020-19295-7
6. Butera D, Passam F, Ju L, Cook KM, Woon H, Aponte-Santamaria C, Gardiner E, Davis AK, Murphy DA, Bronowska A, Luken BM, Baldauf C, Jackson S, Andrews R, Grater F, Hogg PJ (2018) Autoregulation of von Willebrand factor function by a disulfide bond switch. Sci Adv 4(2):eaaq1477. https://doi.org/10.1126/sciadv.aaq1477
7. Butera D, Pijning AE, Avery NG, Coxon CH, Metcalfe C, Mimoun A, Lacroix-Desmazes S, Spiegel PC, Hogg PJ (2025) Patient anti-FVIII drug antibodies bind preferentially to a subset of FVIII covalent states. Blood Adv 9(15):3706–3715. https://doi.org/10.1182/bloodadvances.2025016474
8. Butera D, Wang HJ, Woon HG, Zhao YC, Ju LA, Hogg PJ (2023) Mechano-covalent protection of coagulation factor VIII by von Willebrand factor. Blood Adv 7(10):2117–2128. https://doi.org/10.1182/bloodadvances.2022008650
9. Passam F, Chiu J, Ju L, Pijning A, Jahan Z, Mor-Cohen R, Yeheskel A, Kolsek K, Tharichen L, Aponte-Santamaria C, Grater F, Hogg PJ (2018) Mechano-redox control of integrin de-adhesion. Elife 7:e34843. https://doi.org/10.7554/eLife.34843
10. Pijning AE, Blyth MT, Coote ML, Passam F, Chiu J, Hogg PJ (2021) An alternate covalent form of platelet alphaIIbbeta3 integrin that resides in focal adhesions and has altered function. Blood 138(15):1359–1372. https://doi.org/10.1182/blood.2021012441
11. He L, Diedrich J, Chu YY, Yates 3rd JR (2015) Extracting accurate precursor information for tandem mass spectra by rawconverter. Anal Chem 87(22):11361–11367. https://doi.org/10.1021/acs.analchem.5b02721

Chapter 8

Quantification of Site-Specific Disulfide Bond Redox States in Proteins by Parallel Reaction Monitoring–Mass Spectrometry (PRM–MS)

Aizhen Yang, Yi Wu and Fengwu Chen

Abstract

Conventional non-targeted approaches using data-dependent acquisition (DDA) with isotopic labeling mass spectrometry (MS) have demonstrated limited effectiveness in characterizing site-specific disulfide bond redox states, primarily due to suboptimal coverage and inconsistent reproducibility. Here, we introduce a targeted approach employing differential cysteine alkylation coupled with parallel reaction monitoring (PRM)–MS to identify the redox states of specific disulfide bond sites in proteins, which significantly improves the coverage and reproducibility of mass spectrometry data.

Keywords Disulfide bond, Cysteine alkylation, Redox state, Parallel reaction monitoring (PRM), Mass spectrometry (MS)

1 Introduction

Disulfide bond formation and allosteric changes are critical for protein function [1]. Differential cysteine alkylation coupled with mass spectrometry (MS) provides robust approach to assess the redox status of disulfide bonds in proteins [2]. Currently, the non-targeted DDA-based isotopic labeling MS approach is commonly used for resolving and quantifying the redox states of targeted disulfide bonds [3–7]. However, in the non-targeted MS approaches, such as data-dependent acquisition (DDA), data-independent acquisition (DIA), and label-free MS, precursor ion selection is a stochastic process where ions with higher intensities are more likely to be chosen for subsequent MS/MS acquisition. Under such circumstances, valuable ions may not be selected for fragmentation when they co-elute with abundant high-intensity ions. Therefore, these methods are primarily aimed at obtaining a global quantitative proteomic profile and is commonly applied in discovery studies. However, when analyzing complex samples, it exhibits limitations such as poor reproducibility and low coverage. Whereas in targeted MS approaches, such as

Philip J. Hogg (ed.), *Functional Disulphide Bonds*, Methods in Molecular Biology, vol. 3016,
https://doi.org/10.1007/978-1-0716-5158-2_8,

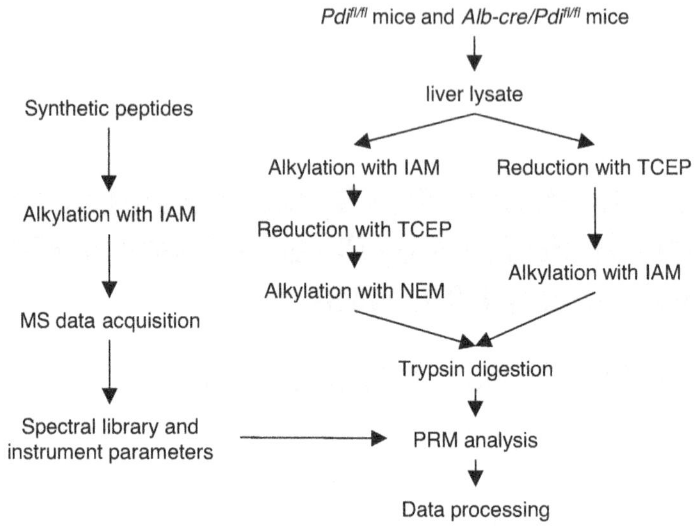

Fig. 1 Schematic of the PRM–MS workflow. (Reproduced from [12] with slight modifications)

parallel reaction monitoring (PRM), a quadrupole is used to select and isolate ions of a predetermined m/z value [8]. The selected precursor ions are then fragmented to generate fragment ions, which are analyzed at high resolution and mass accuracy in the Orbitrap, thereby enabling highly sensitive and specific quantification of known target molecules, and this approach is widely applied in validation studies, despite their lower throughput. PRM–MS has been successfully applied in various studies of posttranslational modifications (PTMs) [9–11].

Here, we employed the PRM–MS approach to identify the redox states of cysteines in the microsomal triglyceride transfer protein (MTP) within liver lysates from hepatocyte-specific protein disulfide isomerase (PDI) knockout ($Alb\text{-}cre/Pdi^{fl/fl}$) and control ($Pdi^{fl/fl}$) mice (Experimental workflow shown in Fig. 1) [12]. As a result, nine cysteines in MTP were quantified, except Cys298 was missed, which is located in a peptide containing only three amino acids acquired by trypsin digestion, too short to be detected by MS. This coverage represents a significant improvement over a preliminary TMT-based mass spectrometry result, which identified only three cysteines in MTP. Collectively, these findings demonstrate that PRM–MS has huge advantages in improving coverage and reproducibility and is a practical method for targeting the redox states of site-specific disulfide bonds in proteins.

2 Materials

2.1 Differential Alkylation of Cysteine in MTP from Mouse Liver

1. Target peptides were synthesized by GenScript Biotechnology Corporation (*see* **Note 1**).

2. 200 mM tetraethylammonium bromide (TEAB) buffer (Promega) (*see* **Note 2**).
3. Freshly prepared 500 mM dithiothreitol (DTT) in water (*see* **Note 3**).
4. Freshly prepared 150 mM iodoacetamide (IAM) in water, keep in dark (*see* **Note 4**).
5. Prechilled acetone.
6. Peptide reconstitution solution: 0.1% formic acid in water.
7. Lysis buffer: 1% SDS containing 1% protease inhibitor cocktail.
8. Freshly prepared 100 mM tris (2-carboxyethyl) phosphine (TCEP) solution in water.
9. Freshly prepared 150 mM *N*-ethylmaleimide (NEM) in ethanol.
10. Ultrasonic processor.
11. BCA Protein Assay Kit.

2.2 Trypsin Digestion and LC–MS/MS Analysis

1. Trypsin, MS grade: 1 mg/mL in 50 mM acetic acid.
2. HPLC solvent A: 0.1% formic acid and 2% acetonitrile in pure water.
3. HPLC solvent B: 0.1% formic acid and 90% acetonitrile in pure water.
4. EASY-nLC 1000 UPLC system.
5. Q Exactive™ Plus mass spectrometer.

2.3 Analysis Software for Mass Spectrometry Data

MaxQuant 1.6.15.0.

Skyline 20.1 software.

3 Methods

3.1 Generation of Spectral Library

1. Target peptides of mouse MTP were synthesized, with their sequences detailed in Fig. 2.
2. Dissolve 20 μg of synthetic target peptide in 200 μL of 200 mM TEAB buffer, Add DTT to a final concentration of 5 mM and incubate for 1 h at 37°C.
3. Add iodoacetamide (IAM) solution to the peptide solution to a final concentration of 15 mM. Incubate at room temperature for 1 h in the dark (*see* **Note 5**).

A

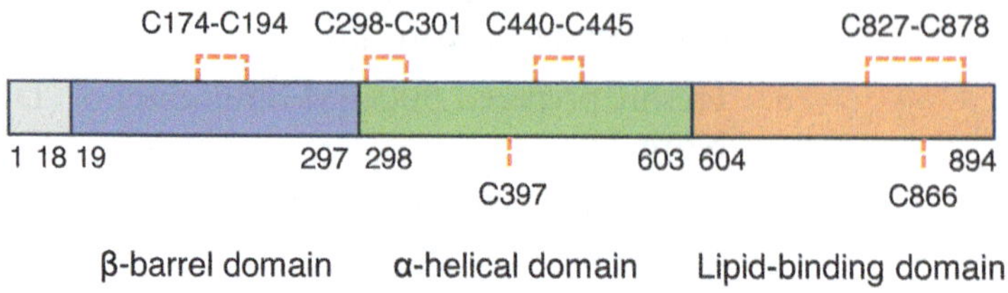

B

Cysteine	Synthetic peptides
Control	EFYSYENEPVGIENLK
C174	GLASLFQMQLSSGTTNEVDISGD**C**K
C194	ALDT**C**K
C298	V**C**K (unsynthesized)
C301	G**C**PSLAEHWK
C397	FLYA**C**GFATHPDEELLR
C440-C445	L**C**QNEG**C**K
C827	AETEAGLEFISTVQFSQYPFLV**C**MQMDK
C866, C878	ESLVAG**C**ELPLHQENSEM**C**NVVFPPQPESDNSGGWF

Fig. 2 Positions (**A**) and sequence (**B**) of the eight target peptides in the MTP protein. (Reproduced from [12] with slight modification)

4. Precipitate the peptides by adding four volumes of precooled acetone and incubate at −20°C for 2 h. Centrifuge at 4500× *g* for 5 min at 4°C (*see* **Note 6**), and discard the supernatant. Wash the pellet three times with precooled acetone and vacuum-dry it. Reconstitute the peptides in 0.1% formic acid at a concentration of 0.5 μg/μL and were ready for MS analysis.
5. Analyze a mix of all IAM-labeled standard peptides on a Q Exactive™ Plus mass spectrometer to determine elution order. Search the raw files using a quantitative proteomics software package (MaxQuant 1.6.15.0), and import the processed data into Skyline 20.1 software to construct a spectral library (Table 1).

3.2 Differential Alkylation of Cysteine in MTP from Mouse Liver

1. Livers from three pairs of *Pdi*$^{fl/fl}$ and *Alb-cre/Pdi*$^{fl/fl}$ mice were ground in liquid nitrogen and lysed with four volumes of lysis buffer (1% SDS containing 1% protease inhibitor cocktail).
2. Sonicate the samples three times on ice using a Fisher Scientific 550 Sonic Dismembrator (probe diameter of 3.2 mm, 40% amplitude, and 10 s pulse on and 10 s off).

Table 1

Spectral library of IAM-alkylated synthesized peptides. Eight peptides of MTP were synthesized and alkylated with IAM, and subsequently analyzed using DDA mode of mass spectrometry. The raw files were searched by using a quantitative proteomics software package (MaxQuant 1.6.15.0), and the processed data were imported into Skyline 20.1 software to build a spectral library as follows. *NCE* normalized collision energy; *CS* charge state

Cysteine	Alkylated synthetic peptides	Mass [m/z]	CS [z]	Start [min]	End [min]	NCE
None (control)	EFYSYENEPVGIENLK	965.957088	2	18.181	24.181	27
C174	GLASLFQMQLSSGTTNEVDISGDC[IAM]K	886.751037	3	19.844	25.844	27
C194	ALDTC[IAM]K	354.173264	2	3.4986	9.4986	27
C301	GC[IAM]PSLAEHWK	395.522085	3	8.596	14.596	27
C397	FLYAC[IAM]GFATHPDEELLR	680.327892	3	24.941	30.941	27
C440–C445	LC[IAM]QNEGC[IAM]K	504.715501	2	2.3451	8.3451	27
C827	AETEAGLEFISTVQFSQYPFLVC[IAM]MQMDK	1090.511629	3	23.289	29.289	27
C866, C878	ESLVAGC[IAM]ELPLHQENSEMC[IAM]N VVFPPQPESDNSGGWF	1354.264394	3	19.4	25.4	27

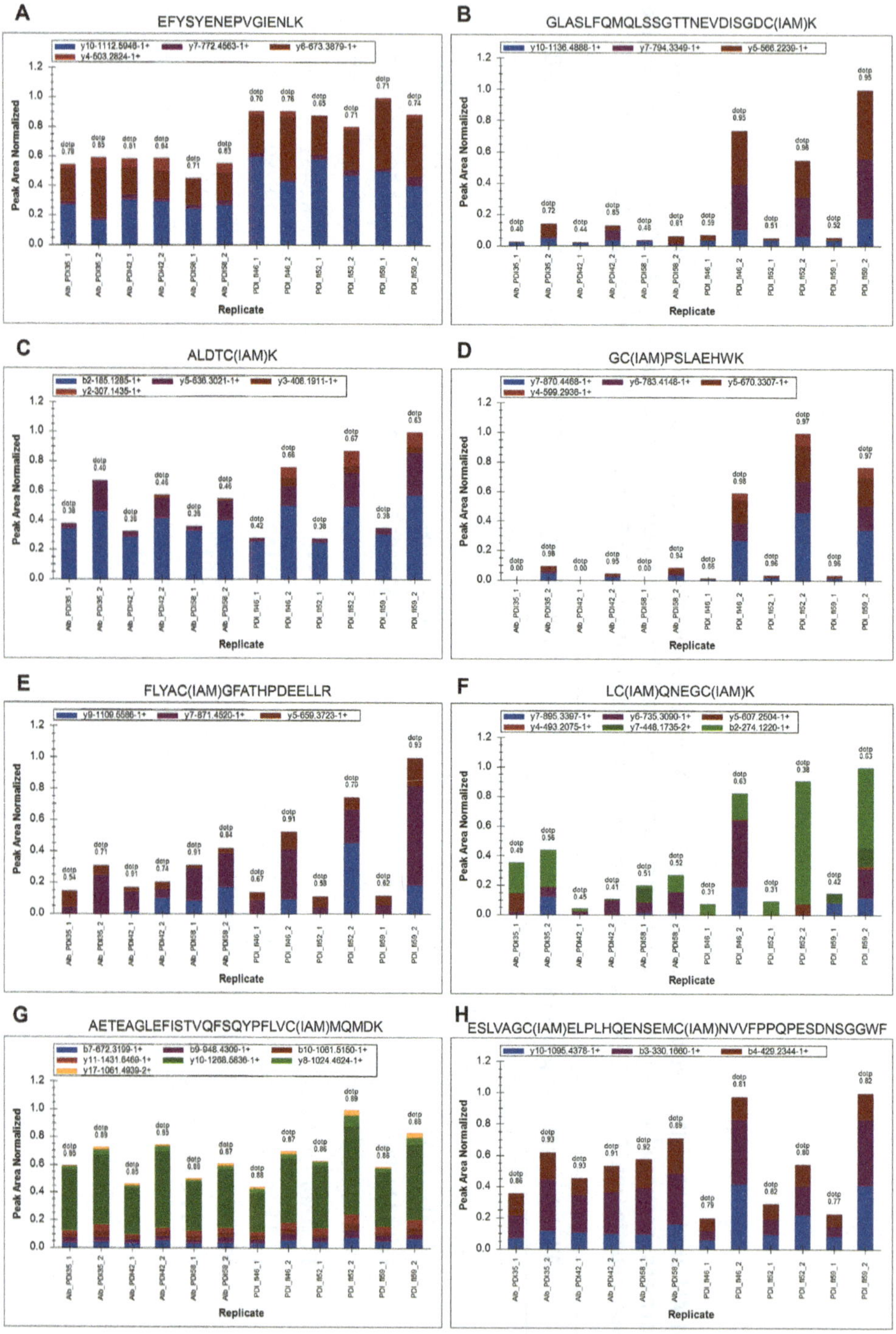

Fig. 3 Peak area distribution of the qualified fragment ions of targeted peptides. Liver lysates of *Pdi*$^{fl/fl}$ and *Alb-cre/Pdi*$^{fl/fl}$ mice were alkylated with IAM and analyzed using PRM–MS, Eight peptides of MTP, namely **(A)** EFYSYENEPVGIENLK, **(B)** GLASLFQMQLSSGTTNEVDISGDC[IAM]K, **(C)** ALDTC[IAM]K, **(D)** GC[IAM]PSLAEHWK, **(E)** FLYAC[IAM]GFATHPDEELLR, **(F)** LC[IAM]QNEGC[IAM]K, **(G)** AETEAGLEFISTVQFSQYPFLVC[IAM]MQMDK, and **(H)** ESLVAGC[IAM]ELPLHQENSEMC[IAM]NVVFPPQPESDNSGGWF, were systematically detected by spectral

3. Remove debris by centrifugation at 12,000 ×g for 10 min at 4°C.
4. Collect the supernatant, and determinethe protein concentration using a BCA Protein Assay Kit.
5. Divide equal amounts of total proteins from each sample into two aliquots. Incubate one aliquot (group 1) with 15 mM of IAM for 1 h at room temperature in the dark to alkylate the native reduced cysteines that did not form disulfide bonds, Then precipitate the proteins with four volumes of precooled acetone at −20°C for 2 h.
6. Centrifuge at 4500× g for 5 min at 4°C, discard the supernatant.
7. Wash the protein pellet three times with precooled acetone and redissolve in lysis buffer.
8. Treat the sample with 10 mM of TCEP for 1 h at 37°C to fully reduce the oxidized disulfide bonds.
9. Alkylate the newly reduced cysteines with 15 mM NEM for 1 h at room temperature in the dark (*see* **Note 7**).
10. Treat the other aliquot (group 2) with 10 mM TCEP for 1 h at 37°C to reduce all the disulfide bonds, Then alkylate the cysteines with 15 mM IAM for 1 h at room temperature in dark.

3.3 Trypsin Digestion and LC–MS/MS Analysis

1. For the final two samples (group 1 and group 2), add one volume of prechilled acetone and vortex. Then add four volumes of prechilled acetone and incubated at −20°C for 2 h.
2. After centrifugation at 4500 ×g for 5 min, discard the supernatant, and wash the pellet three times with prechilled acetone.
3. Air-dry the pellets to remove acetone.
4. Resuspend the air-dried pellet in 200 mM TEAB via ultrasonication.
5. Add trypsin at a 1:50 (trypsin-to-protein, m/m) ratio and digest overnight .
6. Dissolve the digested peptides in solvent A (0.1% formic acid and 2% acetonitrile in pure water) and separate them using an

Fig. 3 (continued) matching using a previously prepared reference library. The traces of the qualified b- and y-type fragment ions of each peptide were extracted, and the peak areas were determined. The abundance of endogenous peptides was determined by summing the peak areas of fragment ions of the corresponding peptide. In the x-axis, the extension 1 after the mouse's name refers to partially IAM alkylated peptide (reduced Cys only), and the extension 2 refers to fully IAM alkylated peptide (total Cys). (Reproduced from [12])

EASY-nLC 1000 UPLC system. The liquid gradient was comprised of an increase from 8% to 30% solvent B (0.1% formic acid in 90% acetonitrile) over 16 min, 30%–40% in 6 min, and up to 80% in 4 min, then holding at 80% for the last 4 min, at a constant flow rate of 500 nL/min.

7. Inject the separated peptides into a nanospray ionization (NSI) source for ionization and analyze them on a Q Exactive™ Plus mass spectrometer with the previously prepared acquisition method (spectral library) loaded. The electrospray voltage was set to 2.1 kV. Precursor peptides and the secondary fragment ions were detected and analyzed by a high-resolution Orbitrap mass spectrometer. The range for full MS scan was set to 350–1405 m/z. The precursor peptides were detected in the Orbitrap at a resolution of 70,000. Peptides were then selected for MS/MS using NCE set to 27/35, and the fragments were detected in the Orbitrap at a resolution of 17,500. Automatic gain control (AGC) was set to 3E6 for full MS and 1E5 for MS/MS. The maximum injection time was set at 50 ms for full MS and 200 ms for MS/MS. The isolation window for MS/MS was set at 1.6 m/z.

3.4 Data Analysis

1. The MS data were processed using the Skyline software. Peptide settings: enzyme was set as trypsin [KR/P], Max missed cleavage was set as 0. The peptide length was set as 7–25, and the variable modification was set as carbamidomethyl on Cys. Transition settings: precursor charges were set as 2, 3, ion charges were set as 1, and ion types were set as b, y. The product ions were set as from ion 3 to last ion, the ion match tolerance was set as 0.02 Da. The identities of the peptides (spectral matching) were confirmed by post-acquisition data processing. The traces of the qualified fragment ions of the peptides were extracted, and the peak areas were determined and used for quantifying peptide abundance (Fig. 3 and Tables 2) (*see* **Note 8**). All targeted peptides were identified by this methodology.

2. The percentage of reduced form of each cysteine was calculated by comparing the peak area of the native alkylated peptide (area of group 1, A_1) to that of the fully alkylated peptide (area of group 2, A_2) (*see* **Note 9**). The peak area of the cysteine-free peptide ($A_{control}$) was used as an internal standard to normalize the quantification of target peptides.

$$\text{Percentage of reduced cysteine} = \left(A_1 / A_{\text{contron1}}\right) / \left(A_2 / A_{\text{control2}}\right).$$

3. The redox state of cysteines in MTP from *Pdi*$^{fl/fl}$ and *Alb-cre/Pdi*$^{fl/fl}$ mouse liver is shown in Fig. 4 (*see* **Notes 10 and 11**).

Table 2

Mass spectrometry peak area of eight targeted peptides in MTP. Liver lysates from three pairs of PDI-knockout (*Alb-cre/Pdi^fl/fl^*) and control (*Pdi^fl/fl^*) mice were differentially cysteine-alkylated as described in Sect. 3, and the peak areas of eight target peptides in MTP protein were quantified by PRM–MS

	*Pdi^fl/fl^*						*Alb-cre/Pdi^fl/fl^*					
	52#		59#		35#		42#		58#			
Targeted peptides in MTP	Native peptide	Fully reduced	Native peptide	Fully reduced	Native peptide	Fully reduced	Native peptide	Fully reduced	Native peptide	Fully reduced	Native peptide	Fully reduced
EFYSYENEPVGIENLK	3.52E+05	3.53E+05	3.41E+05	3.11E+05	3.88E+05	3.43E+05	2.12E+05	2.29E+05	2.27E+05	2.28E+05	1.75E+05	2.14E+05
GLASLFQMQLSSGT TNEVDISGDC(IAM)K	3.09E+05	3.23E+06	2.25E+05	2.40E+06	2.47E+05	4.36E+06	1.17E+05	6.25E+05	1.15E+05	5.85E+05	1.64E+05	2.91E+05
ALDTC(IAM)K	6.09E+06	1.64E+07	6.03E+06	1.88E+07	7.52E+06	2.16E+07	8.22E+06	1.45E+07	7.06E+06	1.24E+07	7.81E+06	1.19E+07
GC(IAM)PSLAEHWK	5.08E+05	1.79E+07	1.10E+06	3.03E+07	1.11E+06	2.33E+07	#N/A	2.97E+06	#N/A	1.47E+06	#N/A	2.61E+06
FLYAC(IAM)GF ATHPDEELLR	1.40E+04	5.21E+04	1.12E+04	7.40E+04	1.18E+04	9.95E+04	1.47E+04	3.08E+04	1.70E+04	2.03E+04	3.09E+04	4.18E+04
LC(IAM)QNEGC(IAM)K	2.54E+04	2.70E+05	3.06E+04	2.97E+05	4.81E+04	3.27E+05	1.16E+05	1.44E+05	1.39E+04	3.52E+04	6.64E+04	8.94E+04
AETEAGLEFISTV QFSQYPFLVC(IAM) MQMDK	2.38E+06	3.76E+06	3.40E+06	5.34E+06	3.16E+06	4.47E+06	3.20E+06	3.91E+06	2.50E+06	4.02E+06	2.71E+06	3.29E+06
ESLVAGC(IAM)ELP LHQENSEMC(IAM) NVVFPPQPESD NSGGWF	2.43E+05	1.18E+06	3.51E+05	6.55E+05	2.76E+05	1.21E+06	4.30E+05	7.51E+05	5.49E+05	6.46E+05	6.96E+05	8.61E+05

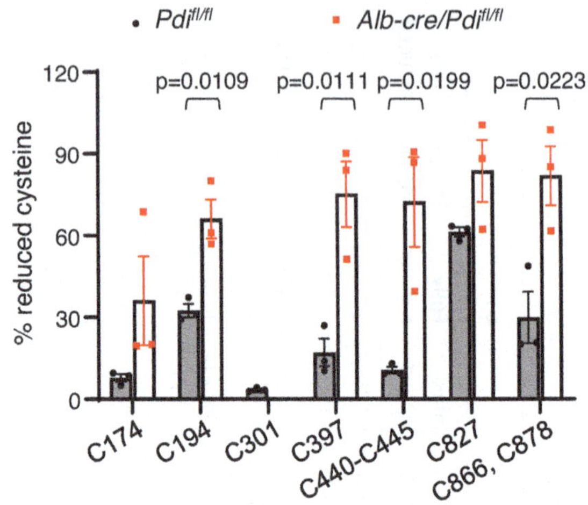

Fig. 4 PDI deficiency increases the levels of reduced cysteines in MTP. Peak area of eight target peptides in MTP protein was quantified by PRM–MS ($n = 3$ biological replicates in each group), and the percentage of reduced form of each cysteine was calculated by comparing the peak area of the native alkylated peptide to that of the fully alkylated peptide. (Reproduced from [12] with slight modification)

4 Notes

1. To generate a spectral library, we synthesized seven peptides covering nine of the ten cysteines in mature mouse MTP protein, and a cysteine-free peptide that was most abundant in liver lysate detectable by a preliminary DDA–MS experiment. Because the peptide with Cys298 acquired by trypsin digestion contained only three amino acids that were too short to be detected by MS, this peptide was not included.
2. All chemicals and reagents should be of LC grade. All operations must be conducted with extra care to prevent unnecessary contamination, and laboratory masks and protective hoods must be worn.
3. TCEP, DTT, IAM, and NEM solutions need to be made fresh immediately prior to use.
4. IAM and NEM are light-sensitive, and related storage and experimental operations should be kept in the dark.
5. We chose to use IAM instead of iodo-TMT labeling for peptides mainly because the spectra of IAM-labeled peptides are generally better than those of iodo-TMT-labeled ones (due to the larger molecular weight of iodo-TMT), and IAM is also more cost-effective.
6. This step is best performed using a swing-bucket centrifuge rather than a fixed-angle centrifuge.

7. The continued use of NEM alkylation serves both to reduce the non-specific labeling from residual IAM and to prevent disulfide bond rearrangement, eliminating background interference.
8. There are two peptides containing two Cys residues, but the case where one Cys is alkylated by IAM and the other Cys is alkylated by NEM was not detected due to low abundance, resulting in no quantitative value. "#N/A" indicates low abundance with no quantitative value.
9. The peak areas of the same peptide labeled with NEM and IAM cannot be directly summed, as they may exhibit ionization efficiency variation.
10. In the absence of PDI, the reduced ratio of three disulfide bonds in MTP, Cys174–Cys194, Cys440–Cys445, and Cys827–Cys878, significantly increased, suggesting that PDI catalyzes the formation of at least these three disulfide bonds in MTP.
11. The reduced ratio of Cys301 in the *Alb-cre/Pdi$^{fl/fl}$* lysate was decreased, which was probably due to the low amount of the peptide in the *Pdi$^{fl/fl}$* and *Alb-cre/Pdi$^{fl/fl}$* liver lysates, which is near the detection threshold boundary and may cause edge effects.

References

1. Cook KM, Hogg PJ (2013) Post-translational control of protein function by disulfide bond cleavage. Antioxid Redox Signal 18(15):1987–2015. https://doi.org/10.1089/ars.2012.4807
2. Chiu J (2019) Quantification of the redox state of protein disulphide bonds. In: Hogg P (ed) Functional Disulphide Bonds. Methods in Molecular Biology, vol 1967. Humana Press, New York, NY, U.S.A. https://doi.org/10.1007/978-1-4939-9187-7_4
3. Passam F et al (2018) Mechano-redox control of integrin de-adhesion. eLife 7:e34843. https://doi.org/10.7554/eLife.34843
4. Pijning AE et al (2021) An alternate covalent form of platelet αIIbβ3 integrin that resides in focal adhesions and has altered function. Blood 138(15):1359–1372. https://doi.org/10.1182/blood.2021012441
5. Lv KY et al (2024) Protein disulfide isomerase cleaves allosteric disulfides in histidine-rich glycoprotein to regulate thrombosis. Nat Commun 15(1):3129. https://doi.org/10.1038/s41467-024-47493-0
6. Dupuy A et al (2023) Mechano-redox control of Mac-1 de-adhesion by PDI promotes directional movement under flow. Circ Res 132(9):E151–E168. https://doi.org/10.1161/Circresaha.122.321926
7. Butera D, Hogg PJ (2020) Fibrinogen function achieved through multiple covalent states. Nat Commun 11(1):5468. https://doi.org/10.1038/s41467-020-19295-7
8. Bezstarosti K, van der Wal L, Doff WAS, Demmers JAA (2024) Parallel reaction monitoring targeted mass spectrometry as a fast and sensitive alternative to antibody-based protein detection. Front Anal Sci 4:1397810. https://doi.org/10.3389/frans.2024.1397810
9. Gallien S, Domon B (2015) Detection and quantification of proteins in clinical samples using high resolution mass spectrometry. Methods 81:15–23. https://doi.org/10.1016/j.ymeth.2015.03.015
10. Bourmaud A, Gallien S, Domon B (2016) Parallel reaction monitoring using quadrupole-Orbitrap mass spectrometer: principle and applications. Proteomics 16(15–16):2146–2159. https://doi.org/10.1002/pmic.201500543
11. Erickson BK et al (2017) A Strategy to combine sample multiplexing with targeted proteomics assays for high-throughput protein signature characterization. Mol Cell 65(2):361–370. https://doi.org/10.1016/j.molcel.2016.12.005
12. Chen FW et al (2025) Differential transport pathways of saturated and unsaturated fatty acid esters in male mouse hepatocytes. Nat Commun 16(1):1344. https://doi.org/10.1038/s41467-025-56620-4

Chapter 9

Identification of Target Disulfide Bonds Using Mechanism-Based Kinetic Trapping Combined with Differential Cysteine Labeling

Jinyu Wang , Yaofeng Li and Chao Fang

Abstract

The formation and cleavage of disulfide bonds, tightly controlled processes, are critical to maintain protein structures and functions in many physiological and pathological scenarios. Thiol isomerases catalyze the rearrangement of disulfide bonds and therefore are important in the regulation of protein functions. ERp72, a member in the thiol isomerase family, directly activates integrin Mac-1 by cleaving functional disulfide bonds. Using a mechanism-based kinetic trapping mutant of ERp72, stepwise labeling of cysteines by different alkylating agents allows the identification of specific disulfide bonds targeted by ERp72. The method described here may also be applied to other oxidoreductases to map potential target disulfides.

Keywords Endoplasmic reticulum protein 72 (ERp72), Disulfide bonds, Cysteine alkylation, Mass spectrometry, Differential labeling

1 Introduction

Disulfide bonds are crucial covalent linkages formed between sulfur atoms of two cysteine residues in proteins, playing a fundamental role in stabilizing protein structures to ensure proper functionality. Their formation is dynamically regulated throughout the life of the protein by surrounding redox states with the assistance of chaperones and oxidoreductases, such as protein disulfide isomerase (PDI) family [1]. Mass spectrometry has been applied to identify disulfide bonds through labeling thiol groups by alkylating agents, such as *N*-ethylmaleimide (NEM) or iodoacetamide (IAM) [2]. Differential labeling by NEM and IAM allows for the determination of which specific cysteine pair is the target of oxidoreductases [3]. Here, we described a strategy to use a mechanism-based kinetic trapping mutant [4] of an oxidoreductase combined with mass spectrometry to identify its target disulfide bonds. ERp72 is a member in the PDI family and has three catalytic domains with each containing a Cys–Gly–His–Cys

Philip J. Hogg (ed.), *Functional Disulphide Bonds*, Methods in Molecular Biology, vol. 3016,
https://doi.org/10.1007/978-1-0716-5158-2_9,

motif (Fig. 1A) [5]. To identify its target disulfide bonds, we mutated the C-terminal Cys in all three active motifs to Ala (ERp72-CA), which forms a stable disulfide bond with its substrate that allows for the stepwise labeling [3] (Fig. 1B).

Mac-1 (integrin αMβ2), composed of CD11b and CD18 subunits, is essential for neutrophil recruitment [6]. ERp72 directly regulates Mac-1 conformation and activity by catalyzing disulfide cleavage on CD11b [3]. In this chapter, the CD11b subunit of Mac-1 is used as an example to demonstrate how target disulfide bonds are mapped for ERp72. The trapping mutant ERp72-CA formed a stable disulfide bond with CD11b (Fig. 2), leading to

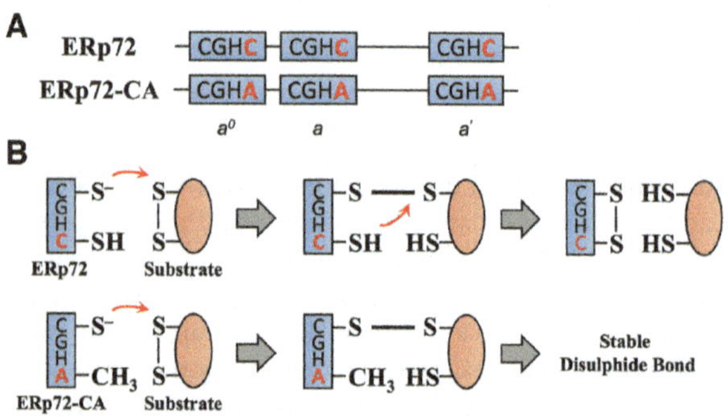

Fig. 1 Mechanism-based kinetic trapping mutant of ERp72. (**A**) Catalytic motifs of ERp72 are mutated from CGHC to CGHA. (**B**) ERp72-CA mutant forms a stable disulfide bond with its substrate that allows for stepwise labeling with different alkylation reagents

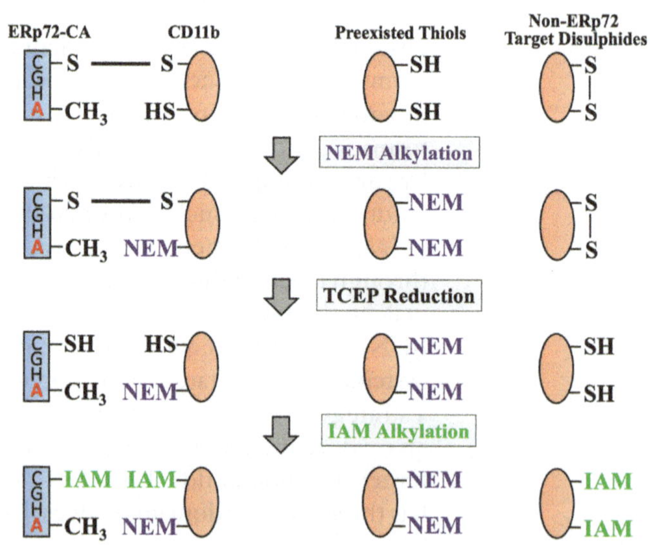

Fig. 2 Sequential labeling of cysteines by different alkylating agents, NEM, and IAM. Only the paired disulfide bonds that are labeled by NEM and IAM, respectively, are considered as the targets

the generation of free thiols that were originally covalently paired. The newly formed or preexisting free thiol groups in CD11b were alkylated by NEM. Upon reduction by TCEP (tris (2-carboxyethyl) phosphine), the disulfide bonds between ERp72-CA and CD11b were broken and the paired cysteines released to be labeled by IAM. The sequentially-labeled CD11b was analyzed by mass spectrometry to map NEM and IAM positions. Based on this strategy, the preexisting free thiols are labeled by NEM only, while the disulfide bonds that do not interact with ERp72 are labeled by IAM only. Only the paired cysteines that are labeled by NEM and IAM, respectively, are considered as the targets for ERp72. Cys711 was labeled by NEM and the paired position Cys654 by IAM, indicating that C654–C711 is the target of ERp72. Differential labeling and mass spectrometry combined with mechanism-based kinetic trapping established a robust and easily accessible approach to map disulfide bonds attacked by oxidoreductases.

2 Materials

1. Erp72-CA recombinant protein (*see* **Note 1**).
2. Dithiothreitol (DTT), dissolved in ultrapure water.
3. Zeba™ spin desalting column 0.5 mL (Thermo).
4. Mac-1 recombinant protein (Kactusbio, itg-hm1mb), dissolved in ultrapure water.
5. D-Hank's buffer, no calcium, and no magnesium.
6. $MgCl_2$, dissolved in D-Hank's buffer.
7. NEM, dissolved in D-Hank's buffer.
8. TCEP-Agarose (Sigma).
9. IAM, dissolved in D-Hank's buffer.
10. Premade PAGE gel (Beyotime, BeyoGel™ Plus PAGE premade gel, Tris-Gly, 4%–20%).
11. 1× NuPAGE™ MOPS running buffer (Thermo).
12. Coomassie blue staining solution.
13. Ammonium bicarbonate, dissolved in ultrapure water.
14. Acetonitrile.
15. Destaining solution: 50% (v/v) 50 mM ammonium bicarbonate, 50% (v/v) acetonitrile.
16. Trypsin solution: 10 ng/μL trypsin in 25 mM ammonium bicarbonate.
17. Formic acid.

3 Methods

3.1 Disulfide Formation Between ERp72-CA and Mac-1

1. For 100 μg of ERp72-CA recombinant protein, add DTT to a final concentration of 20 mM.
2. Incubate on a tube rotator at 4 °C for 20 min.
3. Prepare Zeba™ spin desalting column. Place the column in a collection tube and centrifuge at 1000 × *g* for 2 min to remove storage solution.
4. Mark the top of the slant of the compacted resin after centrifugation on the side of the column. Make sure the mark is facing outward in all subsequent centrifugation steps.
5. Wash the compacted resin with 200 μL ultrapure water.
6. Centrifuge at 1000 × *g* for 2 min to remove ultrapure water.
7. Repeat steps 5 and 6 twice.
8. Place the column in a new collection tube. Remove the cap and slowly add ERp72-CA protein sample from step 2 to the center of the compact resin bed (*see* **Note 2**).
9. Add 40 μL ultrapure water to the resin bed after the protein sample is fully absorbed.
10. Centrifuge at 1000 × *g* for 2 min to collect the sample.
11. Add 20 μg of Mac-1 recombinant protein to desalted ERp72-CA protein sample. Add $MgCl_2$ to a final concentration of 1 mM into the reaction.
12. Incubate on a shaker at room temperature for 20 min.

3.2 NEM Labeling

1. Add NEM to a final concentration of 50 mM into the reaction from Sect. 3.1, step 12.
2. Incubate in dark on a shaker at room temperature for 1 h.
3. Desalt as in Sect. 3.1, steps 3–10.

3.3 Reduction of Disulfide Bonds

1. In a new 1.5 mL tube, add 50 μL TCEP-Agarose gel slurry and wash with D-Hank's buffer.
2. Centrifuge at 1000 × *g* for 2 min to remove supernatant (*see* **Note 3**).
3. Add the desalted proteins from Sect. 3.2, step 3 to the washed TCEP-Agarose gel.
4. Vortex and incubate on a shaker at room temperature for 15 min.

3.4 IAM Labeling

1. Add IAM to a final concentration of 50 mM into the reduced proteins from Sect. 3.3, step 4.
2. Incubate in dark on a shaker at room temperature for 20 min (*see* **Note 4**).
3. Quench the reaction by incubation in light for 10 min.

3.5 Collection of CD11b by Gel Electrophoresis

1. Add reducing loading buffer to the reaction from Sect. 3.4, step 3 (*see* **Note 5**).
2. Boil at 95°C for 5 min.
3. Save 5 μL for Western blotting to locate the position of protein bands.
4. Load the rest of the boiled sample in a premade PAGE gel.
5. Run the gel in 1× MOPS running buffer at 150 V for 65 min.
6. Wash the gel twice with ultrapure water.
7. Stain with Coomassie blue staining solution.
8. Remove the staining solution and wash with ultrapure water until the bands are clearly visible.
9. Collect the gel slice corresponding to CD11b protein band according to its position on Western blot from step 3.
10. Destain the gel slice using 1 mL destaining solution in a 1.5 mL tube. Incubate at 4 °C overnight.
11. If necessary, repeat step 10 until the gel slice becomes transparent (*see* **Note 6**).
12. Remove the destaining solution.

3.6 In-Gel Reduction with DTT and Alkylation with IAM

1. Add 100% acetonitrile and rock at 37 °C for 10 min to dehydrate the gel slice from Sect. 3.5, step 12.
2. Remove acetonitrile.
3. Add DTT solution to a final concentration of 10 mM (*see* **Note 7**).
4. Incubate at 56 °C for 1 h (*see* **Note 8**).
5. Remove the supernatant.
6. Add 100 μL 100% acetonitrile.
7. Let it sit for 5 min.
8. Remove the supernatant. If the gel slice is not fully dehydrated, repeat the dehydration from steps 6–8.
9. Add 1 M IAM solution to a final concentration of 40 mM.
10. Incubate in dark at room temperature for 1 h.
11. Remove the supernatant.
12. Add 100 μL ultrapure water to wash.
13. Let it sit at room temperature for 15 min.
14. Repeat the wash once.
15. Remove the supernatant.
16. Add 100 μL 100% acetonitrile.
17. Let it sit for 5 min.

18. Remove the supernatant. If the gel slice is not fully dehydrated, repeat the dehydration from steps 16–18.
19. Freeze-dry the gel slice in a vacuum concentrator.

3.7 In-Gel Digestion and MS Analysis

1. Add trypsin solution to the gel slice from Sect. 3.6, step 19 (*see* **Notes 9 and 10**).
2. Incubate at 4 °C for 30–60 min to rehydrate the gel slice.
3. Add 20–30 μL 25 mM ammonium bicarbonate (without trypsin) at pH 7.8–8.0 (*see* **Note 11**).
4. Incubate at 37 °C for 12–18 h.
5. Add 5% formic acid to a final concentration of 1% to stop the digestion (pH < 4).
6. Vortex and mix well. Centrifuge.
7. Transfer the supernatant to a new tube.
8. Add 100 μL 60% acetonitrile/0.1% formic acid to the original tube from step 6.
9. Sonicate for 15 min.
10. Transfer the supernatant to the tube from step 7 and combine the supernatants.
11. Add 50 μL 60% acetonitrile/0.1% formic acid.
12. Let it sit for 4 min.
13. Centrifuge at maximum speed for 1 min.
14. Transfer the supernatant to the tube from step 7 and combine the supernatants.
15. Repeat 4 times from steps 11–14.
16. Put the combined supernatants in a vacuum concentrator (*see* **Note 12**).
17. Load the protein sample in Orbitrap Exploris 480 to identify cysteine modification by NEM or IAM.
18. Analyze the mass spectrometry data using BioPharma Finder with the peptide mapping analysis module (*see* **Note 13**).
19. Analyze the protein structure of CD11b using AlphaFold and identify the cysteine pairs in CD11b. Only the paired cysteines that are labeled by NEM and IAM, respectively, are the specific targets of ERp72.

4 Notes

1. The recombinant protein was expressed in bacteria and purified by Ni-NTA His Bind Resin [3]. The eluted protein is in Tris buffer (20 mM Tris, 150 mM NaCl, and pH 7.4) and its concentration is 1–2 mg/mL.

2. The optimal volume is 30–130 μL for Zeba™ spin desalting column 0.5 mL. If necessary, other types of Zeba™ spin desalting columns are available to load more protein samples. The volume of the wash buffer and stacker needs to be adjusted accordingly.
3. After centrifugation, the volume of the settled gel is 25 μL.
4. The reaction should be kept in dark as IAM is sensitive to light.
5. Either DTT or beta-mercaptoethanol is fine.
6. Extend the destaining process if necessary. Protein degradation is not an issue as it is unlikely to happen under this condition.
7. The volume of added DTT solution is dependent on the size of gel slice. To make sure the gel slice is fully rehydrated, add 100–200 μL more DTT solution after the gel slice is rehydrated.
8. If the gel slice is incubated at room temperature, extend the incubation time to 1–2 h.
9. If trypsin solution remains after the gel slice fully rehydrates, remove the excess solution. If the gel slice is not fully rehydrated due to insufficient solution, add more trypsin solution.
10. If more protein sample is loaded in the gel from Sect. 3.5 **step 4**, the amount of trypsin should increase accordingly. The mass ratio of trypsin to the analyzed protein should be 1:30 to 1:100.
11. The volume of ammonium bicarbonate is dependent on the size of gel slice. The gel slice should be fully covered by the solution with at least 20 μL excess volume.
12. The protein sample is ready for mass spectrometry analysis. If necessary, store the samples at −80 °C.
13. The molecular mass shifts by NEM and IAM alkylation are 125 Da and 57 Da, respectively.

Acknowledgments

This work was supported in part by the National Natural Science Foundation of China (92169114 and 81802932), the Natural Science Fund of Hubei Province (2022CFA054 and 2024AFB077), the National Key R&D Program of China (2018YFE0113600), the Global Research Award and the Supplement Award from the American Society of Hematology, the Undergraduate Teaching and Research Project (2024130), and Graduate Teaching and Research Project (2025YB015) of Huazhong University of Science and Technology.

References

1. Xu X, Chiu J, Chen S et al (2021) Pathophysiological roles of cell surface and extracellular protein disulfide isomerase and their molecular mechanisms. Br J Pharmacol 178(15):2911–2930. https://doi.org/10.1111/bph.15493
2. Lakbub JC, Shipman JT, Desaire H (2018) Recent mass spectrometry-based techniques and considerations for disulfide bond characterization in proteins. Anal Bioanal Chem 410(10):2467–2484. https://doi.org/10.1007/s00216-017-0772-1
3. Li Y, Xu X, Wang HJ et al (2024) Endoplasmic reticulum protein 72 regulates integrin Mac-1 activity to influence neutrophil recruitment. Arterioscler Thromb Vasc Biol 44(3):e82–e98. https://doi.org/10.1161/ATVBAHA.123.319771
4. Lv K, Chen S, Xu X et al (2024) Protein disulfide isomerase cleaves allosteric disulfides in histidine-rich glycoprotein to regulate thrombosis. Nat Commun 15(1):3129. https://doi.org/10.1038/s41467-024-47493-0
5. Kozlov G, Maattanen P, Thomas DY et al (2010) A structural overview of the PDI family of proteins. FEBS J 277(19):3924–3936. https://doi.org/10.1111/j.1742-4658.2010.07793.x
6. Smith CW, Marlin SD, Rothlein R et al (1989) Cooperative interactions of LFA-1 and Mac-1 with intercellular adhesion molecule-1 in facilitating adherence and transendothelial migration of human neutrophils *in vitro*. J Clin Investig 83(6):2008–2017. https://doi.org/10.1172/JCI114111

Check for updates

Chapter 10

Quantification of the Redox State of Integrin Disulfide Bonds

Aster Pijning and Philip J. Hogg

Abstract

We have developed a mass spectrometry-based method to quantify the redox state of protein disulfide bonds using a pair of isotopic alkylators. This technique can be applied in several contexts to investigate disulfide bond regulation in integrins. It may be used to identify disulfide-bond targets of vascular thiol isomerases, such as the two disulfide bonds in the β2 subunit of the integrin Mac-1 that are modulated by protein disulfide isomerase (PDI), or to evaluate the redox state of wild-type and mutant αIIbβ3 integrins recombinantly expressed in mammalian cells. Additionally, the method enables direct assessment of disulfide bond redox states in platelet integrins under native conditions. These streamlined protocols eliminate the immunoprecipitation steps included in previous workflows, significantly reducing both the time and cost of sample preparation and removing the dependency on high-affinity antibodies. In particular, the exclusion of immunoprecipitation in the platelet protocol ensures that no selection bias is introduced into the protein pool, allowing for a more accurate quantification of endogenous disulfide redox states in platelet proteins.

Keywords Disulfide bond, Platelet, Integrin, Thiol isomerases

1 Introduction

Recent investigations employing a pair of isotopic thiol alkylators—also referred to as differential cysteine alkylation mass spectrometry—have demonstrated that select blood proteins exist in multiple partially disulfide-bonded states [1–3]. This heterogeneity in disulfide bond formation in some cases reflects variations in protein function, underscoring the importance of identifying and characterizing distinct redox variants within protein populations [4].

Vascular thiol isomerases (VTIs) are a group of oxidoreductases that catalyze thiol–disulfide exchange reactions and play a key role in regulating thrombosis and hemostasis by modifying disulfide bonds in platelet receptors and blood coagulation proteins [5, 6]. New VTIs, including ERp29 and ERp46, continue to be identified, and their platelet disulfide bond targets are actively being investigated.

Philip J. Hogg (ed.), *Functional Disulphide Bonds*, Methods in Molecular Biology, vol. 3016,
https://doi.org/10.1007/978-1-0716-5158-2_10,

Characterizing these interactions requires sensitive techniques capable of detecting redox heterogeneity without introducing bias.

This chapter describes three related protocols for preparing samples prior to differential cysteine alkylation and mass spectrometry, each tailored to interrogate different aspects of integrin disulfide bond redox state. The first section outlines the preparation of recombinant Mac-1 integrin incubated with purified PDI, which allows the investigation of specific disulfide bond targets in the integrin (Fig. 1). The second section focuses on sample preparation for studying the disulfide redox states of wild-type and mutant forms of αIIbβ3 integrin recombinantly expressed in BHK-21 cells. This system uses immunoprecipitation to isolate integrins prior to alkylation and enables detailed analysis of how specific cysteine mutations affect the disulfide bonding pattern.

The third section presents a method for analyzing the redox state of platelet αIIbβ3 integrin in its native state. To avoid introducing selection bias, this protocol deliberately omits immunoprecipitation, as it is becoming clear that antibody binding can preferentially enrich certain disulfide states, and thereby skew redox quantification of disulfide bonds. This streamlined protocol allows for the accurate quantification of disulfide redox states directly from washed platelet lysates (Fig. 2).

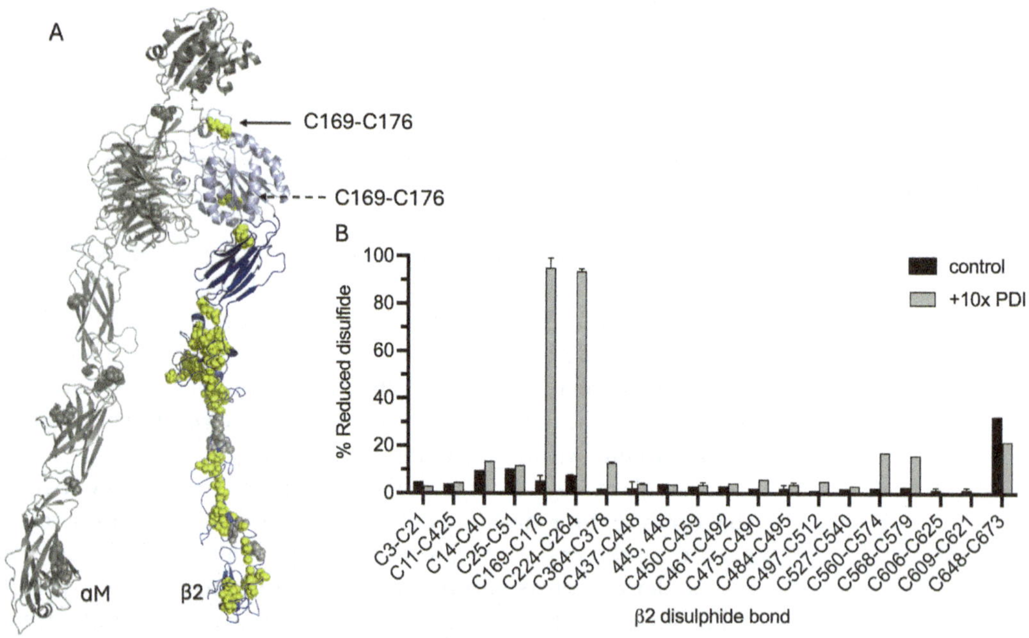

Fig. 1 Determining the disulfide bond targets of oxidoreductases in recombinant integrin. (**A**) A structural model of macrophage-1 antigen (Mac-1) integrin in extended conformation. The ɑM subunit of Mac-1 is shown in gray, and the β2 subunit in blue. Disulfide bonds are indicated as spheres, with bonds that were quantified shown in yellow and bonds that were not quantified shown in gray. The β2 disulfide targets of protein disulfide isomerase (PDI) are indicated. (**B**) The redox state of 20 of the 28 β2 disulfide bonds was quantified with and without pretreatment of ten-fold molar excess PDI. Data shown are mean ± SD. Figure adapted from Dupuy et al. [5]

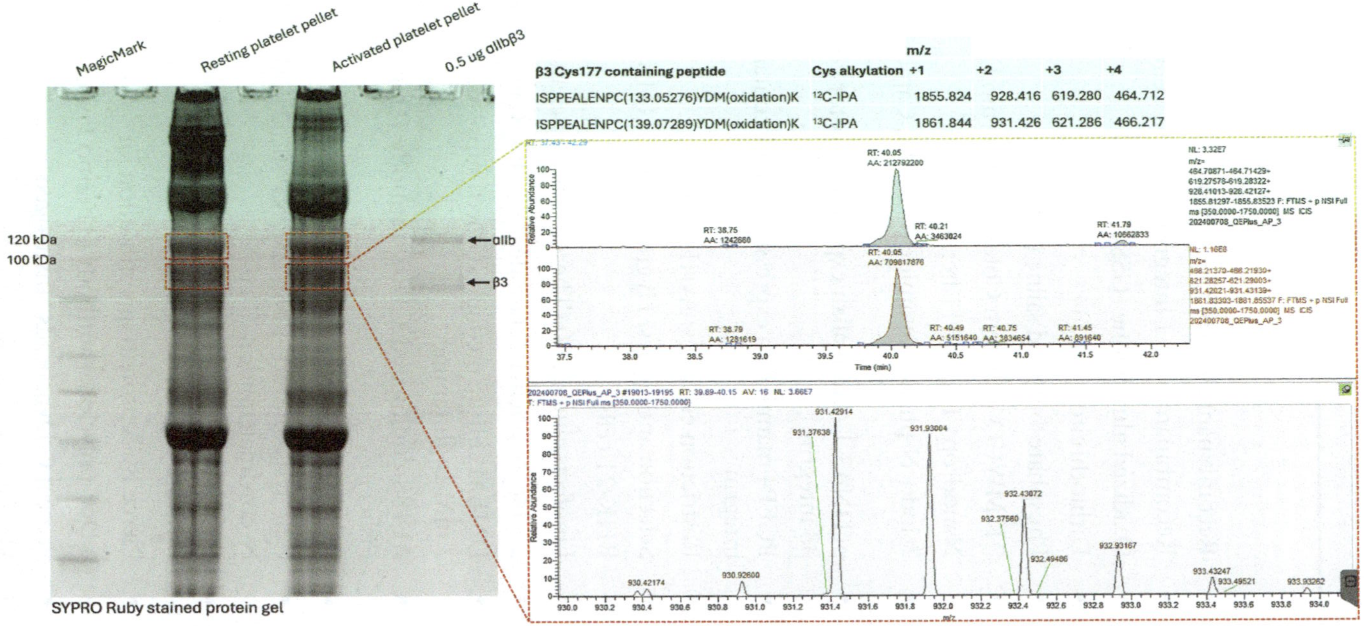

Fig. 2 Example of a SYPRO Ruby-stained protein gel containing a resting and activated platelet sample. Bands were cut at the indicated red dotted lines and processed for mass spectrometry (see Chap. 7). αIIb and β3 cysteine-containing peptides could be identified in the samples, one example of which is shown on the right

Together, these protocols provide a comprehensive framework for exploring disulfide bond regulation in integrins, thus advancing our understanding of redox control in hemostatic processes.

2 Materials

2.1 Measuring Disulfide Targets of PDI in Recombinant Mac-1 Integrin

- Recombinant PDI (purified from *E. Coli*)
- Recombinant Mac-1 integrin (R&D systems)
- Oxidized glutathione (GSSG)
- Dithiothreitol (DTT)
- Phosphate-buffered saline (PBS)
- 7 K MWCO Zebaspin Columns (Thermo Fisher Scientific)
- Nanodrop (Thermo Fisher Scientific)
- 2-iodo-N-phenylacetamide (^{12}C-IPA, Cambridge Isotopes)

2.2 Measurement of Disulfide Status in Recombinantly Expressed αIIbβ3 in BHK-21 Cells After Immunoprecipitation

- pcDNA3.1 mammalian expression vector containing cDNA of β3 integrin
- PCEP4 mammalian expression vector containing cDNA of αIIb integrin
- Restriction enzymes: AvrII, PVUI
- Selection antibiotics: Hygromycin B, G418
- BHK-21 cells
- Lipofectamine 2000 (Thermo Fisher Scientific)
- 2-iodo-N-phenylacetamide (^{12}C-IPA, Cambridge Isotopes)
- DMEM culture medium (Gibco) supplemented with 10% FCS, 2 mM L-glutamine
- AF488 conjugated CD61 antibody (clone VI-PL2, Thermo Fisher)
- Flow cytometer
- Cell lysis buffer: PBS, 2% NP40, 2 mM EDTA, 50 mM N-ethylmaleimide
- Proteinase inhibitor cocktail (cOmplete™ mini EDTA-free Protease Inhibitor cocktail)
- Pierce BCA protein quantification kit
- CD61 polyclonal antibody (Thermo Fisher Scientific)
- Protein G Dynabeads (Invitrogen)
- Magnetic separation rack

2.3 Measurement of Disulfide Bond Status in Native Platelet Mouse or Human αIIbβ3

- Water bath
- 2-iodo-N-phenylacetamide (^{12}C-IPA, Cambridge Isotopes)
- Citrate 4% concentrated solution or citrated blood collection tubes
- Pasteur pipettes or transfer pipettes
- Acid dextrose solution (ACD): 39 mM citric acid, 75 mM sodium citrate, 135 mM dextrose, pH 7.4
- HEPES Tyrode's buffer: 20 mM HEPES, 137 mM NaCl, 4 mM KCl, 0.5 mM Na_2HPO_4, 0.1 mM $CaCl_2$, pH 7.4
- Prostaglandin E1 (PGE1) dissolved in 100% ethanol to make a 1 mM stock
- Optional:
  - Recombinant protein of interest for size indication on SDS-PAGE gel
  - Hemocytometer
  - Thrombin
  - Equine tendon collagen (type I) (HORM Takeda)
  - GPRP peptide (Sigma)
  - PPACK (Sapphire Bioscience)

2.4 SDS-PAGE, Protein Stain, and Band Excision

- LDS loading buffer
- NuPAGE Bis-Tris Mini Protein Gels, 4%–12%, 10-well, 1.5 mm and compatible gel tank
- NuPAGE MOPS SDS running buffer (Invitrogen)
- Molecular marker such as MagicMark XP Western Protein Standard (Invitrogen)
- UV transilluminator
- SYPRO Ruby fixing solution: Deionized water, 50% (v/v) methanol, 7% (v/v) acetic acid)
- SYPRO Ruby Protein Gel Stain (Invitrogen)
- SYPRO Ruby destaining solution: Deionized water, 10% (v/v) methanol, 7% (v/v) acetic acid
- Scalpel

2.5 Integrin Processing for Mass Spectrometry Analysis

- Thermomixer (Eppendorf)
- Dithiothreitol (DTT)
- Stable carbon-13 isotope of IPA (^{13}C-IPA, Cambridge Isotopes)
- Dimethyl sulfoxide (DMSO)
- SpeedVac (Thermo Fisher Scientific)

- Gel wash buffer: 25 mM ammonium bicarbonate, 50% (v/v) acetonitrile
- Proteomics-grade deglycosylation enzyme PNGase F (Sigma). Reconstituted to 500 units/mL with high purity water
- Sequencing-grade bovine chymotrypsin (Roche). Reconstituted to 1 mg/mL stock in 1 mM HCl and diluted 80-fold to 12.5 μg/mL in 25 mM ammonium bicarbonate with 10 mM $CaCl_2$
- Mass spectrometry-grade Trypsin Gold (Promega). Reconstituted to 1 mg/mL in 50 mM acetic acid and diluted 80-fold to 12.5 μg/mL in 25 mM ammonium bicarbonate
- Protease stop buffer: 5% (v/v) formic acid
- Peptide elution solution: 5% (v/v) formic acid, 50% (v/v) acetonitrile
- Peptide reconstitution solution: 0.1% (v/v) formic acid

3 Methods

3.1 Measuring Disulfide Targets of Oxidoreductases in Recombinant Integrin

1. To determine disulfide targets of PDI in Mac-1 integrin, incubate 2 μg of recombinant integrin with ten-fold excess of reduced/oxidized PDI in a total of 30 μL for 30 min at 37°C (*see* **Notes 1 and 2**).
   a. Reduce PDI by incubating with 10 mM DTT in PBS for 30 min at 25°C.
   b. Oxidize PDI by incubating with 200 μM GSSG in PBS for 16 h at 25°C.
   c. Use a 7 K MWCO Zeba spin column according to instructions to remove the DTT/GSSG and perform a buffer exchange into fresh PBS.
   d. Quantify the reduced/oxidized PDI concentration (e.g., Nanodrop) and use immediately.
2. Prepare 50 mM ^{12}C-IPA (13 mg/mL) in DMSO in an amber colored microcentrifuge tube (or tube covered in tin foil to keep light out).
3. Slowly add 4 mM ^{12}C-IPA to the protein, incubate samples for 1 h at 25°C away from light.
4. Add 10 μL 4× LDS loading buffer with 4 mM ^{12}C-IPA freshly added. Incubate for 30 min at 60°C (away from light).

3.2 Measurement of Disulfide Status in Recombinantly Expressed Integrins in BHK-21 Cells After Immunoprecipitation

1. Cotransfection of linearized integrin expression vectors into BHK-21 cells:
   a. Linearize mammalian expression vectors of αIIb and β3 integrin (WT and/or Cysteine mutants) using AvrII and PVUI restriction enzymes, respectively.
   b. Cotransfect linearized constructs into BHK-21 cells using Lipofectamice 2000, following manufacturer's instructions.
   c. Initiate antibiotic selection of transfected cells 48 h after transfection.
   d. Select stably transfected cells in DMEM containing 10% FCS, 2 mM L-glutamine, 500 μg/mL hygromycin B, and 500 μg/mL G418.
2. Verification of integrin expression by flow cytometry:
   a. Harvest 1×10^5 transfected cells, wash in PBS, and resuspend in 100 μL PBS.
   b. Stain with 10 μg/mL Alexa Fluor 488–conjugated anti-CD61 antibody for 30 min at 4°C.
   c. Wash cells, resuspend in PBS, and analyze by flow cytometry.
   d. Compare expression levels to untransfected and vector-only controls.
   e. *Optional:* If expression levels are low, fluorescence activated cell sorting (FACS) can be used to select highly expressing cells.
3. Culture cells to 80% confluency or above in a T75 culture flask.
4. Detach cells using 5 mM EDTA in PBS and wash cells in PBS (centrifuge at 300 g to pellet cells, then resuspend in PBS).
5. Lyse cells in cell lysis buffer with freshly added protease inhibitor. Incubate for 20 min on ice.
6. Remove debris by centrifugation (10,000 g) and determine total protein concentration using the Pierce BCA kit.
7. Immunoprecipitation of integrin (*see* **Note 3**):
   a. Incubate 30 μL of Protein G Dynabeads with 5 μg of polyclonal CD61 antibody.
   b. Wash labeled Dynabeads twice in PBS using a magnetic rack.
   c. Dilute 1 mg of total protein two-fold in PBS and incubate with labeled and washed Dynabeads, using 360° rotation, for 1 h at room temperature.
   d. Wash beads four times in PBS.

8. Prepare 50 mM ^{12}C-IPA (13 mg/mL) in DMSO in an amber colored microcentrifuge tube (or tube covered in tin foil to keep light out).
9. Dilute ^{12}C-IPA to 4 mM in PBS. Resuspend Dynabeads in the diluted ^{12}C-IPA, cover with tin foil and incubate with rotation for 1 h at room temperature.
10. Discard supernatant and resuspend beads in 40 μL LDS loading buffer with 4 mM ^{12}C-IPA freshly added.
11. Incubate at 60°C for 30 min.
12. Continue with Sect. 3.4.

3.3 Measurement of Disulfide Bond Status in Native Platelet Mouse or Human αIIbβ3

1. Collect fresh blood into 1:10 Citrate solution (or a citrated blood collection tube) to prevent aggregation.
   Mouse: As quickly as possible after euthanasia, perform cardiac puncture to collect maximum blood volume, at least 1.5 mL per mouse needed.
   a. Human: Collect 5 mL per sample via venipuncture.
2. Weigh the blood collection tube and prepare a balance tube within 10 mg of the same weight.
3. Spin the blood in a balanced centrifuge for 20 min at 200 g with *deceleration and acceleration turned off* at room temperature.
4. Use a Pasteur pipette to slowly take platelet-rich plasma (upper layer) and place into a fresh 15 mL conical tube.
   Take care to avoid disturbing the buffy coat and red cells
5. Add 20% of plasma volume of prewarmed ACD and gently roll the tube for mixing (no inversion).
6. Allow platelets to rest for 30 min at 37°C in a water bath.
7. Add Prostaglandin E1 to a final concentration of 1 μM. Gently roll to mix (no inversion).
8. Pellet the platelets by centrifugation at 800 g (human) or 1000 g (mouse) for 15 min with medium acceleration and no deceleration at room temperature in a balanced centrifuge.
9. Remove platelet poor plasma with a Pasteur pipette, without disturbing the pellet.
10. Using a Pasteur pipette, gently resuspend the platelet pellet in a volume of Hepes Tyrode's buffer equal to the starting blood volume.
    Hepes Tyrode's buffer must be warmed to 37°C and 5 mM glucose added on the day of use
11. *Optional:* Take an aliquot of washed platelets for counting on a hemocytometer.
12. Rest platelets in a 37°C water bath for 10 min.

13. *Optional:* Activate resting platelets by adding 1 U/mL thrombin, 5 μg/mL equine tendon collagen (type I), 10 mM $CaCl_2$, 10 mM $MgCl_2$, and 1 mM GPRP peptide. Inactivate thrombin after 6 min by adding 50 μM PPACK.
14. Prepare 50 mM ^{12}C-IPA (13 mg/mL) in DMSO in an amber colored microcentrifuge tube (or tube covered in tin foil to keep light out).
15. Slowly add 4 mM ^{12}C-IPA to platelets, incubate all samples for 1 h at room temperature with gentle rotation away from light (wrapped in tin foil).
16. Spin platelets at 1000 g for 15 min with medium acceleration and no deceleration at room temperature, in a balanced centrifuge.
17. Carefully remove supernatant without disturbing the pellet.
18. *Optional:* Flash-freeze the platelet pellet in liquid nitrogen and store at −80°C.
19. Resuspend platelet pellet in 40 μL LDS loading buffer with 4 mM ^{12}C-IPA freshly added. Incubate for 30 min at 60°C (away from light).
20. *Optional:* Also prepare 0.5 μg of recombinant protein with LDS loading buffer to facilitate cutting the protein bands at the right height.

3.4 SDS-PAGE, Protein Stain, and Gel Band Excision

1. Load samples onto NuPAGE Bis-Tris Mini Protein gel (*see* **Note 3**), run with MOPS buffer at 150 V for 65 min. When loading multiple samples, keep an empty lane between each sample. Also, keep an empty lane between a molecular weight marker and any sample (*see* **Note 4**).
2. Fix gel in SYPRO Ruby fixing solution for 1 h.
3. Stain gel in SYPRO Ruby overnight or for 1 h.
4. Wash gel in SYPRO destaining solution for 1 h.
5. Excise gel band using a UV transilluminator and clean scalpel.
6. Cut the gel bands into 1–2 mm^2 pieces and place them in a microcentrifuge tube.
7. Continue with Sect. 3.5.

3.5 In-Gel Disulfide Bond Reduction with DTT and Alkylation with ^{13}C-IPA

1. Dice gel band into small pieces of 1–2 mm^2 and place gel pieces into a clean microcentrifuge tube.
2. Add 150 μL gel wash buffer and vortex gel pieces for 10 min at 25°C using a thermomixer (Eppendorf). Remove supernatant by pipetting.
3. Repeat Step 2 for another two washes of the gel pieces.

4. Dry gel pieces by adding 100 μL of 100% acetonitrile. Remove acetonitrile by pipetting.
5. Cover gel pieces with 100 μL of 40 mM DTT in 25 mM ammonium bicarbonate and incubate at 56°C for 30 min.
6. Remove supernatant and wash gel pieces three times with 150 μL gel wash buffer. Vortex each time at 1400 rpm for 10 min at 25°C.
7. Dry gel pieces by adding 100 μL of 100% acetonitrile. Remove acetonitrile by pipetting.
8. Prepare 50 mM ^{13}C-IPA (13.4 mg/mL) in DMSO in an amber colored microcentrifuge tube (or tube covered in tin foil to keep light out).
9. Prepare a 1:10 dilution of ^{13}C-IPA in 25 mM ammonium bicarbonate, use immediately and keep away from light.
10. Cover gel pieces with 100 μL of 5 mM ^{13}C-IPA in 25 mM ammonium bicarbonate and incubate in the dark for 1 h at 25°C.
11. Remove supernatant and wash three times with 150 μL gel wash buffer. Vortex each time at 1400 rpm for 10 min at 25°C.
12. Dry gel slices by adding 100 μL of 100% acetonitrile. Remove acetonitrile by pipetting.
13. Dry gel pieces in SpeedVac for 15–20 min to remove traces of acetonitrile prior to enzymatic processing of gel pieces.

3.6 In-Gel Deglycosylation and Digestion of Integrin

1. Add 10 μL (five units) of reconstituted PNGase F to dried gel pieces and incubate at 37°C for 30 min.
2. Add 25 μL of water to the sample such that the gel pieces are just covered by the liquid and incubate at 37°C overnight.
3. Remove supernatant by pipetting (*see* **Note 6**) and wash gel pieces three times with 250 μL deionized water. Each time, vortex at 1400 rpm for 20 mins at 25°C.
4. Remove supernatant and dry gel pieces in SpeedVac until completely dry.
5. Add 35 μL of freshly diluted chymotrypsin (*see* **Note 7**) to the gel pieces and incubate at 37°C for 4 h.
6. Add 30 μL of freshly diluted trypsin to gel pieces and incubate at 25°C overnight.
7. Stop the digestion by adding 10 μL of 5% formic acid.
8. Transfer the peptide containing solution to a new microcentrifuge tube.
9. Elute remaining peptides from gel slices by adding 30 μL of 5% formic acid/50% acetonitrile. Vortex at 100 rom for 20 min at 25°C and transfer the eluate to the tube containing peptides.

10. Repeat Step 9 for another elution of peptides from gel pieces.
11. Dry peptides with SpeedVac and reconstitute samples in 12 μL 0.1% formic acid.
12. Proceed to LC-MS/MS analysis.

4 Notes

1. Determining the redox potential of a new VTI and comparing it to that of the target protein can help predict whether the VTI is likely to reduce or oxidize target disulfide bonds.
2. Keep the total volume of integrin, VTI, and ^{12}C-IPA below 30 μL taking into account the maximum capacity of the SDS-PAGE well in the next steps.
3. Immunoprecipitation results in a selection of molecules from the total population that may have a different disulfide redox state than the total population. We recommend using a polyclonal antibody for IP.
4. Use 10-well, 1.5 mm gels for the highest loading volume.
5. Many molecular weight markers contain reducing agents such as DTT, which need to be avoided prior and during ^{12}C-IPA labeling.
6. Supernatant and washes can be combined for glycan analysis.
7. Modify the protease(s) used based on the amino acid sequence of the integrin of interest. We recommend using the MS-digest tool by ProteinProspector to identify the ideal digestion enzymes and resulting peptides depending on your target protein and the disulfide bonds of interest.

Acknowledgments

This work was funded by the National Health and Medical Research Council of Australia (2026631) and a Senior Researcher Grant from the NSW Cardiovascular Research Capacity Program to PJH.

References

1. Pijning AE, Blyth MT, Coote ML, Passam F, Chiu J, Hogg PJ (2021) An alternate covalent form of platelet alphaIIbbeta3 integrin that resides in focal adhesions and has altered function. Blood 138(15):1359–1372. https://doi.org/10.1182/blood.2021012441
2. Butera D, Hogg PJ (2020) Fibrinogen function achieved through multiple covalent states. Nat Commun 11(1):5468. https://doi.org/10.1038/s41467-020-19295-7
3. Butera D, Wang HJ, Woon HG, Zhao YC, Ju LA, Hogg PJ (2023) Mechano-covalent protection

of coagulation factor VIII by von Willebrand factor. Blood Adv 7(10):2117–2128. https://doi.org/10.1182/bloodadvances.2022008650
4. Pijning AE, Butera D, Hogg PJ (2022) Not one, but many forms of thrombosis proteins. J Thromb Haemost 20(2):285–292. https://doi.org/10.1111/jth.15567
5. Dupuy A, Aponte-Santamaria C, Yeheskel A, Hortle E, Oehlers SH, Grater F, Hogg PJ, Passam FH, Chiu J (2023) Mechano-redox control of Mac-1 de-adhesion by PDI promotes directional movement under flow. Circ Res 132(9):e151–e168. https://doi.org/10.1161/CIRCRESAHA.122.321926
6. Flaumenhaft R, Furie B (2016) Vascular thiol isomerases. Blood 128(7):893–901. https://doi.org/10.1182/blood-2016-04-636456

Check for updates

Chapter 11

Identification of Protein Cysteine Modifications Using Unbiased Mass Spectrometry-Based Proteomics

Mark Schreuder and Mark Larance

Abstract

Post-translational modifications (PTMs) can play vital roles in the protein's structure, function, localization, and interactions, enabling dynamic control of cellular processes. As such, understanding these modifications is essential to elucidate key biological mechanisms. This chapter discusses an unbiased, mass spectrometry-based approach using wildcard (open) searches in the Byonic platform to detect and identify nondisulfide cysteine PTMs. By targeting individual cysteines and scanning for mass shifts across a broad range, this strategy enables discovery of novel or unexpected modifications. The method is applicable to both new and existing datasets and can be extended to any amino acid, offering a versatile tool for redox and PTM-focused proteomics.

Keywords Byonic, Post-translational modifications, Cysteine, Prothrombin, Mass spectrometry

1 Introduction

Cysteine is one of the most chemically reactive amino acids, playing pivotal roles in enzyme proteolysis, metal coordination, and redox regulation [1]. While disulfide bond formation between paired cysteines is well-characterized and commonly studied for its role in protein folding, stability and function, cysteine residues are also subject to a wide array of nondisulfide modifications. These include oxidative post-translational modifications (PTMs) such as sulfenylation, sulfinylation, and sulfonylation, as well as regulatory modifications such as S-nitrosylation, S-glutathionylation, acylation, and persulfidation [2, 3]. Many of these modifications are transient, redox-sensitive, and site-specific, making them challenging to identify and quantify using conventional targeted approaches.

This chapter presents an unbiased mass spectrometry-based strategy for the detection and characterization of nondisulfide cysteine modifications. Rather than focusing on disulfide bond quantification between paired cysteines, this approach targets individual cysteine residues to reveal novel or unexpected PTMs with potential functional

Philip J. Hogg (ed.), *Functional Disulphide Bonds*, Methods in Molecular Biology, vol. 3016,
https://doi.org/10.1007/978-1-0716-5158-2_11,

relevance. The method leverages wildcard (open) searches using the Byonic platform (Protein Metrics), which enables the identification of mass shifts corresponding to modifications within a wide range (e.g., ~40 to +1000 Da), without requiring prior knowledge of the specific PTM (Fig. 1).

This strategy can be applied to any proteomic datasets, from simple to complex protein extracts, although extensive protein preparation and optimized sample preparation conditions may be required to enrich low abundance peptides and proteins. This technique broadens the scope of redox proteomics by allowing for exploratory, discovery-based identification of cysteine modifications in a biological

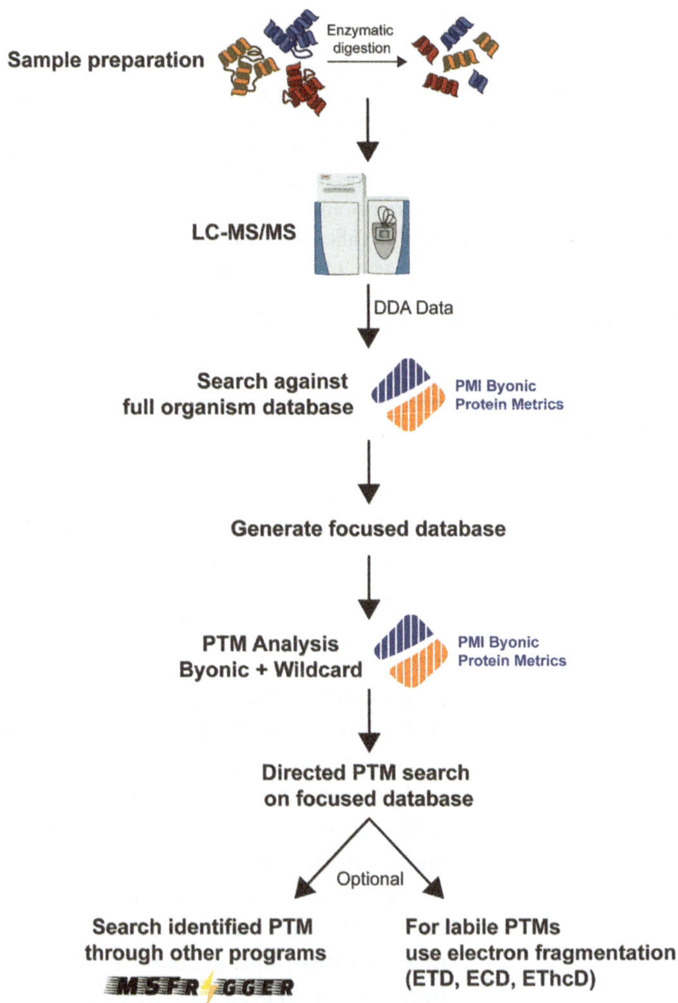

Fig. 1 Overview of the unbiased mass spectrometry-based strategy for the detection and characterization of post-translational modifications. The workflow includes sample preparation and enzymatic digestion, LC-MS/MS data acquisition, and proteomic analysis using the Byonic platform (Protein Metrics). An initial proteome search is followed by the generation of a focused database and a wildcard search to identify unexpected or novel modifications

context. It may provide a key tool in understanding how dynamic cysteine oxidation states contribute to protein function and disease processes beyond traditional disulfide-based regulation. Importantly, this approach can not only be used to identify PTMs of cysteine residues but can be applied to any other amino acid, enabling this to be a very versatile method for the identification of novel PTMs to advance our understanding of protein biology.

2 Materials

This chapter uses the blood coagulation protein prothrombin as a model to demonstrate how to assess post-translational modifications (PTMs) of cysteine residues in proteins. Prothrombin contains twelve disulfide bonds; however, the redox states of the individual cysteines within these bonds are not always equivalent. This discrepancy suggests that one of the cysteines may carry an alternative modification. For instance, in the disulfide bond between Cys191 and Cys231, isotopic labeling with iodoacetamide (IPA) shows an unequal ratio of carbon-12 to carbon-13 (Fig. 2; ~0.12–0.02, respectively), indicating a likely modification on Cys231.

The method described here allows researchers to perform unbiased searches for PTMs—including those of unknown mass—on cysteines or any other amino acid in proteins of interest using mass spectrometry-based approaches. The following data and software are required for the analysis:

- Data-Dependent Acquisition data: mass spectrometry data acquired by selectively fragmenting isolated precursor ions during an experiment.
- Byonic package (Protein Metrics): search engine for peptide and protein identification incorporating open search strategies for unbiased PTM identification.
- UniMod database: a comprehensive online database of protein post-translational modifications that allows identification of unknown modifications based on experimental mass differences.
- Optional, Uniprot browser: a comprehensive and freely accessible database providing detailed and reliable information about proteins from sequence to PTMs and function, across most organisms.

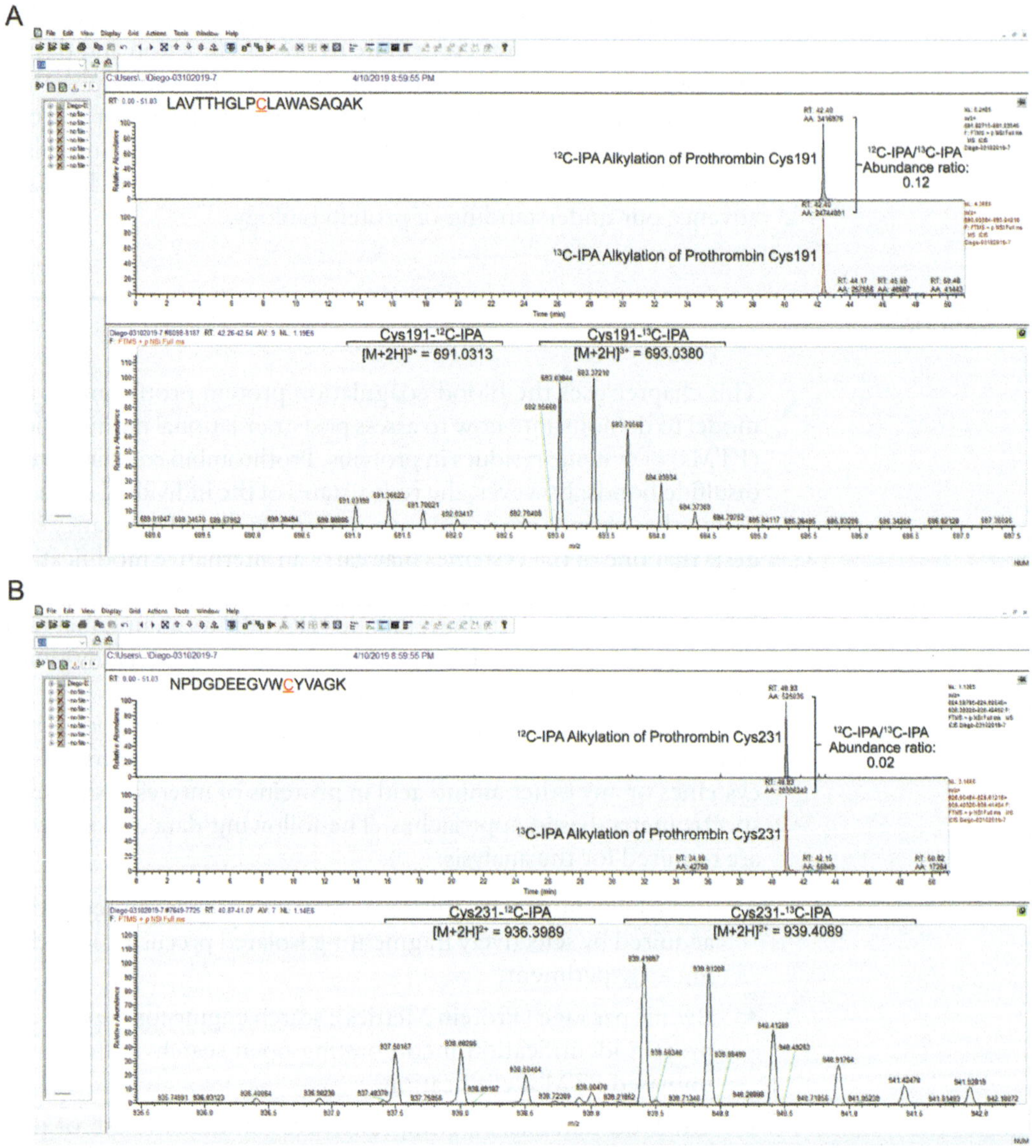

Fig. 2 Screenshots of the XCalibur Qual Browser window showing automated peak detection for the peptides containing alkylated cysteine residues 191 or 231. (**A**) The upper panel shows automated peaks corresponding to the ^{12}C-IPA and ^{13}C-IPA labeled peptide *LAVTTHGLPCLAWASAQAK*, which comprises cysteine 191. The lower panel shows the peptide's *m/z* values of 691.0313 (^{12}C-IPA-Cys191) and 693.0380 (^{13}C-IPA-Cys191), with an isotopic mass distribution of 0.33 (*m/z* +3). (**B**) Automated peaks for the ^{12}C-IPA and ^{13}C-IPA labeled peptide *NPDGDEEGVWCYVAGK*, which comprises cysteine 231, are shown. The isotopic mass distribution is 0.5 (*m/z* + 2) for the ^{12}C-IPA-labeled (936.3989) and ^{13}C-IPA-labeled (939.4089) Cys231-peptide

3 Methods

1. **Protein Sample Preparation**
   Prepare the protein sample using your established laboratory protocol or follow the method described in Harney and Larance [4] (*see* **Notes 1 and 2**).
2. **LC-MS/MS using Data-Dependent Acquisition**
   Analyze the samples using liquid chromatography-tandem mass spectrometry (LC-MS/MS) using data-dependent acquisition (DDA) as per in-house protocols or follow the procedure outlined by Harney and Larance [4] (*see* **Note 3**).
3. **Data Analysis Using Byonic**
   Once data-dependent acquisition (DDA) data has been obtained, analyze the raw MS data using the Byonic software from Protein Metrics. Launch the software, a popup interface with the Byonic search parameters will appear. Enter settings as shown in Fig. 3 that are suitable for the sample and data type (*see* **Note 4**).
   a. Initial proteome search (minimal modification search to identify proteins present from unmodified or minimally modified peptides): Here, we assume that the protein with an unknown modification can also be identified using unmodified peptides from the adjacent protein sequence. Perform an initial database search against the complete proteome of the source organism (e.g., human, mouse) including minimal modifications (e.g., fixed carbamidomethylation of Cys and oxidation of Met; Fig. 3). Set the error tolerances for precursors and fragments that are suitable for the mass analyzer used (e.g., 6 and 20 ppm, respectively, for orbitraps). This step is used to identify the core set of proteins present in the sample. Running the search with minimal variable modifications significantly reduces analysis time (*see* **Note 5**).

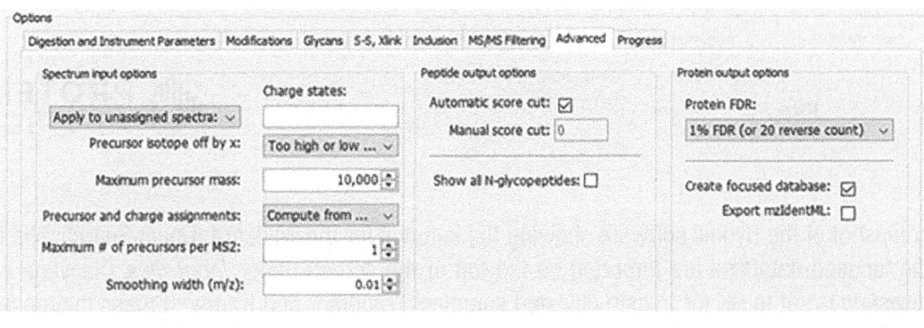

Fig. 3 Screenshot of the Byonic software advanced tab showing the option to "Create focused database" tick box that needs be used for the initial search

b. Focused Database Generation: From the results of the initial unmodified search, a focused database will be generated consisting only of the proteins identified in the sample. This refined database will be used in subsequent open searches for PTM discovery, increasing specificity, and reducing computational load (*see* **Note 6**).

4. Wildcard (Open) Search: Using the focused database generated in Step 3, perform a wildcard search in Byonic to identify mass additions to peptides included in the database.

   a. Set the "Select Protein database file" to be the focused database generated in the previous search (Fig. 4). Turn off the tick boxes for adding decoys and contaminants as these are already contained in the focused database.

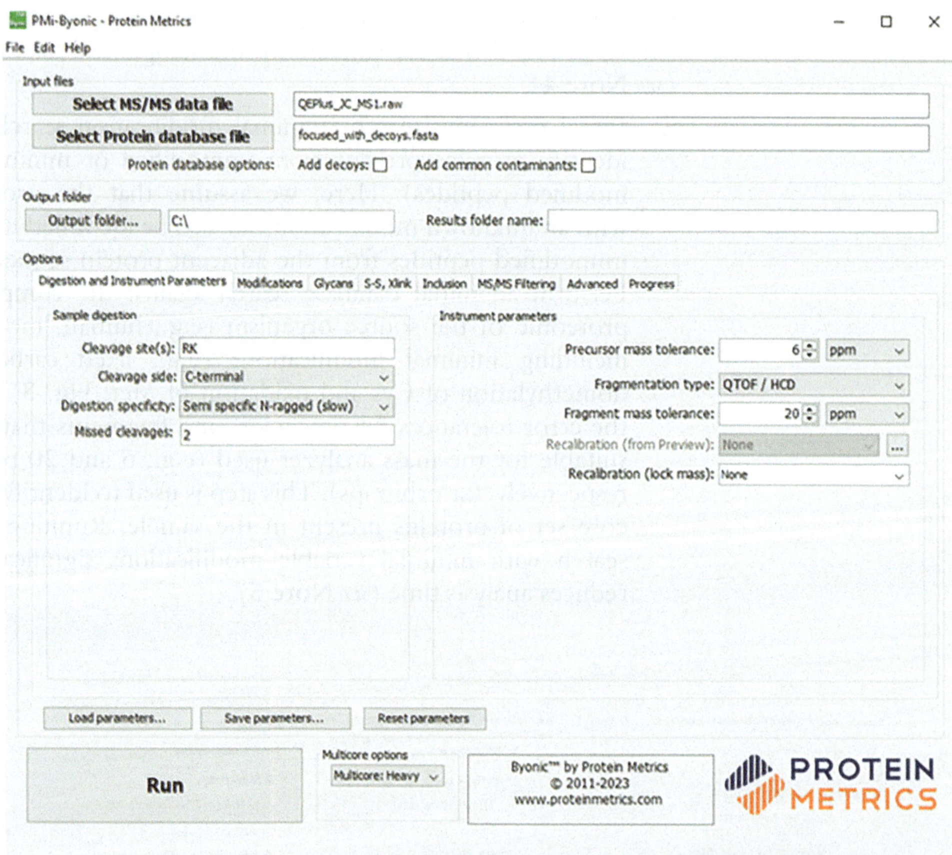

Fig. 4 Screenshot of the Byonic software showing the settings for the Wildcard (Open) Search. The DDA data file and the focused database are imported on the top of the screen under *Input files*. Cleavage site under *Sample digestion* is set to RK for trypsin digested samples. Precursor and fragment mass tolerance, and the fragmentation type are set under *Instrument parameters*

b. Set the protease specificity appropriate to your sample. Many secreted proteins are proteolytically processed and produce semitryptic peptides. If this is the case, use one of the "Semispecific" options such as "Semispecific N-ragged."

c. In the "Common" and "Rare" modification settings, include common modifications such as oxidation (Met), and protein N-terminal acetylation, while keeping the total number of common and rare modifications low (i.e., 1) to reduce search space complexity.

d. Set the wildcard mass range (e.g., ~40 to +300 Da) and restrict the wildcard modification to either cysteine (C) for focused thiol-modification studies, or all amino acids for non-specific PTM searches (Fig. 5) (*see* **Note 6**).

5. Review the wildcard search results for peptides with statistically significant spectral matches such as those with a PEP 2D score <0.001 (colored green in Byonic Viewer PSM table).

6. Identify the mass shift associated with each modified peptide (Fig. 6). Use the UniMod database to match the observed mass shift to known PTMs. This will identify the specific PTM (*see* **Note 7**).

7. Targeted PTM search. Using the same focused database, rerun a targeted Byonic search that includes the identified PTM(s) as a defined variable modification. This step helps to confirm the presence of the identified PTM and reduces the chance that the identified PTM is a false positive hit.

Optional Steps:

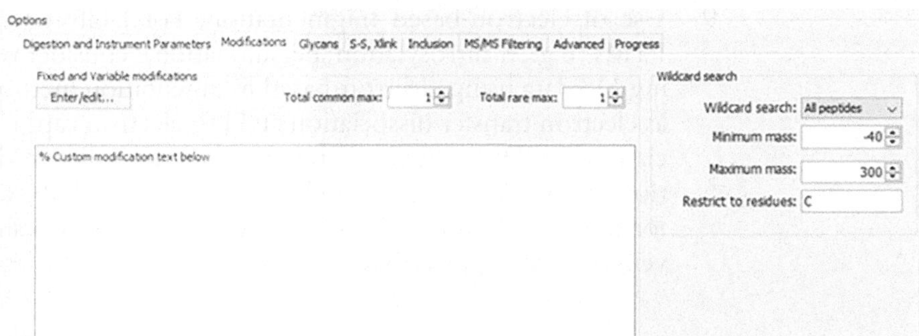

Fig. 5 Screenshot of the Byonic software showing the modifications settings in the Wildcard (Open) Search. The common (total common max) and rare (total rare max) modifications are set to 1. Under the panel *Wildcard search*, the wildcard mass range (e.g., ~40 to +300 Da) and restrictions to specific amino acids can be entered

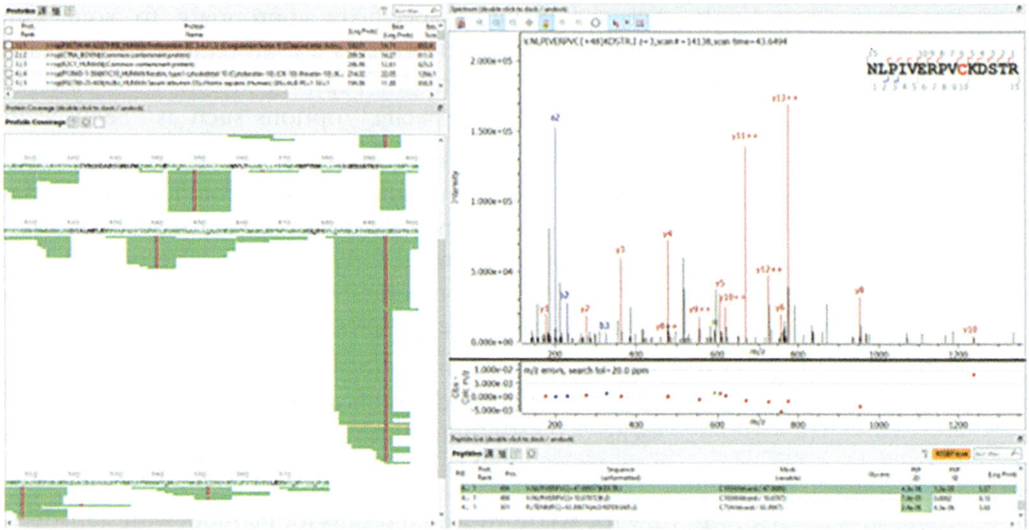

Fig. 6 Screenshot of the Byonic software showing the results of the wildcard (Open) search. The left side shows the detected proteins and the protein coverage highlighted in green. The right side displays the mass additions found in the wildcard search and the corresponding MS/MS spectrum (e.g., the wildcard search observed a mass addition of +47.98857 (sulfonylation) for the cysteine in peptide NLPIVERPVCKDSTR

8. Cross-validation using alternative analysis programs. To further increase the confidence in the identified PTM or potentially identify PTMs that were missed using the Byonic search, repeat the analysis using alternative programs, e.g., MSFragger, Open-pFind, or MetaMorpheus. These programs offer slightly different algorithmic approaches and may capture different modified peptides due to their search strategy and scoring scheme. Findings are particularly strengthened when PTMs are identified in two or more different program analyses.
9. Use of electron-based fragmentation. For labile or complex PTMs (e.g., S-nitrosylation, persulfidation), consider reacquiring MS data using electron-based fragmentation methods such as electron transfer dissociation (ETD), electron capture dissociation (ECD), or ETD with higher-energy collision dissociation (EThcD). These methods preserve labile PTMs, allowing for the identification of PTMs that may be missed using conventional MS approaches.

4 Notes

1. **Low-Abundance and Unstable Protein Modification Considerations**

   For samples in which the protein of interest is expected to be present in low abundance, enrichment strategies may be required.

These can include fractionation techniques such as size-exclusion chromatography, ion-exchange chromatography, or immunoprecipitation. For example, refer to Harney and Larance [4] for details on fractionation of complex proteomes. For modifications that are unstable under standard sample preparation or LC-MS conditions (e.g., low pH, high temperature), optimization of buffers, and incubations may be needed.

2. **Analysis of Previously Generated Data**
   Previously acquired DDA data may be used for PTM discovery, provided the data is of sufficient quality (i.e., high resolution, accurate mass) and in a format compatible with Byonic (e.g., mzML, MGF, WIFF, or RAW files).
3. **Mass Spectrometer Requirements**
   For optimal PTM identification, high-resolution MS instruments capable of accurate mass measurements of both precursor and fragment ions (e.g., Orbitrap or Q-TOF platforms) are recommended. Ensure appropriate resolution settings are used to achieve high mass accuracy.
4. **Byonic Software**
   This protocol makes use of the standalone Byonic software. The Byonic software can also be integrated in Thermo Fisher Scientific's Proteome Discoverer as a node. The integrated version will use the same steps as described in this protocol; however, the exact click-by-click steps may differ between the standalone and integrated Byonic versions.
5. **Reference Proteome Selection and Digestion Specificity**
   Search against the proteome that matches the biological origin(s) of the sample. For instance, use the human proteome if the sample was generated in human embryonic kidney (HEK293) cells or human-derived samples. If the proteins of interest are secreted or membrane proteins, the open search for the unknown PTMs may benefit from using a semitryptic search or incorporating a database that includes known cleavage sites (e.g., signal peptidase cleavage) [5].
6. **MS1 and MS2 Mass Tolerance**
   For optimal peptide and PTM identification accuracy, it is recommended to set the MS1 and MS2 mass tolerance to <7 ppm when using high-resolution instruments such as Orbitrap, or <10 ppm when using Q-TOF systems. Lower ppm values improve search times and sensitivity for PTM identification.
7. **High Mass Accuracy Requirement**
   As described in **Note 5**, high mass accuracy is critical for the confident identification of PTMs, especially in a wildcard (open) search where the modification mass is not predefined. The mass accuracy directly impacts the ability to assign the correct elemental composition and identify the PTM using resources like

UniMod. The closer the experimental mass shift matches a known modification mass in UniMod, the higher the confidence in the assignment. Poor mass accuracy can lead to missed identifications or false positives, especially when analyzing low-abundance or labile PTMs.

References

1. Jacob C, Giles GI, Giles NM, Sies H (2003) Sulfur and selenium: the role of oxidation state in protein structure and function. Angew Chem Int Ed Engl 42(39):4742–4758. https://doi.org/10.1002/anie.200300573
2. Boutin C, Clement C, Rivoal J (2024) Post-translational modifications to cysteine residues in plant proteins and their impact on the regulation of metabolism and signal transduction. Int J Mol Sci 25(18):9845. https://doi.org/10.3390/ijms25189845
3. Chung HS, Wang SB, Venkatraman V, Murray CI, Van Eyk JE (2013) Cysteine oxidative post-translational modifications: emerging regulation in the cardiovascular system. Circ Res 112(2):382–392. https://doi.org/10.1161/CIRCRESAHA.112.268680
4. Harney DJ, Larance M (2023) The Small-Protein Enrichment Assay (SPEA) for analysis of low abundance peptide hormones in plasma. Methods Mol Biol 2628:265–276. https://doi.org/10.1007/978-1-0716-2978-9_17
5. Harney DJ, Larance M (2022) Annotated protein database using known cleavage sites for rapid detection of secreted proteins. J Proteome Res 21(4):965–974. https://doi.org/10.1021/acs.jproteome.1c00806

Check for updates

Chapter 12

Mapping Protein Disulfide Bonds by Mass Spectrometry

Diego Butera and Philip J. Hogg

Abstract

Disulfide bonds are typically identified from protein tertiary and quaternary structures. By analyzing electron density maps of proteins, investigators can identify disulfide bonds based on the distance and the geometry between sulfur atoms of cysteine residues. However, there are many instances where the disulfide pairing in protein crystal structures differs from the pairing in the soluble protein. In this chapter, we describe how mass spectrometry can be used to identify new or unexpected disulfide bonds in proteins. The method is exemplified by identification of an unexpected disulfide bond linking Cys2528 and Cys2533 in von Willebrand factor. This same approach can apply to any protein.

Keywords Disulfide bond, Cysteine alkylation, Mass spectrometry

1 Introduction

The method described herein is a variation of the method described in Chap. 7 to quantify the redox state of protein disulfide bonds. Briefly, unpaired cysteines thiols in the protein are blocked with 2-iodo-N-phenylacetamide to prevent any disulfide exchange during analysis. The sample is resolved on nonreducing SDS-PAGE, and the band corresponding to the protein of interest is excised from the gel. The gel slice is destained, the protein digested with enzymes and the resulting disulfide-linked peptides analyzed by mass spectrometry. Disulfide bonds can form between two different peptides or between two cysteine residues in the same peptide. To exemplify this technique, we describe an unexpected disulfide bond between Cys2528 and Cys2533 in the C4 domain of von Willebrand factor [1, 2].

Philip J. Hogg (ed.), *Functional Disulphide Bonds*, Methods in Molecular Biology, vol. 3016,
https://doi.org/10.1007/978-1-0716-5158-2_12,

2 Materials

As mass spectrometry analysis is highly sensitive, uttermost care should be taken to ensure clean working reagents and environment to avoid keratin contaminations that originate from skin, hair, and dust. All chemicals and reagents should be of LC-grade.

2.1 Immuno-precipitation of von Willebrand Factor from Plasma

1. Acid-citrate-dextrose (ACD) buffer: 39 mM citric acid, 75 mM sodium citrate 135 mM dextrose, pH 7.4
2. Phosphate-buffered saline (PBS): 137 mM NaCl, 2.7 mM KCl, 10 mM Na_2HPO_4, 1.8 mM KH_2PO_4, pH 7.4
3. Anti-von Willebrand factor rabbit polyclonal antibody (Dako)
4. Dynabeads Protein G (Life technologies)
5. Nanodrop (Thermo Scientific) for protein quantification
6. Freshly prepared 40 mM IPA (Cambridge isotopes) in DMSO (*see* **Note 1**)
7. NuPAGE™ 4× LDS sample loading buffer (Thermo Scientific)
8. NuPAGE™ 4%–12% Bis-Tris gel (Thermo Scientific)
9. 1× NuPAGE™ MOPS running buffer (Thermo Scientific)
10. Coomassie Blue G solution (Sigma)
11. Gel destaining solution: 10% (v/v) methanol, 5% (v/v) acetic acid

2.2 Processing of von Willebrand Factor for Mass SpectrometryAnalysis

1. 25 mM ammonium bicarbonate
2. 25 mM ammonium bicarbonate, 50% (v/v) acetonitrile
3. Protease stop buffer: 5% (v/v) formic acid
4. Peptide elution solution: 5% formic acid, 50% acetonitrile
5. Peptide reconstitution solution: 0.1% formic acid
6. Sequencing grade bovine chymotrypsin (Roche) is reconstituted to 1 mg/mL stock in 1 mM HCl and diluted 80-fold to 12.5 μg/mL in 25 mM ammonium bicarbonate with 10 mM $CaCl_2$
7. Mass spectrometry grade Trypsin Gold (Promega) is reconstituted to 1 mg/mL in 50 mM acetic acid and diluted 80-fold to 12.5 μg/mL in 25 mM ammonium bicarbonate
8. Speedvac SC210P1-15 (Thermo Scientific)

2.3 Mass Spectrometry Analysis

1. HPLC Buffer A: 95% LC-grade water, 5% acetonitrile, 0.1% formic acid
2. HPLC Buffer B: 20% LC-grade water, 80% acetonitrile, 0.1% formic acid

2.4 Analysis Software of Mass Spectrometry Data

1. Mascot Daemon (*see* **Note 2**)
2. Qual Browser from XCalibur (Thermo Scientific)

3 Methods

3.1 Immuno-precipitation of von Willebrand Factor from Plasma

1. Draw blood into tubes containing ACD buffer. Gently mix tubes by inverting tubes slowly.
2. Centrifuge blood at 2000*g* for 20 min at 22°C, no brake.
3. Transfer the supernatant (platelet poor plasma) to a new tube using a transfer pipette without disturbing the middle layer (*see* **Note 3**).
4. Incubate 1 mL of plasma with 1 mg of anti-von Willebrand factor coated magnetic beads (see Chap. 7) for 1 h rotating on a rotating wheel at 22°C.
5. Wash the beads with 1 mL of PBS. Place the tubes on the magnet for 1 min and discard the washing buffer.
6. Repeat Step 5 twice.
7. Add 100 μL of freshly diluted 4 mM IPA (diluted from 40 mM stock) in PBS. Incubate at 22°C for 1 h in the dark.
8. Remove IPA solution by placing the tube on the magnet for 1 min. Discard IPA solution.
9. Add LDS sample loading buffer and 4 mM IPA, heat samples at 60°C for 30 min.
10. Analyze samples on nonreducing 4%–12% Bis-Tris NuPAGE gel in 1× MOPS running buffer at 150 volts for 75 min.
11. Stain protein gel with 40 mL of Coomassie Blue G solution for 30 min. Remove Coomassie Blue G solution.
12. Destain gel in destaining solution until background becomes clear.
13. Excise protein gel bands corresponding to von Willebrand factor (Fig. 1) with a clean scalpel for further processing. Note, because von Willebrand factor is a multimer, a gel piece corresponding to the bands between the parentheses was excised.

3.2 In-Gel Digestion (Chymotrypsin and Trypsin)

1. Dice gel band into small pieces of 1–2 mm^3. Place gel pieces into a clean tube.
2. Add 150 μL of 25 mM ammonium bicarbonate/50% acetonitrile to gel pieces.

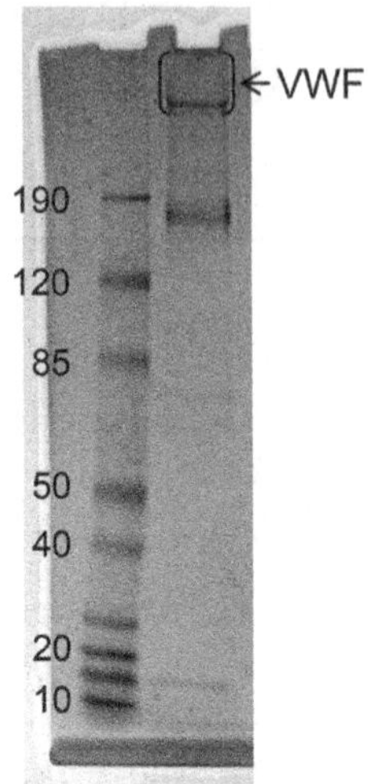

Fig. 1 Analysis of von Willebrand factor immunoprecipitated from plasma by nonreducing SDS-PAGE. Example of IPA-labeled von Willebrand factor resolved on SDS-PAGE and stained with Coomassie blue. Molecular mass standards are shown in the left-hand lane. The band corresponding to von Willebrand factor multimers, indicated with the bracket, is excised for subsequent in-gel processing of samples

3. Vortex gel pieces for 10 min at 22°C using thermomixer (Eppendorf). Remove supernatant by pipetting.
4. Repeat Steps 3 and 4 for another two washes of the gel pieces.
5. Dry gel pieces by adding 100 μL of 100% acetonitrile. Remove acetonitrile by pipetting (*see* **Note 4**).
6. Dry gel pieces in Speedvac for 15–20 min to remove traces of acetonitrile prior to enzymatic processing of gel pieces.
7. Add 35 μL of freshly diluted chymotrypsin to gel pieces and incubate at 37°C for 4 h.
8. Add 30 μL of freshly diluted trypsin to gel pieces and incubate at 30°C overnight.
9. Stop digestion of protein by addition of 10 μL of 5% formic acid.
10. Transfer digest solution which contains peptides to fresh tubes.
11. Elute remaining peptides from gel slices by adding 30 μL of 5% formic acid/50% acetonitrile. Vortex for 30 min at 22°C.
12. Transfer supernatant to tubes containing peptides.
13. Repeat Step 12 for another elution of peptides from gel pieces.
14. Dry peptides with Speedvac.

Table 1
Solvent gradient used in LC

Time (min)	% Buffer A	% Buffer B
0	97	3
20	97	3
22	90	10
44	60	40
47	5	95
50	5	95
50.1	95	5
60	95	5

Acetonitrile gradient for peptide elution is emphasized in bold

3.3 LC-MS/MS analysis

1. For LC MS/MS analysis, reconstitute peptides in 12 μL of 0.1% formic acid (*see* **Note 5**).
2. This protocol is based on the use of a Thermo Scientific Dionex Ultimate 3000. Peptides (3 μL) are injected and resolved on a 35 cm × 75 μm C18 reverse phase analytical column with integrated emitter using a gradient of 2%–35% acetonitrile over 22 min (Table 1) with a flow rate of 300 nL/min (*see* **Note 6**).
3. Peptides are ionized by electrospray ionization at +2.0 kV. Tandem mass spectrometry analysis is performed on a Q-Exactive Plus mass spectrometer (Thermo Scientific) using CID fragmentation. The scan range for full MS is set at 350–1750 m/z. The spectrum data type for full MS is profile mode, whereas for tandem MS is centroid mode.
4. Data-dependent acquisition method is used to obtain MS/MS spectra of the top five most abundant ions at any one point during the gradient.

3.3.1 Database Search and Quantification of Cys-Containing Peptides

1. The monoisotopic mass/charge ratio (m/z) of +2, +3, and +4 are calculated using Peptide/Protein MS Product Utility Program MS-Product available via Protein Prospector Tools (http://prospector.ucsf.edu/prospector/cgi-bin/msform.cgi?form=msproduct). The cysteine thiol bond formation has a mass loss of −2.01565 (Table 2). For example, the bond formation between the two cysteines in the peptide ASPENPCLINECVR will be ASPENPC(−2.01565) LINECVR. Their monoisotopic m/z ratios calculated using MS-Product from Protein Prospector is shown in Table 3. Values for m/z ratios should be at least four decimal places. The second Cys of the peptide is not labeled because it is forming a

Table 2
Monoisotopic mass related to amino acid modification

Amino acid	Modification	Monoisotopic mass*	Composition*
Cys	Iodoacetanilide	133.052764	H(7) C(8) N O
Cys	CCdisulf	−2.01565	H(2)

*Values and composition are obtained from Unimod (www.unimod.org/)

Table 3
Monoisotopic m/z ratios for von Willebrand factor ASPENPCLINECVR peptide containing an intra-peptide disulfide bond

		m/z			
Cys-label	Sequence	+1	+2	+3	+4
CCdisulf	ASPENPC[−2.01565] LINEC[+0.00000]VR	1542.7039	771.8556	514.9062	386.4314

m/z values are calculated using MS-Product tool from Protein Prospector

bond with the first one. For MS-Product, we can either put as modification −2.01565 or CCdisulf (Table 2).

1. To calculate ion abundance of peptides, extracted ion chromatograms were generated using the XCalibur Qual Browser software (v2.1.0; Thermo Scientific). Set autoprocessing parameters and display options for chromatogram and spectra as follows:

   (i) Make the chromatogram cell (top panel) active by selecting the pin  at the upper right corner. To define parameters for autoprocessing for chromatograms, choose **Display** and then **Ranges** from the drop-down menu. When the Ranges dialog box opens, select **Autoprocessing** and click **Use user defined** for Mass Tolerance and set it as 6 ppm. Mass precision is set at 5 Decimals. For display options, choose **Display** and then **Display Options**. Select **Labels** and choose **Retention time** (2 Decimals) and **Area**.

   (ii) Make the spectrum cell (bottom panel) active by selecting the pin  at the upper right corner. To define parameters for auto processing for spectra, choose **Display** and then **Ranges** from the drop down menu. When the Ranges dialog box opens, select **Autoprocessing** and click **Use user defined** for Mass Tolerance and set it as 6 ppm. Mass precision is set at 5 Decimals. For display options, choose **Display** and then **Display Options**. Select **Labels** and choose **Mass** (5 Decimals).

(iii) To save the settings, select **File** and then **Layout** from the drop-down menu, choose **Save as Default**. The above settings for auto processing and display will be used for subsequent quantification of MS peaks.

2. The m/z ratios (+2 and +3) calculated from MS-Product are used to search ions containing the corresponding m/z values. Select the chromatogram cell and choose **Display** the **Ranges** from the drop-down menu. When the Ranges dialog box opens, in the Plot properties, choose Scan filter **FTMS + p NSI Full mass [350.00–1750.00]** from the drop-down menu. Enter the values for +2, +3, and +4 m/z in Ranges.

3. Click the automated peak detection function built into the program to calculate peak area (Fig. 2).

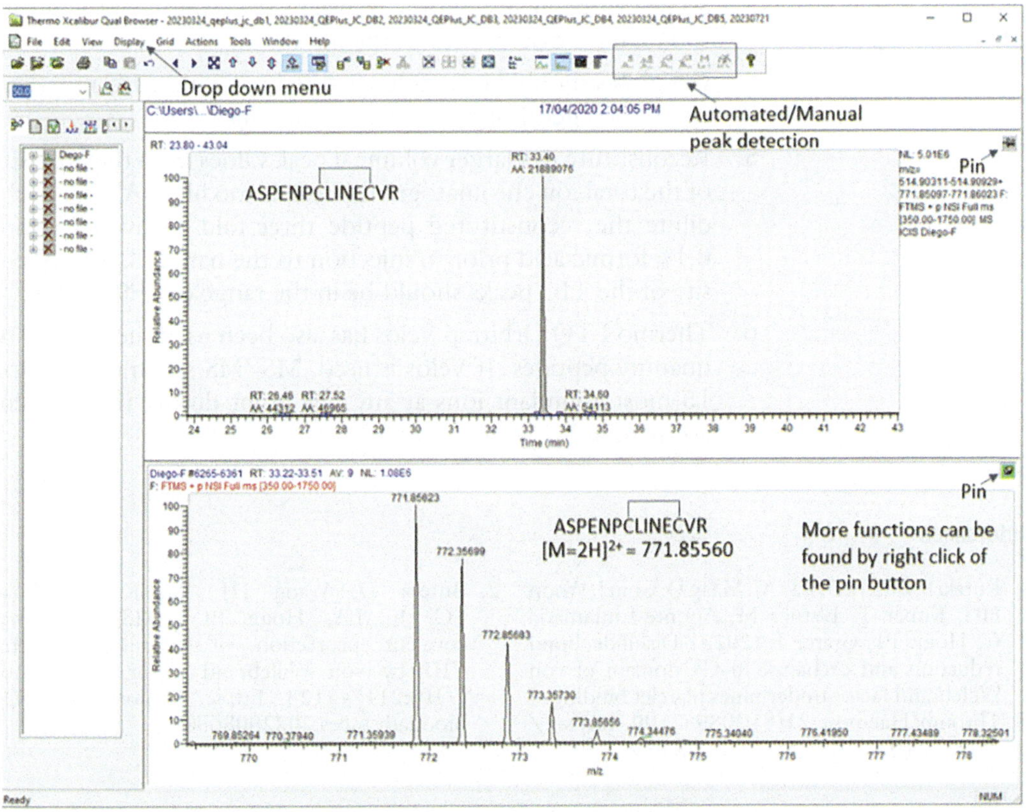

Fig. 2 Screenshot of the XCalibur Qual Browser Window showing automated peaks detected for the peptide APENPCLINECVR where the 2 cysteines form a disulfide bond. The automated area is used for quantification of their abundance. The automated peak corresponds to m/z of 771.85560 with an isotopic mass distribution of 0.5 confirming their m/z ratios as +2. Utilities such as drop-down menu, pin button and automated peak detection are shown by black arrows

4. Automated peaks from Qual Browser are verified by checking the isotopic distribution of the m/z ratios (+2 or +3 or both). For m/z ratio of +2, the isotopic peaks should differ by 0.5 (Fig. 2) and for m/z ratio of +3, the isotopic peaks should differ by 0.33.

4 Notes

1. Thiol alkylators are light sensitive.
2. Mascot Server is not a freely available database search engine. MS-Fit from Protein Prospector (http://prospector.ucsf.edu/prospector/cgi-bin/msform.cgi?form=msfitupload) can be used as alternative.
3. Some protocols suggest the transfer of two thirds of the serum to minimize disturbance of the middle layer.
4. Gel pieces should be completely dried out within 1–2 min. Can store dried gel slices after this step at −30°C.
5. Reconstitute in a larger volume if peak values from the intensity of the total ion chromatogram (TIC) is too high. Alternatively, dilute the reconstituted peptide three-fold to five-fold with 0.1% formic acid prior to injection to the nanoLC. The intensity of the TIC peaks should be in the range of 1E9–1E10.
6. Thermo LTQ Orbitrap Velos has also been used successfully to quantify peptides. If Velos is used, MS/MS spectra of the top 15 most abundant ions at any one point during the gradient are acquired.

References

1. Kutzki F, Butera D, Lay AJ, Maag D, Chiu J, Woon HG, Kubar T, Elstner M, Aponte-Santamaria C, Hogg PJ, Grater F (2023) Disulfide bond reduction and exchange in C4 domain of von Willebrand factor undermines platelet binding. J Thromb Haemost 21(8):2089–2100. https://doi.org/10.1016/j.jtha.2023.03.039
2. Butera D, Wang HJ, Woon HG, Zhao YC, Ju LA, Hogg PJ (2023) Mechano-covalent protection of coagulation factor VIII by von Willebrand factor. Blood Adv 7(10):2117–2128. https://doi.org/10.1182/bloodadvances.2022008650

Chapter 13

Determining the Redox Potential of a Protein Disulfide Bond

Kristina M. Cook

Abstract

The redox potential of a protein disulfide bond is one of the most important factors for determining the role of a disulfide bond. Disulfide bonds could have a stabilizing role for the structure of a protein or they can play a functional role that can regulate protein activity or binding. Determining the redox potential of disulfides can help distinguish the functional from the structural disulfide bonds. In this chapter, two methods for determining the redox potential of a protein disulfide bond are described. The first method uses maleimide-biotin labeling of free cysteine thiols and western blot densitometry to determine the fraction of reduced disulfide bond under various redox buffering conditions. The second method uses differential cysteine labeling and tandem mass spectrometry to determine the redox potential.

Keywords Redox potential, Disulfide bond, Protein, Thiols, Cysteine labeling, Tandem mass spectrometry

1 Introduction

Protein disulfide bonds are widely distributed through the proteome, serving one of two purposes for the protein, structural or functional. While structural disulfide bonds are not actively reduced and do not change in redox state, functional bonds can change redox state and act as an important regulatory mechanism for the protein. A critical factor in deciding whether a disulfide bond is structural or functional is the redox potential. Redox potentials for functional disulfide bonds (including oxidoreductases) range from –89 mV to –330 mV [1–4]. Structural disulfide bonds typically have redox potentials as low as –470 mV [2]. Oxidoreductases will only cleave a disulfide bond with a larger (less negative) redox potential, meaning that oxidoreductases do not cleave structural bonds. For examples of experimentally derived redox potentials of functional disulfide bonds see Tables 1–3.

There are two types of functional bonds: catalytic and allosteric. Catalytic bonds are present within the active sites of oxidoreductases, such as thioredoxin [5, 6] while allosteric disulfide bonds regulate

Philip J. Hogg (ed.), *Functional Disulphide Bonds*, Methods in Molecular Biology, vol. 3016,
https://doi.org/10.1007/978-1-0716-5158-2_13,

Table 1
Examples of redox potentials for human oxidoreductase active site disulfide bonds

Protein	Redox Potential	Reference
ERp57 a and a' domain	−167 mV, −156 mV	[25]
ERp5 a and a' domain	−206 mV, −211 mV	[26]
PDI a' and a domain *(Human)*	−169 mV, −163 mV	[27]
Thioredoxin	−270 mV	[28]

Table 2
Examples of redox potentials for common reducing agents

Reducing agent	Redox potential	Reference
Dithiothreitol	−332 mV, −307 mV	[29, 22]
Glutathione/glutathione disulfide (GSH/GSSG)	−240 mV	[20]

Table 3
Examples of redox potentials for identified human allosteric disulfide bonds

Protein	Redox potential	Reference
Adenosine 5'-phosphosulfate (APS) kinase	−255 mV	[30]
Angiotensinogen	−230 mV	[8]
βII-tryptase	−301 mV	[7]
CD4	−241 mV	[31]
Lon protease	−227 mV	[32]
Methionine aminopeptidase 2	−261 mV	[18]
Tissue factor	−278 mV	[16]
Transglutaminase 2	−184 mV	[33]

the function of the protein in which they reside (Fig. 1). Cleavage or formation of allosteric disulfide bonds can alter enzymatic activity (e.g., β-tryptase substrate hydrolysis and catalytic efficiency [7]) or lead to changes in ligand binding (β2-glycoprotein I and β3 integrin, reviewed in [5]). Altered proteolysis and oligomer formation have also been documented (angiotensinogen [8] and VWF [9]).

Identifying allosteric disulfides and understanding what leads to a change in state (reduced or oxidized) is essential to understanding

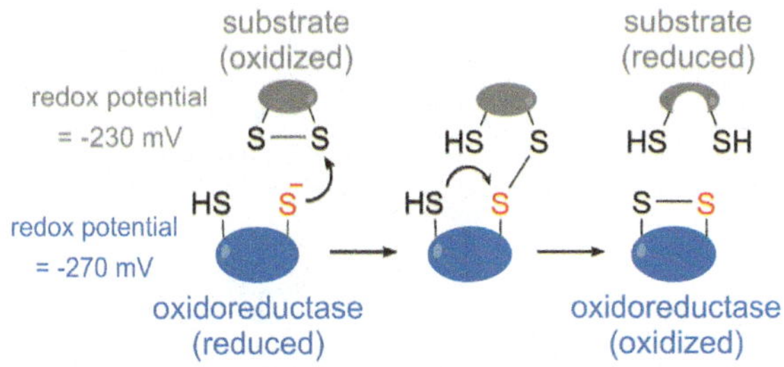

Fig. 1 Cleavage of an allosteric disulfide bond by an oxidoreductase. The redox potential of the allosteric disulfide bond is one of the keys to determining whether the bond will be cleaved by an oxidoreductase. The catalytic disulfide within the oxidoreductase will only cleave an allosteric disulfide bond with a larger (less negative) redox potential. Calculating the redox potential of an allosteric disulfide bond is, therefore, important for understanding how it might be regulated by an oxidoreductase. Reduction of the allosteric disulfide bond can alter the substrate protein structure, thereby controlling protein function or activity [5]

how the disulfide bond controls protein function. Allosteric disulfide bonds could be reduced by thiol-disulfide exchange or by the catalytic disulfide bonds of oxidoreductases. Whether or not an allosteric disulfide bond is cleaved by an oxidoreductase is determined by several aspects, including steric factors, such as accessibility and conformation of the substrate and oxidoreductase disulfide bond. Cellular/extracellular localization of the proteins can also determine whether a substrate is likely to encounter a particular oxidoreductase. Finally, the redox potentials of both the oxidoreductase catalytic disulfide and the substrate allosteric disulfide are critical in determining whether cleavage of the allosteric disulfide will occur (the allosteric bond must be larger/less negative than the oxidoreductase bond). There are a number of human allosteric disulfide bonds for which the redox potential has not been calculated (β2-glycoprotein I [10], Factor XI [11], MICA [12], plasminogen [13], and more [14]), and the protocols described within this chapter can be used to further this area.

The overall structure of a protein, the strain of the disulfide bond, and the electrostatic environment of the protein can all affect the redox potential of the disulfide bond [15]. Several methods have been used to determine the redox potential of disulfide bonds, and the technique is often selected based on the individual properties of the protein of interest. Some of the protein-unique methods to determine a disulfide redox potential include identifying shifts in the size of the tissue factor protein on SDS-PAGE [16] or observing changes in the intrinsic tryptophan fluorescence of DsbA and its variants [17]. Two particular techniques will be highlighted in this methods chapter (Fig. 2). The first technique (*Basic Protocol*) uses maleimide-biotin (MPB) labeling (which specifically tags free thiols

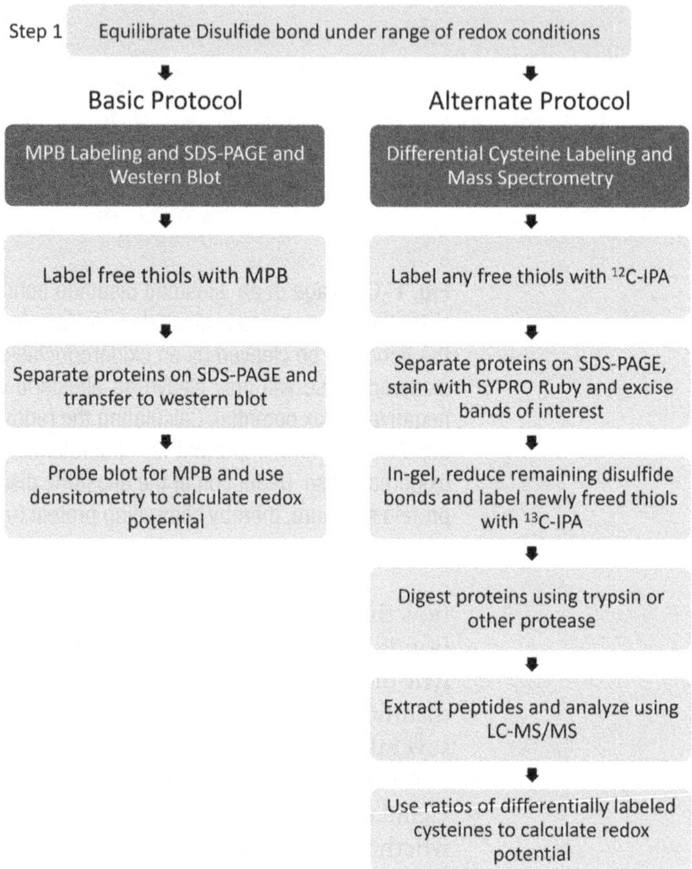

Fig. 2 Simplified flow diagram of the *Basic and Alternate Protocols* for determining the redox potential of a disulfide bond

and not cysteines participating in a disulfide bond) under a range of redox buffering conditions, followed by SDS-PAGE and western blotting. Densitometry of the resulting western blot can be used to calculate the percentage of reduced disulfide bond within the protein. This technique works particularly well when recombinant purified protein is available, and there is a single functional disulfide bond within the protein [7, 16, 18].

The second technique (*Alternate Protocol*) uses differential cysteine labeling [19] and tandem mass spectrometry to quantify oxidized versus reduced cysteines under various redox buffering conditions [9, 20, 21]. The cysteine thiolate-specific labels, iodoacetamide derivatives ^{12}C-IPA and ^{13}C IPA, are used to sequentially label the reduced and oxidized forms of the disulfide bond [19]. This method can be used to calculate the redox potential of a particular disulfide bond in a protein with multiple disulfides. It is also possible to use this technique with proteins that are not purified, though this is considerably more difficult than when using a highly abundant and/or pure protein. The highest chance of success for detecting a particular peptide tends to be with highly abundant, semipure or purified proteins.

2 Materials for SDS-PAGE and Western Blotting Based Redox Determination

Basic Protocol

2.1 Equilibration of Substrate Disulfide Bond Under a Range of Reducing Conditions

1. Phosphate-buffered saline (PBS) containing 0.1 mM EDTA, pH 7.4
2. Protein of interest ("substrate protein")
3. "Substrate stock 1": PBS containing 0.1 mM EDTA, pH 7.4, 0.55 mM oxidized DTT (DTT_{ox}: 4,5-dihydroxy-1,2-dithiane), and 1.1 μM substrate protein (*see* **Note 7.1.1**)
4. 100 mM stock of DTT_{red} (reduced dithiothreitol) in PBS, 0.1 mM EDTA
5. 55 μM stock of MPB (3-(*N*-maleimido-propionyl) biocytin (MPB), (Thermo Scientific)) in DMSO
6. 5 mM glutathione in PBS

2.2 Resolving MPB Labeled Samples via SDS-PAGE and Western Blot

1. NuPAGE or Bolt 4%–12% Bis-Tris gels (Thermo Scientific)
2. 4× NuPAGE or BOLT LDS sample loading buffer
3. MOPS 20× SDS-PAGE running buffer, (Thermo Scientific)
4. PVDF membranes (Millipore)
5. Streptavidin-peroxidase (Thermo Scientific)
6. Tris-buffered saline-Tween (TBST) containing 5% bovine serum albumin (BSA) (w/v)
7. Chemiluminescent substrate

2.3 Converting the Western Blot Images to Numerical Values Using Densitometry

1. ImageJ software (available free at https://imagej.nih.gov/ij/index.html)

2.4 Calculating the Redox Potential from the Densitometry Values

1. Prism software (GraphPad) or other graphing software

3 Methods for SDS-PAGE and Western Blotting Based Redox Determination

All procedures should be performed at room temperature unless otherwise specified. Stocks of DTT_{ox} and DTT_{red} (*see* **Note 7.1.1**) should be made up fresh on the day. All buffers and water should be flushed with argon for at least 1 h prior to experiments to remove dissolved oxygen that can interfere with redox reactions.

3.1 Equilibration of Substrate Disulfide Bond Under a Range of Reducing Conditions

Fifteen separate reactions would be made, each containing a different ratio of reduced DTT to oxidized DTT ($[DTT_{red}]/[DTT_{ox}]$). At the final concentrations, each reaction should contain PBS, 0.1 mM EDTA, 1 μM substrate protein, and 0.5 mM DTT_{ox}. A different concentration of DTT_{red} will also be present in each reaction to create a range of redox buffering conditions (*see* **Note 7.1.2**).

1. Make up 15 reactions in 0.6 mL tubes, with each containing 45 μL of "Substrate stock 1" (1.1 μM substrate protein and 0.55 mM DTT_{ox} in PBS and 0.1 mM EDTA).
2. Separately, 15 stocks of DTT_{red} (A–O) (Table 4) need to be made at 10× final desired concentration. First, make up a 100 mM stock of DTT_{red} in PBS, 0.1 mM EDTA. This stock will be added to the 15 tubes containing "substrate stock 1" to make the various ratios of DTT_{red}/DTT_{ox}. These ratios

Table 4
DTT_{red} stocks for Step 3.1.2 under Sect. 3.1

Reaction	Final 10× concentration of DTT_{red}	← To make this stock, Combine →	Volume of PBS, 0.1 mM EDTA	Volume of 100 mM DTT_{red} (stock to add)
A	0		1000 μL	0
B**	100 μM		990 μL	10 μL of **10 mM DTT_{red}
C	1 mM		990 μL	10 μL
D	2 mM		980 μL	20 μL
E	4 mM		960 μL	40 μL
F	6 mM		940 μL	60 μL
G	8 mM		920 μL	80 μL
H	10 mM		900 μL	100 μL
I	11.75 mM		882.5 μL	117.5 μL
J	13.5 mM		865 μL	135 μL
K	15 mM		850 μL	150 μL
L	17 mM		830 μL	170 μL
M	18.75 mM		812.5 μL	187.5 μL
N	20 mM		800 μL	200 μL
O	25 mM		750 μL	250 μL

Table 5
Reaction mixtures for Step 3.1.3 under Sect. 3.1

Reaction #	1*	2	3	4	5	6	7	8	9
Vol. of "Substrate stock 1"	*See* **Notes 7.1.4 and 7.1.10**	45 μL	45 μL	45 μL	45 μL	45 μL	45 μL	45 μL	45 μL
Vol of 10× DTT_{red} rxn	5 μL of A	5 μL of B	5 μL of C	5 μL of D	5 μL of E	5 μL of F	5 μL of G	5 μL of H	5 μL of I

Reaction #	10	11	12	13	14	15*
Vol. of "Substrate stock 1"	45 μL	45 μL	45 μL	45 μL	45 μL	*See* **Note 7.1.4**
Vol of 10× DTT_{red} rxn	5 μL of J	5 μL of K	5 μl of L	5 μL of M	5 μL of N	5 μL of O

determine the redox buffering environment for the disulfide bond (*see* **Notes 7.1.3** and **7.1.4**).

3. Add 5 μL of each 10× DTT_{red} stock to the corresponding 45 μL of "substrate stock 1" according to Table 5. The final ($[DTT_{red}]/[DTT_{ox}]$) ratios for each reaction are in Table 6.
4. After mixing reactions, microcentrifuge tubes should be quickly flushed with argon gas prior to sealing to minimize oxidation during the incubation period.
5. Incubate tubes for 18 h at room temperature to allow redox equilibrium to be reached.
6. Add 5 μLof a 55 mM stock of MPB to each reaction to react with and label all free thiols. This results in a final reaction concentration of 5 mM MPB.
7. Incubate the reactions containing MPB for 1 h at room temperature to label all free thiols present in the reaction.
8. Excess, unreacted MPB can be quenched by adding 5 mM glutathione and incubating at room temperature for 60 min.

3.2 Resolving MPB Labeled Samples via SDS-PAGE and Western Blot

1. Mix each sample with 18.3 μL of 4× NuPAGE LDS sample buffer and resolve using MOPS buffer and SDS-PAGE (*see* **Note 7.1.5**).
2. Transfer gel to PVDF using semidry or wet transfer methods.
3. The blot should be blocked in Tris-buffered saline-Tween (TBST) containing 5% bovine serum albumin (BSA) or equivalent blocking buffer for 1 h at room temperature.

Table 6

Final ([DTT_{red}]/[DTT_{ox}]) ratios in each reaction for Step 3.1.3 under Sect. 3.1. Additionally, each reaction contains PBS, 0.1 mM EDTA, and 1 μM substrate protein, final concentration. For reactions 1* and 15see* Note 7.1.4**

Reaction #	Final ([DTT_{red}]/[DTT_{ox}])	([DTTred]/[DTTox]) ratio value
1*	([0]/[0]), Native state of the disulfide bond, no redox buffering	NA
2	([10 μM]/[0.5 mM])	0.02
3	([100 μM]/[0.5 mM])	0.2
4	([200 μM]/[0.5 mM])	0.4
5	([400 μM]/[0.5 mM])	0.8
6	([600 μM]/[0.5 mM])	1.2
7	([800 μM]/[0.5 mM])	1.6
8	([1 mM]/[0.5 mM])	2
9	([1.175 mM]/[0.5 mM])	2.35
10	([1.35 mM]/[0.5 mM])	2.7
11	([1.5 mM]/[0.5 mM])	3
12	([1.7 mM]/[0.5 mM])	3.4
13	([1.875 mM]/[0.5 mM])	3.75
14	([2 mM]/[0.5 mM])	4
15*	([2.5 mM]/[0 mM]), fully reduced disulfide, no DTT_{ox}	NA

4. Following blocking, incubate the blot in blocking buffer containing a 1:10,000 dilution of streptavidin-peroxidase for 1 h at room temperature (*see* **Note 7.1.6**).
5. Wash blot in TBST for a minimum of three washes, 10 min each.
6. Expose blot using chemiluminescent substrate and record exposures using radiography film or chemiluminescent digital imager.

3.3 Densitometry Analysis of Western Blot Images

1. Western blot images obtained from the streptavidin-peroxidase exposures are used to determine the total intensity of each band (*see* **Note 7.1.7**) (Fig. 3).
2. Download and open the free program ImageJ (*see* **Note 7.1.8**).
3. Open the Western blot image in ImageJ.
4. Using the rectangle tool, draw a box that fits over the bands of interest in lane 1. Ensure that enough space is included

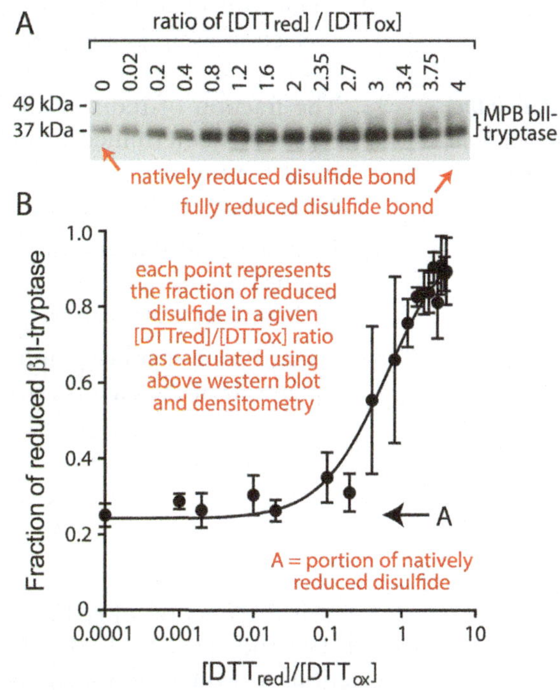

Fig. 3 Example of a western blot for MPB using the *Basic Protocol*. The βII-tryptase Cys[220]-Cys[248] disulfide bond was equilibrated in a range of redox conditions. The band intensity increases as the redox environment becomes more reducing. This indicates the disulfide bond is gradually being reduced, and there is an increase in free cysteine thiols for MPB labeling. Note the background is clean and even, without discoloration and the bands are distinct and clear. Reproduced from Cook et al. [7] with permission from the American Society for Biochemistry and Molecular Biology

both above and below the band so that all bands of interest and background in the adjacent lanes will also be captured in further lanes when the boxes are aligned.

5. On the toolbar, select "Analyze" → "Gels" → "Select Lane 1." A number 1 should appear in the box over the band.
6. Using the mouse, grab the box and place over the band in lane 2. On the toolbar, select "Analyze" → "Gels" → "Select Next Lane."
7. Continue repeating Step 6 until all lanes have been selected.
8. On the toolbar, select "Analyze" → "Gels" → "Plot Lanes." This will generate the lane profile plots.
9. Use the straight line selection tool to draw baselines and drop lines for the peak(s) (*see* **Note 7.1.9**).
10. Measure the area under the curve (AUC) for each peak by clicking the "Wand (tracing) tool" button in the toolbar and then clicking each enclosed AUC.

11. This will generate a new window with the AUC for each selection listed as a numerical value.

3.4 Calculating the Redox Potential from the Densitometry Values

1. The fraction of reduced disulfide bond in a particular redox reaction is determined from the intensity (AUC) of the MPB band for the reaction divided by the intensity (AUC) of the MPB band for the fully reduced protein (reaction 15*).
2. Using GraphPad Prism, the fraction of reduced protein (y-axis) is graphed against the ratio value for ($[DTT_{red}]/[DTT_{ox}]$) (x-axis). Biological replicates can be combined to give standard deviations.
3. Using Prism or other graphing software, a best nonlinear least squares fit of the data is calculated using Equation:

$$R = \frac{\left(\frac{[DTT_{red}]}{[DTT_{ox}]}\right)}{K_{eq} + \left(\frac{[DTT_{red}]}{[DTT_{ox}]}\right)}$$

R is the fraction of reduced protein at equilibrium (*see* **Note 7.1.10**) and ($[DTT_{red}]/[DTT_{ox}]$) is the ratio for the individual reaction. An equilibrium constant (K_{eq}) is generated from the best fit line (*see* Fig. 3).

4. The value for K_{eq} is then used in the Nernst equation to calculate the redox potential of the protein's disulfide bond or $E^{0'}$. Nernst equation:

$$E^{0'} = E^{0'}_{DTT} - \frac{RT}{2F} \ln K_{eq}$$

R is the universal gas constant and equivalent to $8.314 J/K^{-1} mol^{-1}$. T is temperature in Kelvin, while F is the Faraday constant and equal to $9.6485 \times 10^4 C/mol^{-1}$. $E^{0'}_{DTT}$ is the redox potential of the DTT disulfide bond, which has been found to be −307 mV [22]. $E^{0'}$ is the calculated redox potential of the disulfide bond in the protein of interest.

Alternate Protocol

4 Mass Spectrometry Redox Determination

One of the potential downsides to the MPB-labeling and western blot method is that it works best if there is a single disulfide bond (or there is a significant difference in redox potential between multiple bonds) within the protein. MPB labels all free thiols in a protein, including free cysteines not part of a disulfide, which can interfere with determining the redox potential. A method to overcome these

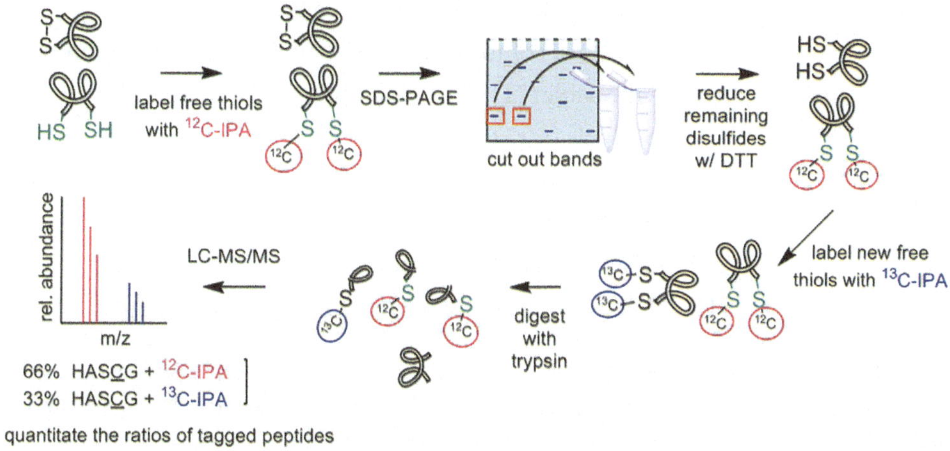

Fig. 4 Differential cysteine labeling and mass spectrometry protocol. The protocol initially labels any free cysteine thiols with ^{12}C-IPA. The labeled protein is isolated using SDS-PAGE and the remainder of the protocol is carried out in-gel. All remaining disulfide bonds are reduced by DTT, then free cysteine thiols are further labeled with ^{13}C-IPA. The protein is digested into smaller peptide fragments by a proteolytic enzyme, such as trypsin or chymotrypsin. The complex mixture of peptides are separated using liquid chromatography and subject to MS/MS analysis for sequencing of peptides

caveats involves differential cysteine labeling [19] and tandem mass spectrometry [7, 9, 20, 21]. Differential cysteine labeling uses one alkylating reagent to label any reduced cysteines, and then following treatment of the protein with a strong reducing agent, such as DTT, a second and different cysteine label is used to label any newly reduced thiols (Fig. 4). "Heavy" and "light" chemically identical labels, such as 2-iodo-*N*-phenylacetamide (^{12}C-IPA) and 2-iodo-*N*-phenylacetamide (^{13}C-IPA) can be used [19]. This results in the two labels having the same retention time for a given peptide in the HPLC component of the mass spectrometry. Ratios of heavy to light are used to determine the degree of disulfide reduction.

5 Materials for Mass Spectrometry Redox Determination

5.1 Equilibration of Substrate Disulfide Bond Under a Range of Reducing Conditions

See materials for Sect. 2.1.

5.2 In-Solution Differential Cysteine Labeling with ^{12}C-IPA and SDS-PAGE

1. 2-iodo-*N*-phenylacetamide (^{12}C-IPA), molecular weight = 261.06 g/mol, Cambridge Isotopes.
2. Gel fixative: 50% methanol, 7% acetic acid solution (mix 50 mL methanol, 7 mL acetic acid, and 43 mL ultrapure water).

3. SYPRO Ruby (ThermoFisher Scientific).
4. Gel wash solution: 10% methanol, 7% acetic acid in water.
5. Ultraviolet (UV) or blue light transilluminator.
6. Scalpels or razorblades.
7. Acetonitrile (HPLC grade).
8. Methanol (HPLC grade).
9. 25 mM ammonium bicarbonate.
10. Ultrapure water (18 MΩ/cm recommended).

5.3 In-Gel Differential Cysteine Labeling with ^{13}C-IPA

1. 25 mM ammonium bicarbonate/50% acetonitrile solution: make by dissolving 1.98 g of ammonium bicarbonate into 500 mL ultrapure water and then mixing with 500 mL of acetonitrile to make a total volume of 1 L.
2. 100 mM DTT.
3. 5 mM2-iodo-*N*-phenylacetamide(^{13}C-IPA),molecularweight= 267.02 g/mol, Cambridge Isotopes: make by dissolving 1.3 mg of ^{13}C-IPA into 500 μL DMSO and then mixing the dissolved ^{13}C-IPA with 500 μL of 50 mM ammonium bicarbonate (*see* **Note 7.2.5**).
4. DMSO.

5.4 Proteolytic Digestion of Labeled Protein

1. Acetic acid (HPLC grade).
2. Vortex/ thermomixer (Eppendorf).
3. Shaker or rocker.
4. Trypsin, Chymotrypsin, and other proteases as needed (Promega and Roche).
5. 50% acetonitrile/5% formic acid: make stock by mixing 500 μ L of acetonitrile with 50 μL of formic acid and 450 μL of ultrapure water.

6 Methods for Mass Spectrometry Redox Determination

6.1 Equilibration of Substrate Disulfide Bond Under a Range of Reducing Conditions

1. Complete Steps 1–6 for Sect. 3.1 Equilibration of Substrate Disulfide Bond under a range of reducing conditions, and move to Sect. 6.2.

6.2 In-Solution Differential Cysteine Labeling with ^{12}C-IPA and SDS-PAGE

1. Make 55 mM stock of 2-iodo-*N*-phenylacetamide (^{12}C-IPA) by dissolving 1.4 mg of ^{12}C-IPA into 100 μL DMSO. Add 5 μL to each reaction to label all free thiols. This gives a final concentration of 5 mM ^{12}C-IPA in all reactions. Incubate for 1 h room temperature in the dark.

2. Mix each sample with 18.3 µL of 4× NuPAGE LDS sample buffer and resolve using SDS-PAGE and MOPS buffer.
3. Add the gel to a clean tray containing 100 mL gel fixative and place on a rocker for 1 h room temperature (*see* **Note 7.2.1**).
4. Remove fixative and add 60 mL SYPRO Ruby stain for 1 h or up to overnight depending on the amount of substrate protein added (*see* **Note 7.2.2**).
5. Remove SYPRO Ruby stain and add 100 mL of gel wash solution. Place on rocker for 30 min.
6. View bands by placing gel on UV or blue light transilluminator (*see* **Note 7.2.3**).
7. Use a clean scalpel or razor blade to excise each band for the protein of interest (*see* **Note 7.2.4**) and place each gel band into individually labeled 1.6 mL microfuge tubes.

6.3 In-Gel Differential Cysteine Labeling with ^{13}C-IPA

1. Add 100 µL of 25 mM Ammonium bicarbonate/50% acetonitrile to each of the microfuge tubes containing gel pieces, vortex for 10 min at room temperature. Remove and discard supernatant.
2. Repeat wash and 10 min vortex twice more, and remove supernatant leaving only the gel band.
3. Dry gel pieces in centrifugal evaporator for 10–15 min or until gel pieces shrink and harden.
4. Add 50 µL of the 100 mM DTT solution to each microfuge tube containing the dehydrated gel pieces. If 50 µL is not sufficient to rehydrate gel pieces, top up reactions with 25 µL as needed.
5. Incubate reactions at 37°C for 1 h, then remove from heat. Allow reactions to cool to room temperature.
6. Centrifuge tubes in a microcentrifuge and aspirate off any excess DTT.
7. Add 100 µL of 25 mM ammonium bicarbonate/50% acetonitrile to each tube and vortex for 10 min at room temperature. Remove and discard supernatant.
8. Repeat Step 10 twice more, and remove supernatant leaving only the gel band.
9. Dry gel pieces in centrifugal evaporator for 10–15 min or until gel pieces shrink and harden.
10. Add 50 µL of the 5 mM ^{13}C-IPA solution (in 50% DMSO/50% 25 mM ammonium bicarbonate) to each microfuge tube containing the dehydrated gels.
11. Incubate for 1 h room temperature in the dark.
12. Centrifuge tubes in a microcentrifuge and aspirate off and dispose of excess ^{13}C-IPA labeling buffer. Add 100 µL of 25 mM

ammonium bicarbonate/50% acetonitrile, vortex for 10 min at room temperature. Remove and discard supernatant.

13. Repeat wash and 10 min vortex twice more, and remove supernatant leaving only the gel band.
14. Dry gel pieces in centrifugal evaporator for 10–15 min or until gel pieces shrink and harden.

6.4 Proteolytic Digestion of Labeled Protein

1. Make a 20 ng/μL mass spectrometry grade trypsin (or other protease as needed, *see* **Note 7.2.6**) stock in 25 mM ammonium bicarbonate.
2. Add 25 μL of the trypsin stock directly to the dehydrated gel pieces and incubate on ice until gel piece is rehydrated. Top up with 25 mM ammonium bicarbonate if needed to completely cover gel pieces.
3. Incubate tubes at 37°C overnight.
4. The following morning, add 5 μL of a 50% acetonitrile/5% formic acid to stop the reaction.
5. Aspirate supernatant to a fresh microfuge tube (this solution contains the peptides for mass spectrometry analysis).
6. Add 30 μL of 50% acetonitrile/5% formic acid to the gel pieces to extract remaining peptides and vortex for 15 min.
7. Centrifuge tubes in a microcentrifuge and aspirate off solution (this solution contains the peptides for mass spectrometry analysis) and add supernatant to microfuge tubes containing corresponding supernatants with peptides.
8. Place microfuge tubes with supernatants in centrifugal evaporators to dry solutions into a pellet containing peptides.

6.5 Mass Spectrometry of Labeled Peptides

Below protocol may vary slightly depending on mass spectrometry equipment availability and specific requirements for the peptides of interest. We recommend optimizing the protocol as needed.

1. Resuspend dehydrated peptide pellets in 10 μL of 0.1% formic acid, desalt samples and inject into a C18 precolumn (this will vary depending on available equipment, we have used a fritless nano column (75 μm × ~10 cm) containing C18 medium (5 μ, 200 Å Magic; Michrom) manufactured as described [23]).
2. Separate peptides using liquid chromatography on an Ultimate 3000 HPLC (Dionex) or equivalent. Peptides should be resolved using an acetonitrile gradient (we have used a linear gradient of H_2O:CH_3CN [98:2, 0.1% formic acid] to H_2O:CH_3CN [64:36, 0.1% formic acid] at 250 nL/min over 30 min).

3. Generate positive ions by electrospray using high voltage (2000 V) applied to a low volume tee (Upchurch Scientific), with the column tip positioned 0.5 cm from the heated capillary (280°C) of an Orbitrap Velos (Thermo Electron) operating in data-dependent acquisition mode.
4. The scan range for full MS is set to m/z 350–1750 (resolution 30,000 at m/z 400, with an accumulation target value of 1,000,000 ions) with lockmass enabled.
5. Data-dependent acquisition is used to obtain up to the ten most abundant ions (5000 counts).

6.6 Data Analysis of Mass Spectrometry Results

1. Convert LC-MS/MS files to Mascot generic format using a free conversion program such as RawConverter (others are also available on the web). This program is available free of charge at http://fields.scripps.edu/rawconv/ [24].
2. Use the program Mascot Daemon (Matrix Science) to analyze data files and display the identified peptides from your protein of interest.
3. Settings for Mascot Daemon should be adjusted for individual needs, but we generally use the following settings: Swissprot database and precursor tolerance of 6 ppm, with product ion tolerances of ±0.6 Da. For modifications, at a minimum ^{12}C-IPA and ^{13}C-IPA cysteine derivatives should be selected to identify the labeled cysteine peptides. Additionally, you may be more likely to find your peptides by selecting common variable modifications such as oxidation of methionine. Up to three missed proteolytic cleavages should be selected to ensure full peptide coverage.
4. The Mascot search results will provide a list of proteins identified based on the number of peptide hits. Within the individual proteins, the peptides identified will be listed, as well as protein sequence coverage. The peptides listed must contain your cysteines of interest. If none of the cysteines of interest are present in the peptides identified, further optimization of peptides can be done by varying the protease used in the protocol (*see* **Note 7.2.6**).
5. Using the identified cysteine peptides from the disulfide of interest, the monoisotopic mass/charge ratio (m/z) of +2, +3, and +4 should be calculated the Protein Prospector Tool: MS-Product (http://prospector.ucsf.edu/prospector/cgi-bin/msform.cgi?form=msproduct) or MS-Isotope (http://prospector.ucsf.edu/prospector/cgi-bin/msform.cgi?form=msisotope). MS-Isotope can be used with the chemical formula of the labeled peptides, while MS-product can be used to give the monoisotopic m/z with the peptide sequence and additional modifications selected. ^{12}C-IPA and ^{13}C-IPA add an

additional monoisotopic m/z mass-adduct for each peptide of 133.05276 and 139.07289, respectively (charge 1+). ^{12}C-IPA and ^{13}C-IPA add an additional C_8H_7NO and $^{13}C_8H_7NO$ to each peptide, respectively (*see* **Note 7.2.7**). m/z ratios should be carried out 4–5 decimal places. These m/z ratios will be used to search the MS spectrums to give the area for the peptide. The areas of the two differentially labelled peptides will then be used to calculate the fraction of each cysteine present as a disulfide bonded form or reduced form in each of the redox conditions.

6.7 Calculating the Ion Abundance of Labeled Peptides and the Fraction of Reduced Disulfide Bond

1. Once the m/z ratios have been calculated, the next step is to calculate the ion abundance of the peptides and the ratio of ^{12}C-IPA to ^{13}C-IPA cysteine labeling for a given cysteine.
2. Use XCalibur™ Qual Browser software (ThermoFisher Scientific) for generating extracted ion chromatograms.
3. Parameters for autoprocessing in XCalibur™ should be as follows: set "Mass tolerance" to 6 ppm and "Mass precision" to five decimals.
4. On the display screen, setup two window cells, with the top displaying the chromatogram (choose settings: display retention time and area labels) and the bottom displaying the spectrum (choose settings: display mass labels).
5. Within the chromatogram cell, open the ranges search function and enter the calculated m/z ratios for the +2 and +3 ratios for the ^{12}C-Cys-IPA labeled peptide (separate values by a comma).
6. On the second line, enter the calculated m/z ratios for the +2 and +3 ratios for the ^{13}C-Cys-IPA labeled peptide. Choose Scan filter "FTMS + p NSI Full mass [350.00–1750.00]."
7. The automated peak function will identify the peaks for the correct m/z and hits can be verified using the spectrum cell (Fig. 5). Results can also be cross checked using the identified retention time (RT) in XCalibur™ and compared to the RT identified by Mascot for the labeled peptide. The two RTs between XCalibur™ and Mascot should be the same. The RTs should also be the same for a given peptide whether it is labeled with ^{12}C-IPA or ^{13}C-IPA.
8. Once the areas for each of the ^{12}C-IPA and ^{13}C-IPA labeled peptides of a particular cysteine has been obtained, the fraction of reduced cysteine can be calculated by taking the area of a particular ^{12}C-IPA-peptide and dividing by the sum of the ^{12}C-IPA and ^{13}C-IPA labeled forms of the same peptide. This gives the fraction reduced for a particular redox buffering condition.

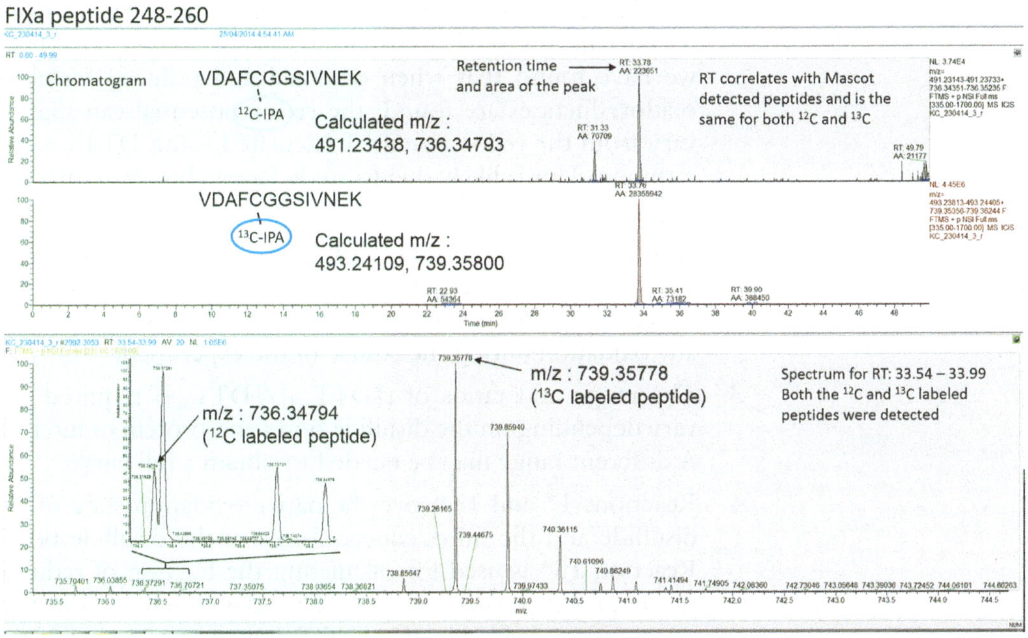

Fig. 5 Example mass spectrometry chromatogram and spectrum of a differentially labelled cysteine peptide. Chromatogram and spectrum generated using XCalibur™ (unpublished data). In this example, the peptide VDAFCGGSIVNEK contains a cysteine that participates in a disulfide bond in the full length protein (from which the peptide was derived). The *Alternate Protocol* described in the text was followed and data analyzed to assess the oxidation state. Both ^{12}C-IPA and ^{13}C-IPA labeled peptides were detected. The following areas were obtained for each peptide: ^{12}C (reduced form) = 223651 and ^{13}C (disulfide bond/ oxidized form) = 28355942. Using the below equation, the fraction of oxidized (disulfide bond) and reduced (free thiol) cysteine can be calculated

9. Steps 6.7.1–6.7.9 should be repeated for the same peptide for each reaction/redox buffering condition.
10. Then Steps 6.7.1–6.7.10 should be repeated for the peptide(s) containing the *other* cysteine that participates in the disulfide bond. The fraction of reduced cysteine for each individual cysteine in the disulfide bond should be similar.

6.8 Calculating the Redox Potential Using the Fractions of Reduced Disulfide Bond

1. Following the *Basic Protocol*, follow Steps 3.4.2–3.4.4 to graph the fraction of reduced disulfide and derive the best nonlinear least squares fit of the data and K_{eq}. The K_{eq} can be used to solve for the redox potential using the Nernst equation.

7 Notes

7.1 Notes on MPB Labeling Protocol

1. Redox buffers other than DTT can be used, including glutathione or oxidoreductase buffers, such as thioredoxin and protein disulfide isomerase (PDI). True redox potentials

should be independent of the reducing agent. This is likely to be true for small molecule reducing agents. However, we have found that when oxidized and reduced forms of oxidoreductases are found, the redox potential can slightly vary from the redox potential calculated using DTT or glutathione. This is likely due to steric factors between interacting proteins.

2. Solutions of both DTT_{red} and DTT_{ox} must be made fresh on the day of the experiment. They should be dissolved in water/buffers that have been flushed with argon to prevent any oxidation during the course of the experiment.
3. The range and ratios of ($[DTT_{red}]/[DTT_{ox}]$) required will vary depending on the disulfide bond and protein of interest. A different range may be needed to obtain a full curve.
4. Reactions 1* and 15* give the native oxidation state of the disulfide and the fully reduced forms of the disulfide bond. Reaction 15* is used for calculating the fraction of reduced protein for the other reactions. The AUC obtained for other reactions are divided by the AUC for reaction 15* to yield the fraction of reduced disulfide. Reaction 15* should be made separately from the other reactions as it should not contain DTT_{ox}. Make up as follows: In a 0.6 mL tube, add 1.1 μM substrate protein and make up to a total volume of 45 μL with PBS and 0.1 mM EDTA.
5. Reactions must be resolved on the same gel, as the calculations are based on relative density. Reactions from the same experiment run on different gels are not comparable without running on the same gel as reaction 15*.
6. We have found that the dilution ratio of streptavidin-peroxidase varies between companies. Different ratios may need to be tested to get optimal exposures depending on source of streptavidin-peroxidase, chemiluminescent substrate and method of imaging used.
7. Images of the exposed blots can be obtained in two ways. If traditional radiography film exposures are used, images can be scanned and converted to a digital form using a scanner. Otherwise, the newer digital imaging chemiluminescent gel docks convert chemiluminescence directly to a digital image and have greater sensitivity. We recommend the second method due to enhanced sensitivity.
8. ImageJ software is available free of cost at https://imagej.nih.gov/ij/index.html. If digital imaging is used, for example using the BioRad ChemiDoc™ imagers, proprietary software may also be used to determine density of bands according to manufacturer's instructions.

9. For best results, blots should be clean and free of background or artefacts, with sharp, even bands. This is usually easily achieved when using recombinant and purified proteins (for an example see Fig. 3). Blots can be optimized further by determining the amount of labeled protein that should be loaded to give sharp, clean bands. If the blot is not clean, interpreting the AUC and calculating the redox potential will be extremely difficult.
10. We have found that the native state of some proteins contains a small fraction of reduced disulfide bond. We have seen this using both gel and mass spectrometry based methods. To account for this, the offset value A has been included in the equation below and is only necessary when natively reduced disulfide is observed. The fraction of reduced disulfide bond in the native protein is calculated using reaction 1*, where no redox buffer is included in the reaction. The fraction of reduced disulfide in the nonredox buffered reaction is the amount of natively reduced disulfide, and is then accounted for in the below equation 1 as value A. The significance of this small percentage of reduced disulfide bond is unknown. If the native state of the disulfide bond contains no amount of reduced disulfide then the standard equation described in the protocol can be used (no offset):

$$R = \frac{A + \left\{(1 - A) * \left(\frac{[DTT_{red}]}{[DTT_{ox}]}\right)\right\}}{K_{eq} + \left(\frac{[DTT_{red}]}{[DTT_{ox}]}\right)}$$

7.2 Notes on Mass Spectrometry Redox Determination

1. Use powder free nitrile gloves to prevent contamination. Latex gloves should be avoided as the gloves can increase keratin and other protein contamination. Furthermore, care should be taken to avoid contamination with dust, hair, and skin that can lead to keratin in samples, which will affect mass spectrometry peptide detection.
2. In our experience, we find that 1 h incubations in SYPRO Ruby are enough to visualize bands containing 1 μg of protein. If less protein is used, we recommend longer staining periods.
3. Ultraviolet radiation is hazardous and correct safety equipment should be used to protect eyes and skin. Blue light transilluminators are a bright light source and specialized safety goggles are required.
4. Clean scalpels or razor blades should be used for each band or should be cleaned between cutting each band to minimize any cross contamination between samples. The minimum amount of gel should be cut to ensure complete excision of protein band. Furthermore, each band should be diced into ~1 mm^3 pieces to maximize surface area and exposure to reagents.

We also recommend using a fresh box of microfuge tubes to minimize contamination risk of dust, hair, and skin.

5. In the phenyl ring of 2-iodo-*N*-phenylacetamide (^{13}C-IPA), all six carbon atoms have a mass of 13. The benefit of using "light" (^{12}C) and "heavy" (^{13}C) analogues of the same molecule to label at different steps means that differentially labeled cysteines in the same equivalent peptide will have identical retention times in liquid chromatography.
6. The protease required for individual proteins will vary. Ideally, you want to find a protease that will cleave at sites within your protein to yield a peptide containing a single cysteine of interest and of a sequence length that will give you an m/z of ~350–1750. A variety of proteases are available including trypsin, chymotrypsin, Glu-C, Asp-N, and Lys-C. All of these proteases target different sequences for protein cleavage. The MS-Digest website (http://prospector.ucsf.edu/prospector/cgi-bin/msform.cgi?form=msdigest) can be useful in identifying the peptides produced by individual proteases and/or combinations of proteases. Different proteases may be used to obtain ideal peptides for different cysteines in the protein.
7. Microsoft Excel can be very useful in organizing the list of possible peptides, the chemical formulas of peptides and modifications, and their monoisotopic m/z for the different charges.

$$\textit{Fraction of reduced disulfide bond} = \frac{\textit{Area of reduced}}{\textit{Area of oxidized} + \textit{Area of reduced}}$$

This results in the following based on the example in Fig. 5:

$$\textit{Fraction of reduced disulfide bond} = 0.7\% = \frac{223651}{28355942 + 223651}$$

As expected, in this example, 99.3% of the disulfide bond is formed in the native protein, with only 0.7% in the reduced state.

References

1. Schmidt B, Ho L, Hogg PJ (2006) Allosteric disulfide bonds. Biochemistry 45(24):7429–7433. https://doi.org/10.1021/bi0603064
2. Gilbert HF (1990) Molecular and cellular aspects of thiol-disulfide exchange. Adv Enzymol Relat Areas Mol Biol 63:69–172
3. Wunderlich M, Glockshuber R (1993) Redox properties of protein disulfide isomerase (DsbA) from *Escherichia coli*. Protein Sci 2(5):717–726. https://doi.org/10.1002/pro.5560020503
4. Wouters MA, Fan SW, Haworth NL (2010) Disulfides as redox switches: from molecular

mechanisms to functional significance. Antioxid Redox Sign 12(1):53–91. https://doi.org/10.1089/ars.2009.2510
5. Cook KM, Hogg PJ (2013) Post-translational control of protein function by disulfide bond cleavage. Antioxid Redox Sign 18(15):1987–2015. https://doi.org/10.1089/ars.2012.4807
6. Butera D, Cook KM, Chiu J, Wong JW, Hogg PJ (2014) Control of blood proteins by functional disulfide bonds. Blood 123(13):2000–2007. https://doi.org/10.1182/blood-2014-01-549816
7. Cook KM, McNeil HP, Hogg PJ (2013) Allosteric control of beta II-tryptase by a redox active disulfide bond. J Biol Chem 288(48):34920–34929. https://doi.org/10.1074/jbc.M113.523506
8. Zhou A, Carrell RW, Murphy MP, Wei Z, Yan Y, Stanley PL, Stein PE, Broughton Pipkin F, Read RJ (2010) A redox switch in angiotensinogen modulates angiotensin release. Nature 468(7320):108–111. https://doi.org/10.1038/nature09505
9. Butera D, Passam F, Ju L, Cook KM, Woon H, Aponte-Santamaria C, Gardiner E, Davis AK, Murphy DA, Bronowska A, Luken BM, Baldauf C, Jackson S, Andrews R, Grater F, Hogg PJ (2018) Autoregulation of von Willebrand factor function by a disulfide bond switch. Sci Adv 4(2):eaaq1477. https://doi.org/10.1126/sciadv.aaq1477
10. Ioannou Y, Zhang JY, Passam FH, Rahgozar S, Qi JC, Giannakopoulos B, Qi M, Yu P, Yu DM, Hogg PJ, Krilis SA (2010) Naturally occurring free thiols within beta 2-glycoprotein I in vivo: nitrosylation, redox modification by endothelial cells, and regulation of oxidative stress-induced cell injury. Blood 116(11):1961–1970. https://doi.org/10.1182/blood-2009-04-215335
11. Giannakopoulos B, Gao L, Qi M, Wong JW, Yu DM, Vlachoyiannopoulos PG, Moutsopoulos HM, Atsumi T, Koike T, Hogg P, Qi JC, Krilis SA (2012) Factor XI is a substrate for oxidoreductases: enhanced activation of reduced FXI and its role in antiphospholipid syndrome thrombosis. J Autoimmun 39(3):121–129. https://doi.org/10.1016/j.jaut.2012.05.005
12. Kaiser BK, Yim D, Chow IT, Gonzalez S, Dai Z, Mann HH, Strong RK, Groh V, Spies T (2007) Disulphide-isomerase-enabled shedding of tumour-associated NKG2D ligands. Nature 447(7143):482–486. https://doi.org/10.1038/nature05768
13. Butera D, Wind T, Lay AJ, Beck J, Castellino FJ, Hogg PJ (2014) Characterization of a reduced form of plasma plasminogen as the precursor for angiostatin formation. J Biol Chem 289(5):2992–3000. https://doi.org/10.1074/jbc.M113.539924
14. Pijning AE, Chiu J, Yeo RX, Wong JWH, Hogg PJ (2018) Identification of allosteric disulfides from labile bonds in X-ray structures. R Soc Open Sci 5(2):171058. https://doi.org/10.1098/rsos.171058
15. Li W, Baldus IB, Grater F (2015) Redox potentials of protein disulfide bonds from free-energy calculations. J Phys Chem B 119(17):5386–5391. https://doi.org/10.1021/acs.jpcb.5b01051
16. Liang Hai Po H, Brophy Teresa M, Hogg Philip J (2011) Redox properties of the tissue factor Cys186–Cys209 disulfide bond. Biochem J 437(3):455–460. https://doi.org/10.1042/bj20110718
17. Huber-Wunderlich M, Glockshuber R (1998) A single dipeptide sequence modulates the redox properties of a whole enzyme family. Fold Des 3(3):161–171. https://doi.org/10.1016/S1359-0278(98)00024-8
18. Chiu J, Wong JWH, Hogg PJ (2014) Redox regulation of methionine aminopeptidase 2 activity. J Biol Chem 289(21):15035–15043. https://doi.org/10.1074/jbc.M114.554253
19. Pasquarello C, Sanchez JC, Hochstrasser DF, Corthals GL (2004) N-t-butyliodoacetamide and iodoacetanilide: two new cysteine alkylating reagents for relative quantitation of proteins. Rapid Commun Mass Spectrom 18(1):117–127. https://doi.org/10.1002/rcm.1286
20. Bekendam RH, Bendapudi PK, Lin L, Nag PP, Pu J, Kennedy DR, Feldenzer A, Chiu J, Cook KM, Furie B, Huang M, Hogg PJ, Flaumenhaft R (2016) A substrate-driven allosteric switch that enhances PDI catalytic activity. Nat Commun 7:12579. https://doi.org/10.1038/ncomms12579
21. Read SA, O'Connor KS, Suppiah V, Ahlenstiel CLE, Obeid S, Cook KM, Cunningham A, Douglas MW, Hogg PJ, Booth D, George J, Ahlenstiel G (2017) Zinc is a potent and specific inhibitor of IFN-λ3 signalling. Nat Commun 8:15245. https://doi.org/10.1038/ncomms15245 . https://www.nature.com/articles/ncomms15245#supplementary-information
22. Rothwarf DM, Scheraga HA (1992) Equilibrium and kinetic constants for the thiol-disulfide interchange reaction between glutathione and dithiothreitol. Proc Natl Acad Sci U S A 89(17):7944–7948
23. Gatlin CL, Kleemann GR, Hays LG, Link AJ, Yates 3rd JR (1998) Protein identification at the low femtomole level from silver-stained gels using a new fritless electrospray interface for liquid chromatography-microspray and nanospray mass spectrometry. Anal Biochem 263(1):93–101. https://doi.org/10.1006/abio.1998.2809

24. He L, Diedrich J, Chu YY, Yates 3rd JR (2015) Extracting accurate precursor information for tandem mass spectra by rawconverter. Anal Chem 87(22):11361–11367. https://doi.org/10.1021/acs.analchem.5b02721
25. Frickel EM, Frei P, Bouvier M, Stafford WF, Helenius A, Glockshuber R, Ellgaard L (2004) ERp57 is a multifunctional thiol-disulfide oxidoreductase. J Biol Chem 279(18):18277–18287. https://doi.org/10.1074/jbc.M314089200
26. Passam F, Chiu J, Ju L, Pijning A, Jahan Z, Mor-Cohen R, Yeheskel A, Kolsek K, Tharichen L, Aponte-Santamaria C, Grater F, Hogg PJ (2018) Mechano-redox control of integrin de-adhesion. eLife 7:e34843. https://doi.org/10.7554/eLife.34843
27. Chambers JE, Tavender TJ, Oka OBV, Warwood S, Knight D, Bulleid NJ (2010) The Reduction potential of the active site disulfides of human protein disulfide isomerase limits oxidation of the enzyme by ero1α. J Biol Chem 285(38):29200–29207. https://doi.org/10.1074/jbc.M110.156596
28. Krause G, Lundstrom J, Barea JL, Pueyo de la Cuesta C, Holmgren A (1991) Mimicking the active site of protein disulfide-isomerase by substitution of proline 34 in *Escherichia coli* thioredoxin. J Biol Chem 266(15):9494–9500
29. Cleland WW (1964) Dithiothreitol, a new protective reagent for SH groups. Biochemistry 3(4):480–482. https://doi.org/10.1021/bi00892a002
30. Ravilious GE, Nguyen A, Francois JA, Jez JM (2012) Structural basis and evolution of redox regulation in plant adenosine-5'-phosphosulfate kinase. Proc Natl Acad Sci U S A 109(1):309–314. https://doi.org/10.1073/pnas.1115772108
31. Matthias LJ, Azimi I, Tabrett CA, Hogg PJ (2010) Reduced monomeric CD4 Is the preferred receptor for HIV. J Biol Chem 285(52):40793–40799. https://doi.org/10.1074/jbc.M110.190579
32. Nishii W, Kukimoto-Niino M, Terada T, Shirouzu M, Muramatsu T, Kojima M, Kihara H, Yokoyama S (2015) A redox switch shapes the Lon protease exit pore to facultatively regulate proteolysis. Nat Chem Biol 11(1):46–51. https://doi.org/10.1038/nchembio.1688
33. Jin X, Stamnaes J, Klock C, DiRaimondo TR, Sollid LM, Khosla C (2011) Activation of extracellular transglutaminase 2 by thioredoxin. J Biol Chem 286(43):37866–37873. https://doi.org/10.1074/jbc.M111.287490

Chapter 14

Oxidative Protein Folding Using *trans*-3, 4-Dihydroxyselenolane Oxide

Kenta Arai and Michio Iwaoka

Abstract

trans-3,4-Dihydroxyselenolane oxide (DHS^{ox}), a water-soluble cyclic selenoxide reagent, is useful for rapid and quantitative formation of disulfide (SS) bonds in a reduced state of SS-containing proteins because the selenoxide is a strong but selective oxidant for thiol substrates (RSH) in a wide range of pH. Due to this advantage over common disulfide reagents, such as oxidized dithiothreitol (DTT^{ox}) and glutathione (GSSG), DHS^{ox} enables clear characterization of oxidative folding pathways of proteins. DHS^{ox} is also useful for facile diagnosis of weakly folded structure, or reactivity (i.e., pK_a) of the thiols, present in a reduced polypeptide chain and the partially oxidized folding intermediates, identification of the key SS intermediates that can be oxidized directly to the native state, and preparation of SS-scrambled misfolded protein species. In this chapter, these diverse utilities of DHS^{ox} in protein folding study are demonstrated.

Keywords Disulfide formation, Disulfide intermediates, Redox potential, Kinetic analysis, Oxidative folding pathways, Oxidation pulse, pH effects

1 Introduction

Formation of a disulfide (SS) bond is an important chemical reaction, which is relevant to various biological processes, such as oxidative folding of nascent polypeptide chains in endoplasmic reticulum (ER) [1] and regulation of oxidative stress in cellular compartments [2]. In vivo, the SS-formation reaction is mediated by a number of oxidoreductases, such as protein disulfide isomerase and its family enzymes, as well as by small redox molecules, such as oxidized glutathione (GSSG) and lipoic acid. These enzymes and small biomolecules (RSSR) exchange their SS bond with a protein substrate ($P(SH)_2$) to oxidize the two thiol (SH) functional groups to a SS bond: $RSSR + P(SH)_2 \rightarrow 2RSH + P(SS)$. This SS-exchange process can be realized in a flask by using small disulfide reagents, such as GSSG, oxidized dithiothreitol (DTT^{ox}), cystine (CysSSCys), cystamine ($H_2NCH_2CH_2SSCH_2CH_2NH_2$), etc. However, these disulfide reagents have low oxidation potentials ($E^{\circ\prime} = -265$ to -236 mV [3]).

Philip J. Hogg (ed.), *Functional Disulphide Bonds*, Methods in Molecular Biology, vol. 3016,
https://doi.org/10.1007/978-1-0716-5158-2_14,

Therefore, the SS-exchange reaction progresses slowly and reversibly, hence an excess amount of the reagent being required to complete the reaction.

On the other hand, selenoxide reagents (>Se=O) have higher oxidation potentials than disulfide reagents. They can oxidize SH groups of a substrate rapidly and quantitatively in a wide range of pH: >Se=O + $P(SH)_2$ → >Se + P(SS) + H_2O. We have recently demonstrated that *trans*-3,4-dihydroxyselenolane oxide (DHS^{ox}), a new water-soluble cyclic selenoxide reagent, is useful for determination of oxidative folding pathways of SS-containing proteins, such as bovine pancreatic ribonuclease A (RNase A) [4], hirudin variant CX-397 [5], hen egg white lysozyme (HEL) [6], α-lactalbumin (αLA) [7], β-lactoglobulin (βLG) [8], insulin [9], relaxin [9], and the component peptides [10]. The oxidation potential of DHS^{ox} ($E^{o\prime}$ = +375 mV [10]) is greatly higher than those of common disulfide reagents. Therefore, when DHS^{ox} is reacted with a SS-reduced protein, the cysteinyl SH groups are rapidly and quantitatively oxidized, forming SS bonds at random. The SS-formation reaction takes place via two steps (Fig. 1). The features of DHS^{ox} as an oxidative protein folding reagent are threefold.

1. In the SS-formation reaction (Fig. 1), the rate-determining step is the first bimolecular step [5]. Therefore, the oxidation rate-constant (k_{ox}) is generally proportional to the number of SH groups present in a substrate [4, 5].

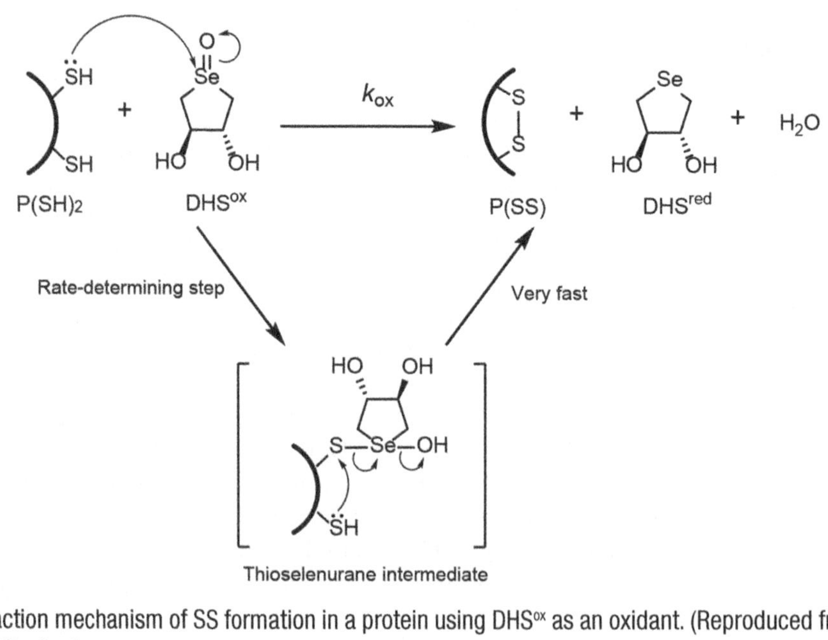

Fig. 1 Reaction mechanism of SS formation in a protein using DHS^{ox} as an oxidant. (Reproduced from [4] with slight modification)

Table 1
Second-order rate constants for SS formation of DTT^{red} with DHS^{ox} as an oxidant at 25°C [10]

pH	4.1	4.7	7.0	8.0	10.0
k_{ox} (mM^{-1} s^{-1})	0.55 ± 0.15	0.45 ± 0.01	1.3 ± 0.1	5.1 ± 0.2	12.0 ± 1.0

2. The SS-formation proceeds rapidly and irreversibly. Therefore, no oxidant (i.e., DHS^{ox}) is left in the solution, hence no oxidation takes place after the reaction. Thus, the SS rearrangement (isomerization) from partially oxidized intermediates to thermodynamically stabilized specific SS intermediates can be clearly observed [4, 6–8, 11].
3. The oxidative protein folding experiments can be carried out in a wide pH range (ca. pH 3–10). Thus, pH effects on the oxidative folding pathways can be analyzed. The value of k_{ox} is variable depending on the solution pH as determined for dithiothreitol (DTT^{red}) (Table 1). These values can be used as standards when one diagnoses weakly folded structure present in the folding intermediates [10].

These features are in significant contrast to disulfide reagents and have been found to be advantageous in several points when DHS^{ox} is applied to protein folding study.

1. Oxidative folding pathways of SS-containing proteins can be clearly characterized by using DHS^{ox}. To regenerate a native state (N) of a protein from the reduced state (R), two chemical processes are necessary; one is SS formation (oxidation), and the other is SS rearrangement. The latter process is important to obtain correct native SS bonds because the former SS formation usually occurs randomly. For the case of RNase A, which has four SS bonds in N (Fig. 2) [12, 13], the SS formation and SS rearrangement processes, i.e., Phases 1 and 2, respectively, could be clearly separated by using DHS^{ox} as a folding reagent (Figs. 3 and 4) [4, 11]. When three equivalents of DHS^{ox} were reacted with R of RNase A, the 3S intermediate ensemble, which has three random SS bonds, was generated as a major product within 1 min (Phase 1) (Fig. 3a). Then, SS rearrangement slowly took place to generate key SS intermediates (i.e., des intermediates) that have three native SS bonds (Phase 2) (Fig. 3b). These des intermediates could be observed clearly because no oxidant was present in the reaction solution after SS formation. The final SS formation of the des intermediates (Phase 3) proceeded much more slowly after Phase 2 (Fig. 4).
2. Weakly folded structure (or the reactivity of the cysteinyl thiols) of a reduced polypeptide chain (R) or the partially oxidized

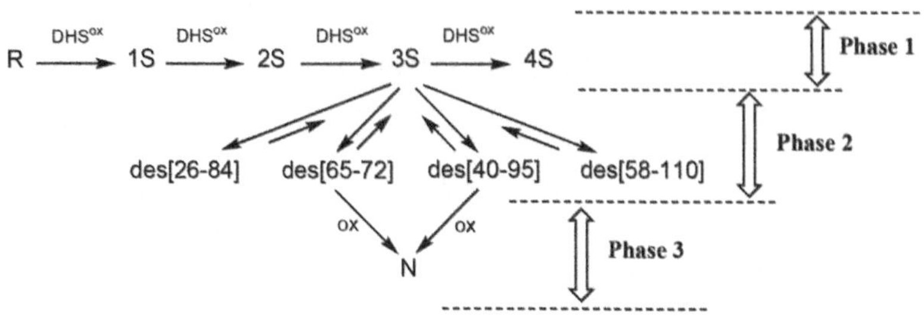

Fig. 2 Oxidative folding pathways of RNase A. Phases 1, 2, and 3 represent a random SS formation process losing a chain entropy, a SS rearrangement process to search for the native SS bonds accompanying conformational folding, and a final oxidation process to generate N, respectively. Des[*n*–*m*] is a specific SS intermediate having three native SS bonds but lacking one native Cys*n*–Cys*m* SS bond. (Reproduced from [4, 11, 14] with slight modification)

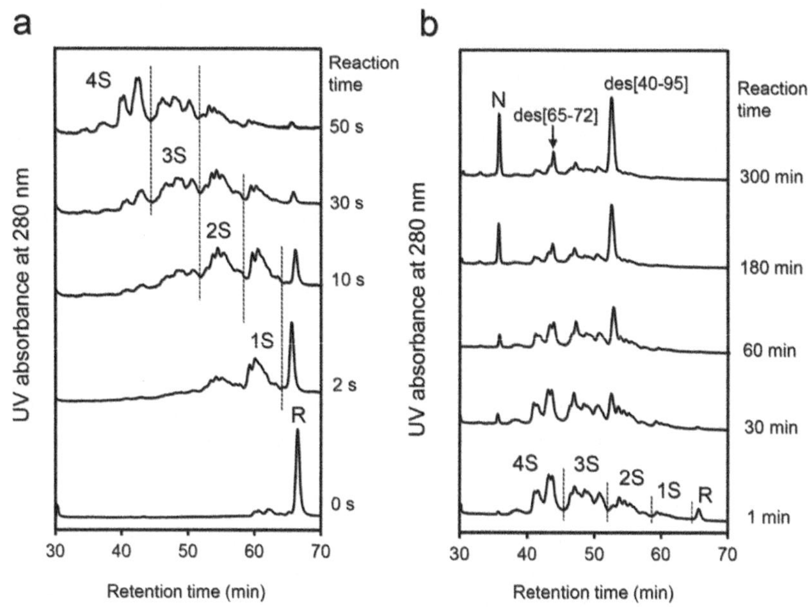

Fig. 3 Typical cation-exchange HPLC chromatograms obtained from the oxidative folding of RNase A. (**a**) A short-term SS-formation experiment using a quench-flow instrument. The reaction conditions were $[R]^0 = [DHS^{ox}]^0/3 = 17.6\,\mu M$ at pH 7.0 and 25°C. (**b**) A long-term oxidative folding experiment by manual mixing. The reaction conditions were $[R]^0 = [DHS^{ox}]^0/3 = 24.6\,\mu M$ at pH 8.0 and 25°C. (Reproduced from [11, 14] for (a, b), respectively, with slight modification)

folding intermediates (1S, 2S, 3S, etc.) can be easily diagnosed by comparing the k_{ox} values for the SS formation with those of DTT^{red} (Table 1) [10]. If the normalized k_{ox} value, which is obtained by dividing k_{ox} by a half number of the SH groups present in the substrate, is smaller than that of DTT^{red}, there would be some structure in the substrate to hinder the SS

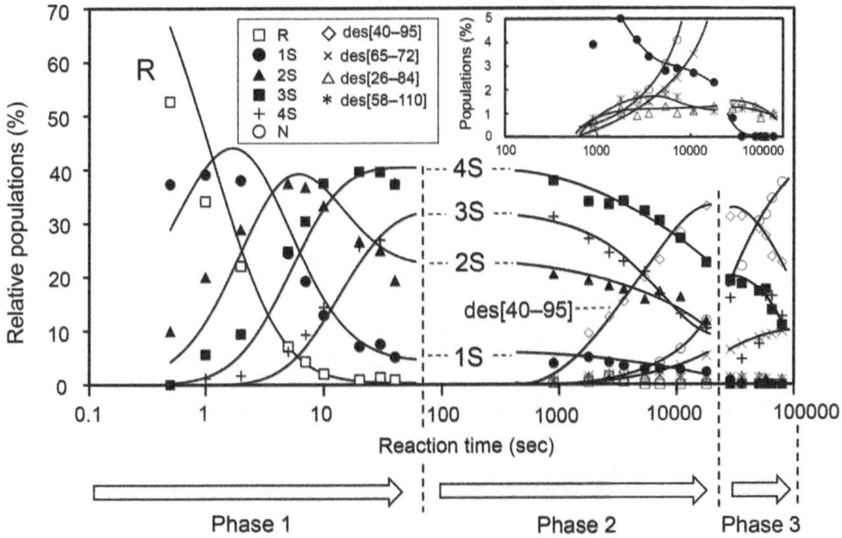

Fig. 4 A whole time-course of relative populations of the SS intermediates generated during the oxidative folding of RNase A using three equivalents of DHS^{ox} at 25°C. Phase 1: reaction conditions were the same as Fig. 3a. Phase 2: reaction conditions were the same as Fig. 3b. Phase 3: reaction conditions were $[R]^0 = [DHS^{ox}]^0/3 = 17.0\,\mu M$ at pH 8.0. (Reproduced from [11] with slight modification)

formation. If the normalized k_{ox} value is larger, on the other hand, the reactivity of the SH groups is statistically enhanced by some reasons, such as the lower pK_a values due to the existence of a basic amino acid residue such as histidine or lysine near the cysteine residues.

3. The key SS intermediates that are direct precursors of N can be identified by the addition of DHS^{ox} to the folding solution. This technique is called an oxidation pulse [11].
4. SS-scrambled misfolded protein species can be prepared by reacting excess DHS^{ox} with R [14]. The same species can be obtained alternatively by reacting hydrogen peroxide with R in the presence of a catalytic amount of a reduced form of DHS^{ox} (i.e., DHS^{red}) [15]. The generated misfolded species is useful as a substrate in the investigation of SS-isomerase activity of PDI-family enzymes and their artificial small-molecule mimics [16].

In this chapter, above-mentioned utilities of DHS^{ox} in protein folding study are demonstrated with the practical protocols for oxidative folding experiments of RNase A using an quench-flow instrument (Sect. 3.3) and by a manual mixing method (Sect. 3.4), the oxidation pulse experiment using DHS^{ox} (Sect. 3.5), HPLC analysis of the generated folding intermediates (Sect. 3.6), determination of the second-order rate constants (k_{ox}) (Sect. 3.7), and preparation of SS-scrambled misfolded species of RNase A (Sects. 3.8 and 3.9).

2 Materials

Prepare all buffer solutions using analytical grade chemicals and deionized water, which is prepared by purifying distilled water. Purge the buffer solutions with nitrogen or argon to remove dissolved oxygen before use. Diligently follow all waste disposal regulations when disposing of waste materials, especially those containing selenium. Prepare *trans*-3,4-dyhydroxyselenolane oxide (DHS^{ox}) [17] and 2-aminoethyl methanethiosulfonate (AEMTS) [18] (*see* **Note 1**) as described in Sect. 3.1. Other reagents with the highest grade are commercially available and can be used without further purification.

2.1 Buffer Solutions

1. 10× Tris stock solution: 1 M Tris–HCl, 10 mM EDTA. Add about 700 mL of water to tris(hydroxymethyl)aminomethane (Tris) (121.1 g, 1 mol) and ethylenediaminetetraacetic acid disodium salt dihydrate (EDTA) (3.72 g, 10 mmol) in a glass beaker. Magnetically stir the aqueous solution at room temperature to dissolve the materials. Make up the volume to 1 L with water. Filter the solution through a nitrocellulose membrane under reduced pressure. Store the filtrate at room temperature.
2. 10× HEPES stock solution: 250 mM HEPES, 10 mM EDTA. Add about 800 mL of water to 4-(2-hydroxyethyl)-1-piperazineethanesulfonic acid (HEPES) (59.6 g, 250 mmol) and EDTA (3.72 g, 10 mmol) in a glass beaker. Magnetically stir the aqueous solution at room temperature to dissolve the materials. Make up the volume to 1 L with water. Filter the solution through a nitrocellulose membrane under a reduced pressure. Store the filtrate at room temperature.
3. Denaturating buffer (pH 8.5): 100 mM Tris–HCl, 4 M GdmSCN, and 1 mM EDTA. Add 5 mL of 10× Tris stock solution and about 15 mL of water to guanidinium thiocyanate (GdmSCN) (23.6 g, 0.2 mol) in a glass beaker. Magnetically stir the aqueous solution at room temperature to dissolve the materials. Adjust pH of the solution to 8.5 by adding 2 M hydrochloric acid (HCl). Make up the volume to 50 mL with water. Filter the solution through a nitrocellulose membrane under a reduced pressure. Store the filtrate at room temperature.
4. Acetate buffer (pH 4.0): 200 mM sodium acetate/acetic acid. Add about 800 mL of water to sodium acetate (13.4 g, 163 mmol) in a glass beaker. Magnetically stir the aqueous solution at room temperature to dissolve the material. Add acetic acid (2.1 mL, 37 mmol) and stir. Adjust pH of the solution to 4.0 by adding 2 M acetic acid or 2 M sodium acetate in water. Make up the volume to 1 L with water. Filter the solution through a nitrocellulose membrane under a reduced pressure. Store the filtrate at room temperature.

5. Tris buffer (pH 7.0): 100 mM Tris–HCl, 1 mM EDTA. Dilute 100 mL of 10× Tris stock solution with about 800 mL of water in a glass beaker. Adjust pH of the solution to 7.0 by adding 5 M HCl. Make up the volume to 1 L with water. Store the solution at room temperature.
6. Tris buffer (pH 8.0): 100 mM Tris–HCl, 1 mM EDTA. Dilute 100 mL of 10× Tris stock solution with about 800 mL of water in a glass beaker. Adjust pH of the solution to 8.0 by adding concentrated HCl. Make up the volume to 1 L with water. Store the solution at room temperature.
7. Sodium bicarbonate buffer (pH 10.0): 25 mM $NaHCO_3$–NaOH, 1 mM EDTA. Add about 800 mL of water to sodium bicarbonate ($NaHCO_3$) (2.1 g, 25 mmol) and EDTA (0.372 g, 1 mmol) in a glass beaker. Magnetically stir the solution at room temperature to dissolve the material. Adjust pH of the solution to 10.0 by adding 1 M sodium hydroxide (NaOH) in water. Make up the volume to 1 L with water. Filter the solution through a nitrocellulose membrane under a reduced pressure. Store the filtrate at room temperature.
8. HPLC eluent A: 25 mM HEPES, 1 mM EDTA. Dilute 100 mL of 10× HEPES stock solution with about 800 mL of water in a glass beaker. Adjust pH of the solution to 7.0 by adding 1 M NaOH. Make up the volume to 1 L with water.
9. HPLC eluent B: 25 mM HEPES, 0.5 M Na_2SO_4, and 1 mM EDTA. Dilute 100 mL of 10× HEPES stock solution with about 700 mL of water in a glass beaker. Add sodium sulfate (Na_2SO_4) (71.0 g, 0.5 mol) to the aqueous solution. Magnetically stir the solution to dissolve Na_2SO_4. Adjust pH of the solution to 7.0 by adding 1 M NaOH. Make up the volume to 1 L with water. Filter the solution through a nitrocellulose membrane under a reduced pressure. Store the filtrate at room temperature.

2.2 Quench-Flow Instrument

Because SS formation is rapidly completed (within 1 min) after mixing a reduced protein and DHS^{ox}, it is necessary to use a quench-flow instrument (Fig. 5) in order to follow the fast SS formation phase. A reduced protein solution, a DHS^{ox} solution, an aqueous AEMTS solution, and a driving buffer solution are loaded in reservoirs A–D, respectively. By pushing up Cylinder 1, the same volumes of the reduced protein and DHS^{ox} solutions are rapidly mixed in Mixer 1. After a certain period of time (0.1–60 s), Cylinder 2 is pushed up to move the reaction solution from Mixer 1 to Mixer 2 by using the driving solution and to quench the folding reaction by blocking the free thiol groups present in the reaction solution with AEMTS. The whole system should be maintained at 25.0 ± 0.1 °C.

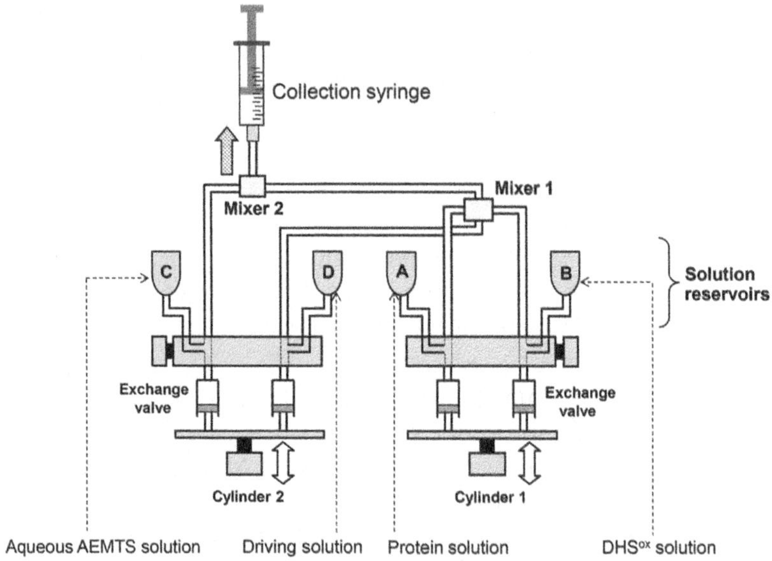

Fig. 5 A set-up of a quench-flow instrument for short-term SS formation experiments. (Reproduced from [4] with slight modification)

3 Methods

3.1 Chemical Syntheses

1. *trans*-3,4-Dyhydroxyselenolane (DHS^{red}): slowly add an aqueous solution (25 mL) of sodium borohydride ($NaBH_4$) (2.15 g, 57.0 mmol) to gray selenium powder (2.25 g, 28.5 mmol) suspended in water (20 mL) via a dropping-funnel with vigorously stirring in a two-neck round-bottom flask on an ice bath (*see* **Note 2**). Stir the resulting solution for 10 min at room temperature to obtain a colorless solution of sodium hydrogen selenide (NaHSe) [19]. To the solution, add 1,3-butadiene diepoxide (1.8 mL, 23.3 mmol) slowly via a glass syringe at room temperature. Stir the mixture for 10 min at room temperature under nitrogen, and then overnight under air (*see* **Note 3**). Filter the resulting mixture under a reduced pressure to remove precipitates. Extract the filtrate with diethyl ether for 3 days by using a liquid/liquid continuous extractor (Soxhlet). Concentrate the obtained ether layer to obtain a yellow solid of the crude product (*see* **Note 4**). Purify the crude product by recrystallization from chloroform to give a colorless crystal of DHS^{red} [yield, 2.7 g (69%); m.p. 79–80 °C; ^{1}H NMR (500 MHz in $CDCl_3$) δ 2.10 (d, J = 5.0, 2 H), 2.87 (dd, J = 4.0 and 11.0, 2 H), 3.16 (dd, J = 4.0 and 11.0, 2 H), 4.27 (m, 2 H); ^{13}C NMR (125.65 MHz) δ 26.8, 78.9; and ^{77}Se NMR (95.35 MHz) δ 65.6].

2. *trans*-3,4-Dyhydroxyselenolane oxide (DHS^{ox}) (*see* **Note 5**): add H_2O_2 (35.8%, 1.13 mL, and 13.2 mmol) to an aqueous solution (20 mL) of DHS^{red} (2.00 g, 12.0 mmol) cooled at 0 °C on an ice bath. Stir the mixture for 1 h at 0 °C and then overnight at room temperature. Lyophilize the resulting mixture to obtain a white solid. Purify the crude product by recrystallization from methanol to give a colorless crystal of DHS^{ox} [yield, 1.72 g (78%); m.p. 124–125 °C; ^1H NMR (500 MHz in CD_3OD) δ 2.87 (d, J = 13.0, ^1H), 3.08 (dd, J = 3.0 and 12.0, ^1H), 3.58 (dd, J = 5.5 and 13.0, ^1H), 3.78 (dd, J = 2.5 and 12.0, ^1H), 4.63 (m, ^1H), 4.72 (dd, J = 3.0 and 5.5, ^1H), 4.80 (s, ^2H); ^{13}C NMR (125.65 MHz) δ 55.0, 58.0, 79.1, 79.5; and ^{77}Se NMR (95.35 MHz) δ 941.8].
3. 2-Aminoethyl methanethiosulfonate (AEMTS) (*see* **Note 6**): slowly add methanesulfonyl chloride (24 mL, 0.31 mol) to an aqueous solution (80 mL) of sodium sulfide nonahydrate ($Na_2S{\cdot}9H_2O$) (74.5 g, 0.31 mol) in a round-bottom flask via a dropping funnel at room temperature. Reflux the mixture solution for 20 h with stirring (*see* **Note 7**). Remove H_2O from the solution under vacuum to obtain an off-white solid. Extract the solid with ethanol using a solid/liquid continuous extractor (Soxhlet) for 15 h. Concentrate the extract to about a half of the volume by evaporation. Store the solution at −20°C to crystalize sodium methanethiosulfonate (CH_3SO_2SNa) as a white solid. Collect the white solid by filtration under a reduced pressure. Purify the crude product by recrystallization from ethanol to obtain a white crystal of CH_3SO_2SNa [yield, 35.2 g (85%). m.p. 272–274°C]. Add 2-bromoethylamine hydrobromide (13.4 g, 65.4 mmol) to an ethanol solution (125 mL) of CH_3SO_2SNa (8.9 g, 66.3 mmol). Reflux the mixture solution for 6.5 h with stirring. Filter the resulting solution to remove a white precipitate. Cool the filtrate to room temperature and then store the solution at 4°C overnight. Filter the solution while it is cold to remove a white precipitate, which is not AEMTS. Concentrate the filtrate to about a half of the volume by evaporation. Store the solution at 4°C to crystalize AEMTS as a white solid. Collect the solid by filtration. Purify the crude product by recrystallization from ethanol to give a white crystal of AEMTS as a hydrobromide salt [yield, 6.6 g (43%). m.p. 112–113°C].

3.2 Preparation of SS-Reduced RNase A (see Note 8)

1. Add an excess amount of DTT^{red} (7–9 mg) to a solution of RNase A (7–10 mg) dissolved in 0.5 mL of denaturating buffer (pH 8.5).
2. Incubate the solution at room temperature for 50 min.
3. Desalt the resulting solution through a Sephadex G25 gel-filtration column (1.5 × 30 cm) into a variable pH buffer

solution (acetate buffer (pH 4.0), Tris buffer (pH 7.0), Tris buffer (pH 8.0), or sodium bicarbonate buffer (pH 10.0)) under a continuous flow condition (3.2–3.8 mL/min).

4. Collect the fraction containing R, which is detected by UV absorbance at 254 or 280 nm (*see* **Note 9**).
5. Determine the concentration of R by UV absorbance at 275 nm based on the molar extinction coefficient ($\varepsilon = 8600\,M^{-1}\,cm^{-1}$ [14, 20]) (*see* **Note 10**).

3.3 Oxidative Folding of RNase A Using a Quench-Flow Instrument (see Note 11)

1. Prepare the solutions of R (30–70 μM) and DHS^{ox} (one to four times higher concentration than R) using the same buffer solution.
2. Prepare an aqueous AEMTS solution (7 mg/mL in water).
3. Load the solutions of R, DHS^{ox}, AEMTS, and a driving buffer in reservoirs A–D of the quench-flow instrument (Fig. 5) and maintain the solutions at 25°C by using a thermostated circulating water bath system.
4. Rapidly mix 50 μL each of R and DHS^{ox} solutions at Mixer 1 by pumping up Cylinder 1.
5. Wait for a certain period of time (0.1–60 s).
6. Quench the reaction by mixing 109 μL of an aqueous AEMTS solution with the reaction solution, which remains in a tube between Mixer 1 and Mixer 2, by pumping up Cylinder 2.
7. Collect the quenched sample solution into a collection syringe by pumping up Cylinder 2 again.
8. Transfer the collected sample solution to a 1.5 mL microtube.
9. Add 5 μL of acetic acid to the solution to adjust the pH to 3–4 (*see* **Note 12**).
10. Store the sample solutions at −30°C until the HPLC analysis (Fig. 3a).

3.4 Oxidative Folding by a Manual Mixing Method (see Note 13)

1. Prepare the solutions of R (30–70 μM) and DHS^{ox} (one to four times higher concentration than R) using the same buffer solution.
2. Prepare an aqueous AEMTS solution (7 mg/mL in water).
3. Mix manually 200 μL each of R and DHS^{ox} solutions in several 1.5 mL microtubes.
4. Stir immediately each reaction solution using a vortex mixer for 5 s.
5. Incubate the reaction solutions in a drythermo bath regulated at any temperature (generally 5–45°C) for 1–1440 min.
6. Quench the reaction by adding 200 μL of an aqueous AEMTS solution to the reaction solution.

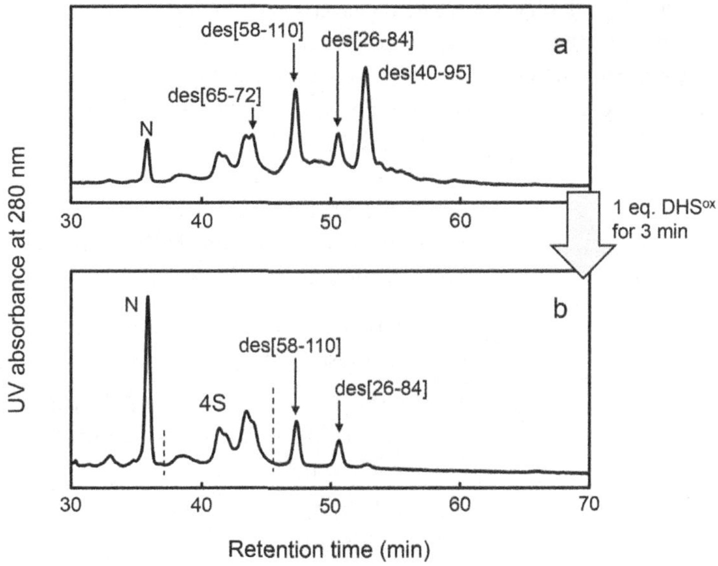

Fig. 6 Typical cation-exchange HPLC chromatograms obtained from the oxidation pulse experiment for the folding intermediates of RNase A. (a) Before the oxidation pulse. Conditions of the oxidative folding were $[R]^0 = [DHS^{ox}]^0/3 = 20.7\,\mu M$ for 180 min at pH 8.0 and 15°C. (b) After the oxidation pulse. Conditions of the oxidation pulse were one equivalent of DHS^{ox} with respect to R for 3 min at pH 8.0 and 15°C. (Reproduced from [11] with slight modification)

7. Stir tJhe mixture using a vortex mixer for 5 s.
8. Add 5 μL of acetic acid to the solution to adjust the pH to 3–4 (*see* **Note 12**).
9. Store the sample solutions at –30 °C until the HPLC analysis (Fig. 3b).

3.5 Oxidation Pulse Using DHS^{ox}

1. Initiate the oxidative folding reaction of RNase A (or the other SS-containing proteins) manually and incubate the reaction solutions by following the procedures 1–5 of Sect. 3.4.
2. Prepare a new DHS^{ox} solution with the same concentration as R, using the same buffer solution.
3. Add 200 μL of the DHS^{ox} solution (oxidation pulse) to the folding solution at the same temperature as the oxidative folding condition.
4. Incubate the mixture for 3 min.
5. Add 200 μL of an aqueous AEMTS solution (7 mg/mL in water) to the reaction mixture.
6. Add 5 μL of acetic acid to the solution to adjust the pH to 3–4 (*see* **Note 12**).
7. Store the sample solution at –30 °C until the HPLC analysis (Fig. 6).

3.6 Cation-Exchange HPLC Analysis and Characterization of SS Intermediates

AEMTS rapidly reacts with cysteinyl thiol (–SH) groups of a substrate to block them to $-SSCH_2CH_2NH_3^+$ moieties, resulting in an increase of a mass of 76 Da and one positive charge per one SH group. Thus, by AEMTS modification, the SS intermediates can be separated through a cation-exchange column, and the number of SH groups (in turn, the number of SS bonds) present in SS intermediates can be determined according to the mass number. A typical protocol for the cation-exchange HPLC analysis and characterization of the SS intermediates is described below.

1. Thaw the sample solution obtained from the oxidative folding of RNase A (Sects. 3.3–3.5).
2. Desalt the solution into 0.1 M acetic acid by using a Sephadex G25 column (1.5 × 30 cm) under a continuous flow condition (3.2–3.8 mL/min).
3. Collect the fraction containing a protein, which is detected by UV absorbance at 254 nm or 280 nm (*see* **Note 14**).
4. Inject the desalted protein solution into the HPLC system equipped with a 5 mL sample solution loop, UV detector (280 nm) (*see* **Note 15**), and a cation-exchange column (7.5 × 75 mm), which is equilibrated with HPLC eluent A at a flow rate of 0.5 mL/min.
5. Apply a concentration gradient of sodium sulfate (Na_2SO_4) by linearly increasing the ratio of HPLC eluent B from 0% to 45% in 50 min.
6. Collect the separated SS intermediates in different tubes.
7. Measure the mass of each SS intermediate by MALDI-TOF-MS.

3.7 Determination of the Second-Order Rate Constants for SS Formation

1. Integrate peak areas of the separated SS-intermediates on the HPLC chromatograms, as shown in Fig. 3a.
2. Calculate relative populations of the SS intermediates (i.e., R, 1S, 2S, 3S, and 4S) based on the peak areas.
3. Perform data fitting of the obtained time-dependent relative populations to the sequential reaction scheme of the equation below.

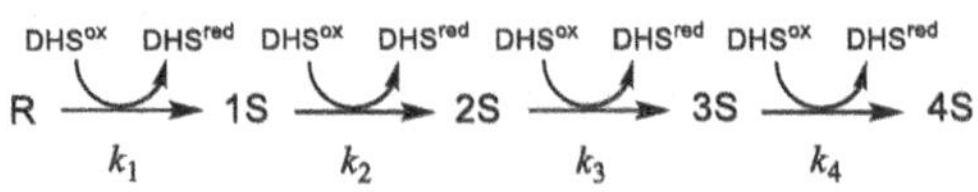

(1)

4. Determine the four second-order rate constants (k_1, k_2, k_3, and k_4).
5. Reproduce the time courses of relative populations of the SS-intermediates by using the obtained rate constants to confirm adequacy of the values obtained for k_1–k_4 and the reaction scheme of Eq. 1 (*see* **Note 16**).

3.8 Preparation of Scrambled RNase A Using DHS^{ox} (see Note 17)

1. Manually mix a R solution (1 mL) and a DHS^{ox} solution (1 mL) with six times higher concentration than R at pH 4.0 in a 15 mL centrifuge tube.
2. Stir the mixture by using a vortex mixer for 5 s.
3. Incubate the solution at room temperature for 5 min.
4. Add 5 μL of acetic acid to the solution.
5. Desalt the solution into 0.1 M acetic acid by using a Sephadex G25 column (1.5 × 30 cm) under a continuous flow condition (3.2–3.8 mL/min)
6. Freeze the desalted sample solution by using liquid nitrogen instantly (*see* **Note 18**).
7. Lyophilize the frozen solution to give a white powder of scrambled RNase A (*see* **Notes 19 and 20**).

3.9 Preparation of Scrambled RNase A Using a Catalytic Amount of DHS^{red} (see Note 21)

1. Prepare a R solution (40 μM) using acetate buffer (pH 4.0).
2. Prepare a DHS^{red} solution (80 μM) using acetate buffer (pH 4.0).
3. Mix 200 μL of the R solution with 100 μL of the DHS^{red} solution.
4. Stir the mixture by using a vortex mixer for 5 s.
5. Add 100 μL of a H_2O_2 solution (8.0 mM) to the solution.
6. Stir the solution by using a vortex mixer for 5 s.
7. Incubate the resulting solution in a drythermo bath regulated at 25°C for 300 min.
8. Add 600 μL of an aqueous AEMTS solution (7 mg/mL) to the solution.
9. Stir the solution by using a vortex mixer for 5 s.
10. Add 5 μL of acetic acid to the solution.
11. Desalt the solution into 0.1 M acetic acid by using a Sephadex G25 column (1.5 × 30 cm) under a continuous flow condition (3.2–3.8 mL/min)
12. Freeze the desalted sample solution by using liquid nitrogen instantly (*see* **Note 18**).
13. Lyophilize the frozen solution to give a white powder of scrambled RNase A (*see* **Notes 19 and 20**).

4 Notes

1. AEMTS is also commercially available.
2. Because harmful hydrogen selenide (H_2Se) is generated, nitrogen should be inhaled gently into the flask via a stainless

needle to lead H_2Se to an aqueous solution of 2 M NaOH. The reaction should be carried out in a fume hood.

3. During the overnight stirring, excess NaHSe is oxidized by air oxygen to red selenium powder.
4. If the crude product is obtained as yellow oil due to contamination of side products, it is better to purify it by silica gel column chromatography using diethyl ether as an eluent before recrystallization.
5. This method is different from the literature method [17], in which DHS^{ox} was synthesized from DHS^{red} by oxidation with *m*-chloroperbenzoic acid in dichloromethane.
6. This method is basically according to the literature [18] but with slight modifications.
7. The color of the solution changes from transparent to red, dark, green, white, and then transparent.
8. A similar protocol can be applied to the preparation of a solution of reduced hirudin variant CX-397 [5], HEL [6], αLA [7], βLG [8], and insulin and relaxin component peptides [10].
9. Collect a fraction with a high concentration of a protein so that the collected volume is about 6–7 mL.
10. The concentration of R would be 30–70 μM. The prepared R solution should immediately be used for the folding experiment because R is slowly oxidized by air oxygen remaining in the solution.
11. For applications to short-term oxidative folding of proteins, see [4, 5, 10, 14].
12. Acidification is necessary to avoid deamidation of the folding intermediates during the storage.
13. For applications to long-term oxidative folding of proteins, see [4, 6, 7, 11, 14].
14. Collect a part of the protein fraction with a high concentration so that the collection volume is about 5.5 mL.
15. The absorption intensity does not significantly change between the folded and unfolded species.
16. The simulation results and the experimental data for the relative populations of SS-intermediates do not match well with each other in a short reaction time (<0.3 s) due to a limit of the reaction rate of AEMTS or a dead mixing time of the quench-flow instrument.
17. For a typical example, see [16].
18. Wear gloves to avoid low-temperature burn by liquid nitrogen.

19. The purity should be confirmed by cation-exchange HPLC.
20. When the white powder of scrambled RNase A is used in an experiment, the powder should be dissolved first in a small amount of 10 mM HCl (<100 μL/mg) before adding a buffer solution because scrambled RNase A, which is an unfolded species, frequently aggregates in a buffer solution at a high concentration. The HCl solution should be subsequently diluted with a buffer solution to 10–20-fold of a volume.
21. For a typical example, see [15].

References

1. Tu BP, Weissman JS (2004) Oxidative protein folding in eukaryotes: mechanisms and consequences. J Cell Biol 164:341–346
2. Espinosa-Diez C, Miguel V, Mennerich D et al (2015) Antioxidant responses and cellular adjustments to oxidative stress. Redox Biol 6:183–197
3. Iwaoka M, Arai K (2013) From sulfur to selenium. A new research arena in chemical biology and biological chemistry. Curr Chem Biol 7:2–24
4. Iwaoka M, Kumakura F, Yoneda M et al (2008) Direct observation of conformational folding coupled with disulphide rearrangement by using a water-soluble selenoxide reagent—a case of oxidative regeneration of ribonuclease A under weakly basic conditions. J Biochem 144:121–130
5. Arai K, Dedachi K, Iwaoka M (2011) Rapid and quantitative disulfide bond formation for a polypeptide chain using a cyclic selenoxide reagent in an aqueous medium. Chem Eur J 17:481–485
6. Arai K, Shibagaki W, Shinozaki R, Iwaoka M (2013) Reinvestigation of the oxidative folding pathways of hen egg white lysozyme: switching of the major pathways by temperature control. Int J Mol Sci 14:13194–13212
7. Shinozaki R, Iwaoka M (2017) Effects of metal ions, temperature, and a denaturant on the oxidative folding pathways of bovine α-lactalbumin. Int J Mol Sci 18:1996
8. Iwaoka M, Mitsuji T, Shinozaki R (2019) Oxidative folding pathways of bovine milk β-lactoglobulin with odd cysteine residues. FEBS Open Bio 9:1379–1391
9. Arai K, Takei T, Shinozaki R et al (2018) Characterization and optimization of two-chain folding pathways of insulin via native chain assembly. Commun Chem 1:26
10. Arai K, Noguchi M, Singh BG et al (2013) A water-soluble selenoxide reagent as a useful probe for the reactivity and folding of polythiol peptides. FEBS Open Bio 3:55–64
11. Arai K, Kumakura F, Iwaoka M (2012) Kinetic and thermodynamic analysis of the conformational folding process of SS-reduced bovine pancreatic ribonuclease A using a selenoxide reagent with high oxidizing ability. FEBS Open Bio 2:60–70
12. Rothwarf DM, Li Y-J, Scheraga HA (1998) Regeneration of bovine pancreatic ribonuclease A: identification of two nativelike three-disulfide intermediates involved in separate pathways. Biochemistry 37:3760–3766
13. Welker E, Narayan M, Volles MJ, Scheraga HA (1999) Two new structured intermediates in the oxidative folding of RNase A. FEBS Lett 460:477–479
14. Arai K, Kumakura F, Iwaoka M (2010) Characterization of kinetic and thermodynamic phases in the prefolding process of bovine pancreatic ribonuclease A coupled with fast SS formation and SS reshuffling. Biochemistry 49:10535–10542
15. Kumakura F, Mishra B, Priyadarsini KI, Iwaoka M (2010) A water-soluble cyclic selenide with enhanced glutathione peroxidase-like catalytic activities. Eur J Org Chem 2010:440–445
16. Arai K, Ueno H, Asano Y et al (2018) Protein folding in the presence of water-soluble cyclic diselenides with novel oxidoreductase and isomerase activities. ChemBioChem 19:207–211
17. Iwaoka M, Takahashi T, Tomoda S (2001) Syntheses and structural characterization of water-soluble selenium reagents for the redox control of protein disulfide bonds. Heteroat Chem 12:293–299
18. Bruice TW, Kenyon GL (1982) Novel alkyl alkanethiolsulfonate sulfhydryl reagents.

Modification of derivatives of L-cysteine. J Protein Chem 1:47–58

19. Klayman DL, Griffin TS (1973) Reaction of selenium with sodium borohydride in protic solvents. A Facile Method for the introduction of selenium into organic molecules. J Am Chem Soc 95:197–1994
20. Rothwarf DM, Scheraga HA (1993) Regeneration of bovine pancreatic ribonuclease A. 1. Steady-state distribution. Biochemistry 32:2671–2679

Chapter 15

Method for Selection of Antithrombin Disulfide-Bonded Subsets

Kenny Lu, Philip J. Hogg and Mark Schreuder

Abstract

Studies of coagulation proteins have demonstrated that they are frequently produced in multiple partially disulfide-bonded states, with each state potentially influencing the protein's function. The ability to isolate distinct disulfide-bonded forms could be valuable for functionally characterizing these protein subsets. Here, we describe protocols for the selection of a disulfide-bonded subset of antithrombin, characterized by its distinct biophysical properties, using a limiting quantity of heparin–Sepharose. Antithrombin is a key anticoagulant protein that regulates the proteolytic activity of several procoagulant proteases, and its deficiency is associated with pathological thrombosis. Antithrombin contains three disulfide bonds, which exist in both formed and unformed states in human plasma, demonstrating that antithrombin circulates in blood in multiple disulfide-bonded states. This method provides a practical approach for isolating and studying partially disulfide-bonded subsets of antithrombin and can be extended to other disulfide-bonded protein states.

Keywords Disulfide bond, Antithrombin, Heparin

1 Introduction

Antithrombin is a key anticoagulant protein that regulates the proteolytic activity of various procoagulant proteases, such as factor Xa and thrombin, and is critical for the maintenance of normal hemostasis. Heparin is an important cofactor for antithrombin and is widely used in anticoagulant therapies. The binding of heparin to antithrombin induces the allosteric activation of antithrombin, significantly enhancing antithrombin's inhibitory activity by up to three orders of magnitude. This underscores the therapeutic importance of clinical heparins, such as enoxaparin and dalteparin in anticoagulant treatment [1–3].

The differential cysteine labeling and mass spectrometry approach developed in our laboratory [4] has demonstrated that various coagulation proteins exist in multiple disulfide-bonded states, including fibrinogen [5], von Willebrand factor [6], and αIIbβ3 integrin [7].

Philip J. Hogg (ed.), *Functional Disulphide Bonds*, Methods in Molecular Biology, vol. 3016,
https://doi.org/10.1007/978-1-0716-5158-2_15,

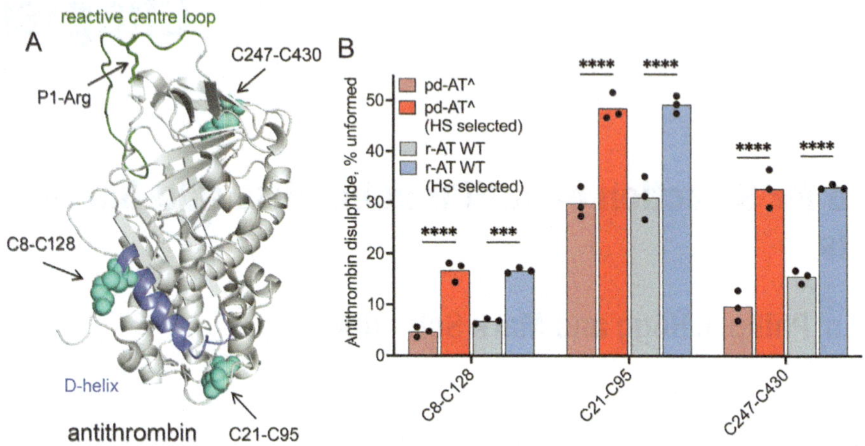

Fig. 1 Limiting heparin–Sepharose selects for a subset of antithrombin disulfide-bonded states. (**A**) Positions of the individual disulfide bonds (cyan spheres) in a ribbon structure of antithrombin (PDB identifier 1e05 [8]) are indicated by the arrows. Green and purple highlighted areas represent the reactive center loop comprising the substrate-specific proteolytic site (P1-Arg), and the heparin-binding site (D-helix), respectively. (**B**) The redox state of human plasma-derived (pd-AT) and recombinant antithrombin (r-AT WT) disulfide bonds captured by a limiting quantity of heparin–Sepharose. Bars represent the mean of three independent experiments. ^Triplicate of same donor (23 y/o, male). ***$p<0.001$, ****$p<0.0001$ according to two-way ANOVA analysis. Disulfide-bonded states were quantified using differential cysteine labeling and mass spectrometry as described previously [4]

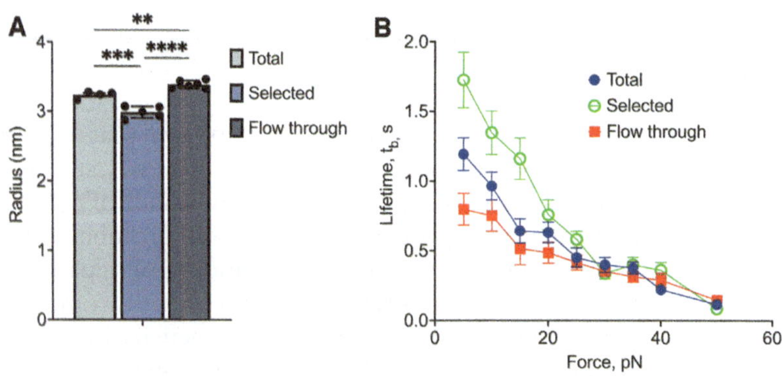

Fig. 2 Biophysical heterogeneity of antithrombin disulfide-bonded subsets. (**A**) The hydrodynamic radii of antithrombin disulfide-bonded subsets were quantified using isothermal dynamic light scattering with Prometheus Panta and analyzed using Panta Analysis software (Nanotemper). Results are presented as mean ± 1 standard deviation of at least five technical acquisitions per point derived from more than four independent experiments. **$p<0.01$, ***$p<0.001$, ****$p<0.0001$ according to one-way ANOVA analysis. (**B**) The lifetime of antithrombin (r-AT WT)-unfractionated heparin (UFH) bonds was measured as a function of clamp force using a biomembrane force probe (BFP) as described previously [9, 10]. r-AT WT was biotinylated and covalently immobilized to streptavidin–maleimide coupled to thiolated glass beads. UFH was adsorbed onto the thiolated glass beads followed by BFP measurements. Results represent mean ± standard error of the mean of at least 20 measurements per data point

Antithrombin is also constitutively produced in partially disulfide-bonded states. The three antithrombin disulfide bonds, C21-C95, C247-C430, and C8-C128, are unformed in ~1 in 3, ~1 in 5, and ~1 in 10 antithrombin molecules, respectively (Fig. 1). Identifying and functionally characterizing specific disulfide-bonded subsets are required for advancing our understanding of this emerging aspect of protein biology.

In the methods described here, a disulfide-bonded subset of antithrombin is isolated using a limiting amount of heparin–Sepharose, selecting for the antithrombin subset that preferentially binds to heparin. Differential cysteine labeling and mass spectrometry revealed that this subset contains significantly more unformed disulfide bonds (Fig. 1). Further evidence for the isolation of a distinct subset was obtained through isothermal dynamic light scattering and force spectroscopy, revealing biophysical differences in hydrodynamic radii and antithrombin–heparin bond lifetimes (Fig. 2).

2 Materials

2.1 Albumin Depletion of Human Plasma

1. Pierce™ albumin depletion kit (Thermo Fisher Scientific).
2. Pierce™ albumin depletion resin bottle.
3. Pierce™ spin column (screw cap) for immunoprecipitation.
4. Binding/wash buffer: 25 mM Tris, 75 mM NaCl, pH 7.5.

2.2 Selection of Partially Disulfide-Bonded Antithrombin Subsets

1. Heparin–Sepharose 6 Fast flow column (GE Healthcare).
2. Phosphate-buffered saline (PBS): 137 mM NaCl, 2.7 mM KCl, 10 mM Na_2HPO_4, 1.8 mM KH_2PO_4, pH 7.4.
3. Purified recombinant antithrombin (r-ATWT): stock 10 mg/mL (172.4 μM) in 10 mM HEPES, 75 mM NaCl, 5 mM EDTA, 50% (v/v) glycerol, pH 7.5. (*see* **Note 1**)
4. Assay buffer: 20 mM $NaPO_4$, 100 mM NaCl, 0.1 mM EDTA, 0.1% PEG 8000, pH 7.4.
5. 10 kDa MWCO Amicon Ultra-0.5 centrifugal filter units (Merck).
6. NanoDrop microvolume spectrophotometer (Thermo Fisher Scientific) for protein quantification.

3 Methods

Carry out all procedures at room temperature unless otherwise specified.

If purified recombinant antithrombin is used, proceed to section "Selection of partially disulfide-bonded antithrombin subsets." If plasma is used, albumin depletion of human plasma is performed prior to the selection of antithrombin subsets to minimize interference caused by the high abundance of albumin in plasma and its known affinity for heparin.

3.1 Albumin Depletion of Human Plasma

1. Shake the Pierce™ albumin depletion resin bottle to resuspend the resin. Using a wide-bore micropipette tip, load 400 µL of the resin (50% slurry) into a Pierce™ spin column and loosely cap the column (*see* **Note 2**).
2. Twist off the bottom closure of the spin column (*see* **Note 3**). Place the spin column into a 1.5–2.0-mL Eppendorf tube. Centrifuge at approximately 12,000 × *g* for 1 min to remove excess liquid. Discard the flow-through by tipping and place the spin column back into the same Eppendorf tube.
3. Add 200 µL of binding/wash buffer to the spin column.
4. Centrifuge at 12,000 × *g* for 1 min. Discard the flow-through and place the spin column into a new Eppendorf tube.
5. Dilute the albumin containing plasma 1 in 3 by adding binding/wash buffer and mix using a pipette. Apply the mixture to the resin and incubate for 1–2 min at room temperature (*see* **Note 4**).
6. Centrifuge at 12,000 × *g* for 1 min. Re-apply the flow-through to the spin column and incubate for 1–2 min at room temperature to ensure maximal albumin binding.
7. Centrifuge at 12,000 × *g* for 1 min. Retain the flow-through which contains albumin-depleted plasma.
8. Wash the resin to release unbound proteins by adding 500 µL of binding/wash buffer.
9. Centrifuge at 12,000 × *g* for 1 min. Retain the flow-through which contains the remaining albumin-depleted plasma.
10. Combine the albumin-depleted plasma from Steps 7 and 9.

3.2 Selection of Partially Disulfide-Bonded Antithrombin Subsets

1. Using a wide-bore pipette tip, pre-equilibrate 8 µL of heparin–Sepharose 6 Fast Flow beads (50% slurry) by adding to 1 mL of PBS in a 1.5 mL Eppendorf tube, mix with a vortex for 30 s, centrifuge at 1000 ×*g* for 1 min and discard the supernatant.
2. Repeat Step 1 for another two times.
3. Skip to Step 6, if using albumin-depleted plasma. For purified proteins: dilute 2 µL of r-AT WT stock (10 mg/mL) ten-fold with PBS (*see* **Notes 5 and 6**).
4. Take out 2 µL of the purified r-AT WT stock and measure protein concentration using Nanodrop at 280 and 320 nm (*see* **Note 7**).

5. Dilute the r-AT WT to 0.4 mg/mL (6.9 μM) (working stock) with PBS.
6. Add 50 μL of albumin-depleted plasma or the r-AT WT working stock to the pre-equilibrate beads.
7. Dilute the samples by adding 300 μL of PBS to the antithrombin and heparin–Sepharose mixture.
8. Incubate with vertical end-over-end rotation at 15–20 rpm for 1 h (*see* **Note 8**).
9. Centrifuge the samples at 1000 × *g* for 30 s and collect the supernatant containing unbound antithrombin (this is the non-selected antithrombin fraction).
10. Wash the antithrombin-bound-beads by adding 1 mL of PBS, mix with a vortex for 30 s, centrifuge at 1000 × *g* for 1 min and discard the supernatant.
11. Repeat Step 10 for another two times.
12. Elute antithrombin from the antithrombin-bound beads by adding 10 μL of 2 M NaCl and mix carefully using a pipette. Centrifuge at 1000 × *g* for 30 s and carefully transfer the supernatant to an Eppendorf tube.
13. Repeat Step 12 for another time to ensure complete elution of antithrombin and combine the eluted antithrombin in the same Eppendorf tube.
14. To desalt the eluted antithrombin, add the antithrombin into an Amicon Ultra-0.5 centrifugal filter unit with 10 kDa molecular weight cutoff.
15. Add 400 μL of assay buffer, centrifuge at 14,000 × *g* for 10 min and discard flow-through while retaining the antithrombin sample in the filter.
16. Repeat Step 15 for another two times.
17. To collect the antithrombin sample in the filter unit, invert the filter unit and place it upside down in a clean Eppendorf tube, and centrifuge at 1000 × *g* for 30 s. The collected material is the selected antithrombin subset.
18. Take out 2 μL of the selected and non-selected antithrombin subsets, dilute it five-fold to ten-fold with PBS. Measure protein concentration as described in Note 7.

4 Notes

1. Purified recombinant antithrombin is recommended for optimal results. Alternatively, albumin-depleted plasma can be used, if high purity is not essential.

2. The amount of resin to use depends on the volume and albumin concentration of the sample being processed. Each Pierce™ albumin depletion resin aliquot of 200 μL settled resin can bind sufficient albumin to process 5–50 μL of serum sample. To process larger serum samples volumes, use more resin volume or multiple spin columns. For serum samples containing high albumin concentrations (e.g., >50 mg/mL), dilute sample with the appropriate binding/wash buffer.
3. Do not snap off the bottom. To remove, twist slightly in one direction followed by twisting it in the other direction.
4. The sample must have <100 mM NaCl for proper albumin binding. Dialyze or dilute the sample as needed.
5. The r-AT WT stock can be prepared at any concentration, provided it allows subsequent dilution to a working concentration of at least 10–20 μM.
6. It is recommended to use recently expressed/purified r-AT WT as we find that commercial stocks are often highly oxidized due to prolonged storage.
7. Using the nanodrop, the protein concentration can be calculated using the following equations:

$$Protein\ concentration\left(\frac{mg}{ml}\right)=\frac{A280corr}{Extinction\ coefficient}$$

In which,

$$A280corr = (10 \times A280)-(10 \times 1.7 \times A320)$$

Molarity of protein can be calculated using the following equation:

$$Molarity\ (mM)=\frac{Protein\ concentration\ \left(\frac{mg}{ml}\right)}{molecular\ weight\ (kDa)}$$

8. Beads can sometimes get stuck to the bottom of the tube, flick it to ensure proper mixing during rotation.

Acknowledgments

This study was funded by the National Health and Medical Research Council of Australia (2026631) and a Senior Researcher Grant from the NSW Cardiovascular Research Capacity Program to PJH.

References

1. Huntington JA, McCoy A, Belzar KJ, Pei XY, Gettins PG, Carrell RW (2000) The conformational activation of antithrombin: a 2.85-Å structure of a fluorescein derivative reveals an electrostatic link between the hinge and heparin binding regions. J Biol Chem 275(20):15377–15383
2. Nugent MA (2000) Heparin sequencing brings structure to the function of complex oligosaccharides. Proc Natl Acad Sci 97(19):10301–10303
3. Shworak NW, Kobayashi T, de Agostini A, Smits NC (2010) Anticoagulant heparan sulfate: to not clot—or not? Prog Mol Biol Transl Sci 93:153–178
4. Chiu J (2019) Measurement of redox states of the β3 integrin disulfide bonds by mass spectrometry. Bio-protocol 9(3):e3156
5. Butera D, Hogg PJ (2020) Fibrinogen function achieved through multiple covalent states. Nat Commun 11(1):5468. https://doi.org/10.1038/s41467-020-19295-7
6. Butera D, Wang HJ, Woon HG, Zhao YC, Ju LA, Hogg PJ (2023) Mechano-covalent protection of coagulation factor VIII by von Willebrand factor. Blood Adv 7(10):2117–2128. https://doi.org/10.1182/bloodadvances.2022008650
7. Pijning AE, Blyth MT, Coote ML, Passam F, Chiu J, Hogg PJ (2021) An alternate covalent form of platelet alphaIIbbeta3 integrin that resides in focal adhesions and has altered function. Blood 138(15):1359–1372. https://doi.org/10.1182/blood.2021012441
8. McCoy AJ, Pei XY, Skinner R, Abrahams JP, Carrell RW (2003) Structure of beta-antithrombin and the effect of glycosylation on antithrombin's heparin affinity and activity. J Mol Biol 326(3):823–833. https://doi.org/10.1016/s0022-2836(02)01382-7
9. Ju L (2019) Dynamic force spectroscopy analysis on the redox states of protein disulphide bonds. In: Functional disulphide bonds: methods and protocols, Springer, New York, pp 115–131
10. Ju L, Dong J-f, Cruz MA, Zhu C (2013) The N-terminal flanking region of the A1 domain regulates the force-dependent binding of von Willebrand factor to platelet glycoprotein Ibα. J Biol Chem 288(45):32289–32301

Check for updates

Chapter 16

Polyclonal Antibody Selection of Partially Disulfide-Bonded Protein States

Diego Butera and Philip J. Hogg

Abstract

Polyclonal antibodies are used extensively to characterize their protein antigens in many experimental settings, including measuring protein concentration and function in vitro, ex vivo and in fixed and living tissues. Disulfide bonds constrain the polypeptide backbone of proteins and reduce the conformational variations of the structure. Several proteins are constitutively produced as multiple partially disulfide-bonded states that result in populations of protein states with a variety of conformational forms, which can influence polyclonal antibody binding to the protein. For two protein antigens described herein, polyclonal antibodies preferentially recognize states with more unformed disulfides and the avidity of the antibodies correlates with extent of unformed disulfide bonds in one of the protein antigens. This selectivity of polyclonal antibodies for their target antigens should be considered when interpreting the results of antibody-based experiments.

Keywords Disulfide bond, Antibody, Cysteine, Fibrinogen, Von Willebrand factor

1 Introduction

The nature of the protein antigen that is recognized by polyclonal antibodies is crucial for correct interpretation of results that rely on this interaction for an output. We show here for two human protein antigens that are constitutively produced as multiple partially disulfide-bonded states, fibrinogen, and von Willebrand factor (VWF), polyclonal antibodies generated against the proteins preferentially bind subsets of the disulfide-bonded states. The same phenomenon has been observed for monoclonal antibodies. Patient anti-blood coagulation factor VIII (FVIII) drug monoclonal antibodies bind preferentially to a subset of FVIII covalent states [1].

Philip J. Hogg (ed.), *Functional Disulphide Bonds*, Methods in Molecular Biology, vol. 3016,
https://doi.org/10.1007/978-1-0716-5158-2_16,

2 Methods

2.1 Proteins and Antibodies

Purified human fibrinogen was obtained commercially (Sigma, catalogue number F4883). Human VWF was purified from Biostate as described in [2]. Polyclonal anti-fibrinogen (catalog number A0080) and anti-VWF (catalog number A0082) antibodies were from Dako.

2.2 Quantification of the Redox States of Fibrinogen and VWF Disulfide Bonds

Dynabeads (2 mg, Life Technologies) were coated with 16 μg of polyclonal anti-fibrinogen or anti-VWF antibodies in 1 mL of phosphate-buffered saline on a rotating wheel for 1 h at 22°C and the excess antibody removed by washing three times with 1 mL of phosphate-buffered saline. Purified fibrinogen (8 μg) or VWF (8 μg) was incubated with 2 mg of coated beads on a rotating wheel for 1 h at 22°C. On some occasions, we varied the incubation times from 10 min to up to 24 h. The beads were collected using a magnet, excess plasma aspirated to reduce the volume and incubated in 0.3 mL of 5 mM 2-iodo-N-phenylacetamide (^{12}C-IPA) in phosphate-buffered saline containing 10% DMSO for 1 h at 22°C in the dark to alkylate unpaired Cys thiols in the proteins. The supernatants were aspirated, and the beads incubated with NuPAGE LDS sample buffer (Life Technologies) containing a further 5 mM ^{12}C-IPA for 30 min at 60°C. On some occasions, the proteins were eluted with glycine buffer at pH 2.5 before LDS incubation. The supernatants were resolved on SDS-PAGE. As controls for the immunoprecipitation, purified fibrinogen (5 μg) or VWF (5 μg) was incubated with 5 mM ^{12}C-IPA in phosphate-buffered saline containing 10% DMSO for 1 h at 22°C in the dark and the protein resolved on SDS-PAGE. The SDS-PAGE gels were stained with colloidal coomassie (Fig. 1) and the protein bands excised, destained, dried, incubated with 40 mM dithiothreitol, and washed. The gel slices were incubated with 5 mM ^{13}C-IPA for 1 h at 22°C in the dark to alkylate the disulfide bond Cys, washed and dried before digestion of proteins with 12.5 ng/μL of trypsin (Promega) in 25 mM NH_4CO_2 overnight at 25°C. Peptides were eluted from the slices with 5% formic acid, 50% acetonitrile. Nano-liquid chromatography (nano-LC) was performed using an Ultimate 3000 HPLC and autosampler system (Dionex). Samples were injected into a fritless nano-LC column (75 μm × ~12 cm) containing C18 media (1.9 μm, 120 Å ReproSil-Pur 120 C18-AQ, Dr Maisch GmbH) and heated to 45°C for all runs. Peptides were eluted using a linear gradient over 52 min, at a flow rate of 0.2 μL/min. Mobile phase A consisted of 0.1% formic acid in H_2O, while mobile phase B consisted of acetonitrile:H_2O (8:2) with 0.1% formic acid. The gradient was: 0 min 2% B; 4 min 2% B, 36 min 45% B, 37 min 80% B, 37.5 min 80% B, 39 min 2% B, and 52 min 2% B.

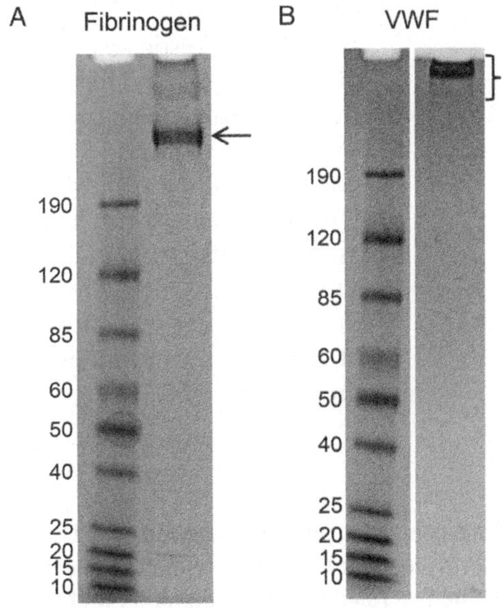

Fig. 1 Example of ^{12}C-IPA-labeled purified fibrinogen and VWF resolved on SDS-PAGE. Fibrinogen (**A**) and VWF (**B**) have molecular masses of 340 and 500–20,000 kDa, respectively. Molecular mass standards are shown in the left-hand lane and size indicated

High voltage (2000 V) was applied to a low volume tee (Upchurch Scientific) and the column tip positioned ~0.5 cm from the heated capillary (T=275°C) of a LTQ Orbitrap Velos (Thermo Electron) mass spectrometer. Positive ions were generated by electrospray ionization and the mass spectrometer operated in data-dependent acquisition mode. Full-scan MS spectra were acquired (m/z 350–1750) by the Orbitrap at a resolution of 30,000. The 15 most abundant ions (>5000 counts) with charge states ≥+2 were sequentially isolated and fragmented within the linear ion trap using collisionally induced dissociation with an activation q=0.25 and activation time of 10 ms at a target value of 30,000 ions. M/z ratios selected for MS/MS were dynamically excluded for 35 s. Ion trap zoom AGC target: 3000.00. Ion trap full AGC target: 30,000.00. Ion trap SIM AGC target: 10,000.00. Ion trap MSn AGC target: 10,000.00. Data were analyzed using Mascot Daemon (version 2.5.0, Matrix Science) against Swissprot database. Search parameters were precursor tolerance of 6 ppm and product ion tolerances of ± 0.6 Da, and iodoacetanilide derivative (Cys), iodoacetanilide ^{13}C derivative (Cys), oxidation (Met) selected as variable modifications with full tryptic cleavage of up to three missed cleavages. With the list of cysteine-containing peptides identified using Mascot, their monoisotopic mass/charge ratio (m/z) of +2, +3, and +4 is calculated using MS-Product (Version v 6.2.2, Protein Prospector Tools).

Cys labeled with ^{12}C-IPA or ^{13}C-IPA has a mass of 133.05276 or 139.07289, respectively. Cys-containing peptides were analyzed (Tables 1 and 2). The different redox forms of the Cys residues were quantified from the relative ion abundance of peptides labeled with ^{12}C-IPA and/or ^{13}C-IPA. To calculate ion abundance of peptides, extracted ion chromatograms were generated using XCalibur Qual Browser (v2.1.0; Thermo Scientific). The area was calculated using the automated peak detection function built into the software. The data were routinely searched for peptides containing free Cys thiols and these were not detected, which indicates that alkylation of unpaired Cys residues by ^{12}C-IPA or ^{13}C-IPA was complete in the protein.

Table 1

Fibrinogen cysteine containing peptides analysed by HPLC and mass spectrometry. Cysteine numbering is according to UniProt identifier P02671 for human fibrinogen α chain, P02675 for the β chain and P02679 for the γ chain. The Cys residues of the disulfide that were measured are underlined in the peptide

Disulfide bond	Cys measured	Peptide
αC47-αC47	αC47	HQSACK
αC55-βC95	αC55	DSDWPFCSDEDWNYK
αC64-γC49	αC64	CPSGCR
	γC49	FGSYCPTTCGIADFLSTYQTK
αC68-βC106	αC68	CPSGCR
βC110-γC45	γC45	FGSYCPTTCGIADFLSTYQTK
αC180-γC161	γC161	VAQLEAQCQEPCK
αC184-βC223	βC223	LESDVSAQMEYCR
αC461-αC491	αC491	EVVTSEDGSDCPEAMDLGTLSGIGTLDGFR
βC227-γC165	βC227	TPCTVSCNIPVVS
	γC165	VAQLEAQCQEPCK
βC231-βC316	βC231	TPCTVSCNIPVVSGK
	βC316	NYCGLPGEYWLGNDK
βC241-βC270	βC241	ECEEIIR
	βC270	VYCDMNTENGGWTVIQNR
βC424-βC437	βC437	CHAANPNGR
γC179-γC208	γC179	DCQDIANK
	γC208	ANQQFLVYCEIDGSGNGWTVFQK
γC352-γC365	γC352	FEGNCAEQDGSGWWMNK
	γC365	CHAGHLNGVYYQGGTYSK

Table 2

VWF cysteine containing peptides analysed by HPLC and mass spectrometry. Cysteine numbering is according to UniProt identifier P04275 for human VWF. The Cys residues of the disulfide that were measured are underlined in the peptide

Disulfide bond	Cys measured	Peptide
767–808	767	SLSCRPPMVK
776–804	776	LVCPADNLR
788–799	788	AEGLECTK
810–821	821	CVALER
914–921	914	GCSHPSVK
889–1031	1031	VSSQCADTR
1046–1089	1046	KVPLDSSPATCHNNIMK
1060–1084	1060	QTMVDSSCR
1071–1111	1071	ILTSDVFQDCNK
1879–1904	1879	ICMDEDGNEK
1927–2088	1927	GLRPSCPNSQSPVK
1972–2123	1972	LTGSCSYVLFQNK
1993–2001	1993	EQDLEVILHNGACSPGAR

3 Results and Discussion

3.1 Polyclonal Anti-Fibrinogen Antibodies Select for Fibrinogen States with More Unformed Disulfides

Fibrinogen is an abundant blood protein that plays a critical role in arresting bleeding in hemostasis and obstruction of blood vessels in thrombosis by cross-linking platelets at sites of vessel injury. Fibrinogen comprises two Aα, two Bβ, and two γ polypeptide chains, where two AαBβγ units are linked head-to-head to form a structure with globular central E and two peripheral D regions. Seven disulfide bonds are located in the E region and 10 in the D regions [3]. The redox states 14 of the 17 disulfide bonds have been measured by differential cysteine alkylation and mass spectrometry [4]. Briefly, the unpaired cysteine thiols in fibrinogen were alkylated with 2-iodo-N-phenylacetamide (^{12}C-IPA), the protein isolated by SDS-PAGE, the disulfide bonds reduced with dithiothreitol, and the disulfide-bonded cysteine thiols alkylated with a stable carbon-13 isotope of IPA (^{13}C-IPA) [5]. The fibrinogen was digested with trypsin and peptides quantified by HPLC and identity established by mass spectrometry. The levels of the different redox forms of the cysteines were determined from the relative abundance of peptides labeled with ^{12}C-IPA and/or ^{13}C-IPA.

Our previous studies demonstrated that 13 of the 17 fibrinogen disulfides are unformed in ~10 to ~50% of fibrinogen molecules immunoprecipitated from human plasma [4], and that this is a constitutive property of the protein as recombinant fibrinogen produced in non-human cells and immunoprecipitated from the conditioned medium has the same distribution of partially disulfide-bonded states. Analysis of purified, lyophilized fibrinogen that does not require isolation by immunoprecipitation for analysis indicated that the 13 disulfide bonds were globally more formed than the plasma-derived protein, ranging from ~5 to ~30% unformed [4]. This result was confirmed herein (Fig. 2).

Fourteen disulfides in purified, lyophilized fibrinogen (five E region and nine D region disulfides) were unformed in ~5 to ~30% of fibrinogen molecules (Fig. 2). This result was compared with the same fibrinogen captured on anti-fibrinogen polyclonal antibody-coated beads. The 14 disulfide bonds in bead-bound fibrinogen were globally more unformed than the unselected fibrinogen. Eight of the 14 fibrinogen disulfides were significantly ($p < 0.05$) more unformed in the bead-bound fibrinogen in three repeat experiments (Fig. 2).

We next determined whether time of incubation of fibrinogen with the polyclonal antibody-coated beads influenced the selection of fibrinogen disulfide-bonded states.

3.2 Polyclonal Anti-Fibrinogen Antibodies Rapidly Select for Fibrinogen States with More Unformed Disulfides

Purified human fibrinogen was incubated with anti-fibrinogen polyclonal antibody-coated beads for 10 min to 24 h and the disulfide bond redox states of the bead-bound fibrinogen determined. The 14 disulfide bonds in bead-bound fibrinogen were globally more unformed than the unselected fibrinogen at all the incubation times (Fig. 3). Also, the results for the 10 min incubation were comparable to the 24 h incubation timepoint, indicating that the polyclonal antibody is rapidly selecting fibrinogen states with more unformed disulfides.

We also determined if we could separate different populations of fibrinogen disulfide states bound to the polyclonal antibody-coated beads.

3.3 Polyclonal Anti-Fibrinogen Antibodies Bind with Higher Avidity to Fibrinogen with More Unformed Disulfide Bonds

Purified human fibrinogen was incubated with anti-fibrinogen polyclonal antibody-coated beads for 25 min and the bead-bound fibrinogen eluted with pH 2.5 buffer, and then with buffer containing sodium dodecyl sulfate (SDS) at 60°C. Acidic pH if often used to dissociate antigens from antibodies, while denaturation of the proteins in SDS at high temperatures is a more stringent means of dissociating antigens from antibodies. Nine of the 11 measured disulfides in fibrinogen eluted under mild (pH 2.5) conditions were more unformed than the total unselected fibrinogen. Eleven of the 11 measured disulfides in fibrinogen then eluted under stringent (SDS) conditions were more unformed than fibrinogen eluted under mild

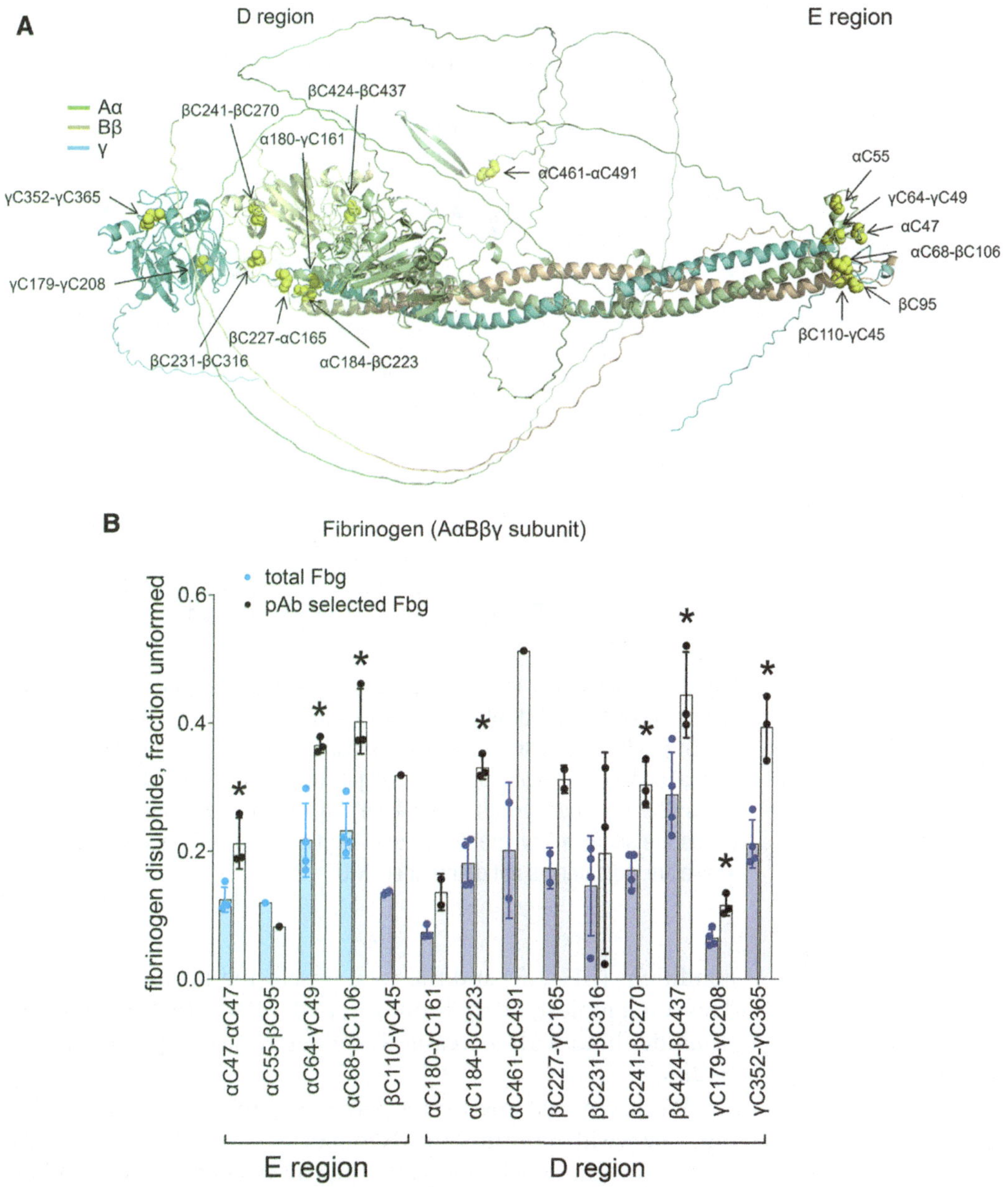

Fig. 2 Polyclonal anti-fibrinogen antibodies select for fibrinogen states with more unformed disulfides. (**A**) Ribbon representation of the AlphaFold structure of human fibrinogen subunit consisting of Aα, Bβ, and γ chains. The disulfide bonds measured in this study are shown as yellow spheres and residue numbers indicated. Cysteine numbering is according to UniProt identifier P02671 for human fibrinogen α chain, P02675 for the β chain and P02679 for the γ chain. Note that the E region αC55-βC95 disulfide bond is unformed in the structure. αC47 bonds with αC47 in another subunit. (**B**) Redox states of the five E region and nine D region disulfides in purified human fibrinogen and the fibrinogen captured on anti-fibrinogen polyclonal antibody-coated beads. The bars and errors are mean ± SD of three to four separate experiments. Parametric unpaired *t* test was used to evaluate differences between groups where there is a minimum of three data points per group (**p* <0.05). The different redox states of the Cys residues were quantified from the relative ion abundance (AUC) of peptides labeled with ^{12}C-IPA and/or ^{13}C-IPA. Missing data points are situations where IPA-labeled peptides had AUC values of <100,000

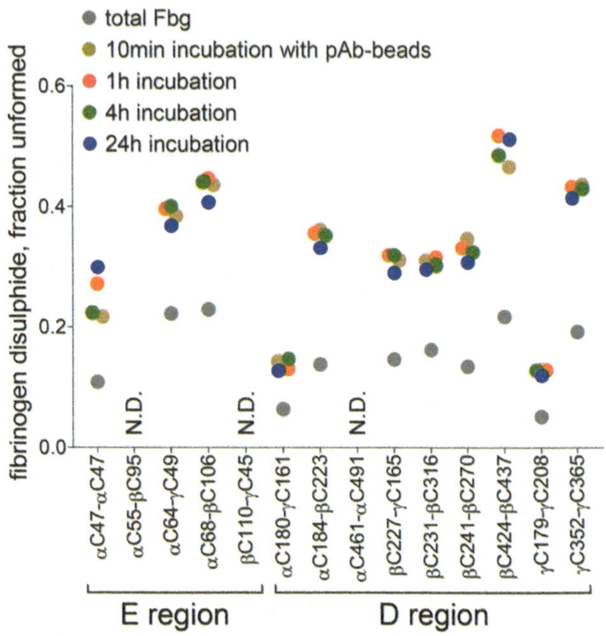

Fig. 3 Polyclonal anti-fibrinogen antibodies rapidly select for fibrinogen states with more unformed disulfides. Redox states of the five E region and nine D region disulfides in purified human fibrinogen captured on anti-fibrinogen polyclonal antibody-coated beads following incubation of the reactants for discreet times. The black data points are the total fibrinogen for comparison. Missing data points are situations where IPA-labeled peptides had AUC values of <100,000. These are indicated as not determined, N.D

conditions (pH 2.5) and the total unselected fibrinogen (Fig. 4). This result indicates that fibrinogen molecules with more unformed disulfides bind with increasingly higher avidity to polyclonal antibodies.

We determined whether the polyclonal antibody selection of fibrinogen states with more unformed disulfides is reflected in other proteins. We chose to study the blood protein, von Willebrand factor (VWF).

3.4 Polyclonal Anti-VWF Antibodies Select for VWF States with More Unformed Disulfides

VWF is a plasma protein produced by vascular endothelial cells and bone marrow megakaryocytes that chaperones blood coagulation cofactor factor VIII and tethers platelets to the damaged vessel wall [6]. VWF circulates in blood as a series of multimers containing variable numbers of 500 kDa dimeric units. VWF was purified from a human plasma concentrate and we quantified the redox state of four disulfide bonds in the VWF TIL' domain (C767-C808, C776-C804, C788-C799, and C810-C821), two disulfides in the D3 domain (C889-C1031 and C914-C921), three in the C8-3 domain

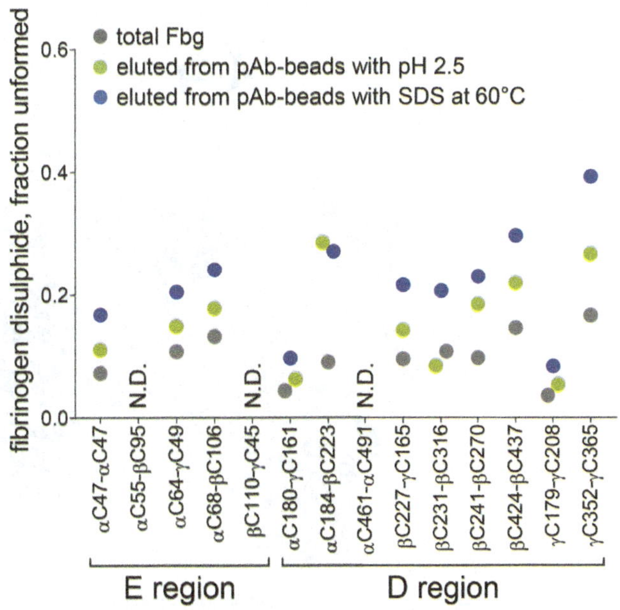

Fig. 4 Polyclonal anti-fibrinogen antibodies bind with higher avidity to fibrinogen with more unformed disulfide bonds. Redox states of the five E region and nine D region disulfides in purified human fibrinogen captured on anti-fibrinogen polyclonal antibody-coated beads and eluted with pH 2.5 buffer, and then with buffer containing SDS at 60°C. The black data points are the total fibrinogen for comparison. Missing data points are situations where IPA-labeled peptides had AUC values of <100,000. These are indicated as not determined, N.D

(C1046-C1089, C1060-C1084, and C1071-C1111), and four in the D4 domain (C1879-C1904, C1927-C2088, C1972-C2123, and C1993-C2001). These disulfide bonds have been defined in structures of VWF fragments [7, 8].

Our previous studies demonstrated that 13 disulfide bonds in the TIL' through E3 domains of VWF are unformed in ~4 to ~21% of VWF molecules immunoprecipitated from human plasma, and, as for fibrinogen, this is a constitutive property of VWF as recombinant protein has comparable distribution of partially disulfide-bonded states [2]. Here, we measured the disulfide-bonded states of purified human VWF that does not require isolation by immunoprecipitation for analysis. The 13 disulfide bonds in the TIL' through D4 domains indicated in the previous paragraph were unformed in ~2 to ~10% of the purified VWF molecules (Fig. 4). This result was compared with the same VWF captured on anti-VWF polyclonal antibody-coated beads. The 13 disulfide bonds in bead-bound VWF were globally more unformed than the unselected VWF (Fig. 5). Some disulfides were up to 30% unformed in the polyclonal antibody-bound VWF.

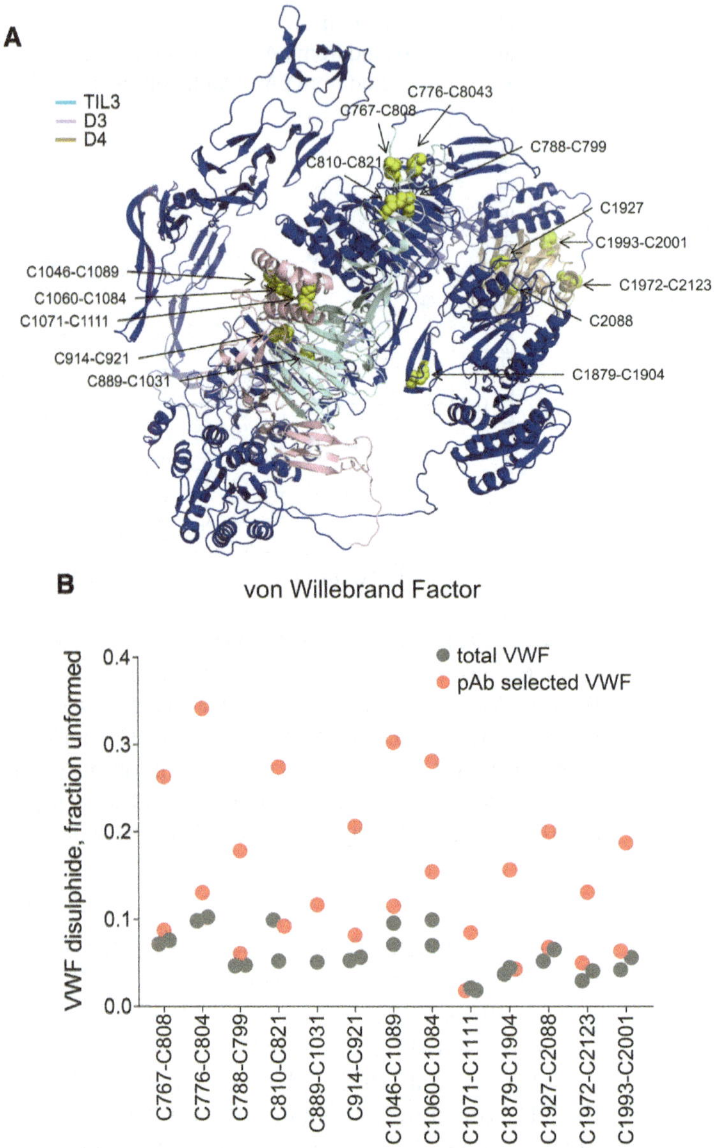

Fig. 5 Polyclonal anti-VWF antibodies select for VWF states with more unformed disulfides. (**A**) Ribbon representation of the AlphaFold structure of human VWF subunit with the TIL3, D3, and D4 domains highlighted. The disulfide bonds measured in this study are shown as yellow spheres and residue numbers indicated. Cysteine numbering is according to UniProt identifier P04275 for human VWF. Note that the D4 domain C1927-C2088 disulfide bond is unformed in the structure. (**B**) Redox states of 13 disulfides in purified VWF and the VWF captured on anti-VWF polyclonal antibody-coated beads. The data points are from two separate experiments. The different redox states of the Cys residues were quantified from the relative ion abundance (AUC) of peptides labeled with ^{12}C-IPA and/or ^{13}C-IPA. Missing data points for some disulfides are situations where IPA-labeled peptides had AUC values of <100,000

4 Summary

Anti-fibrinogen and anti-VWF polyclonal antibodies selectively bind fibrinogen and VWF states where their disulfide bonds are more unformed. For fibrinogen, the polyclonal antibody selection is complete within minutes of incubation and the bound antigen subsets are heterogeneous, where the bound molecules disrupted by low pH have a different distribution of disulfide-bonded states than the bound molecules disrupted by denaturation with SDS. It is important to note that the results for fibrinogen and VWF may not be representative of all antigens that are produced as partially disulfide-bonded states. Some polyclonal antibodies, for instance, may selectively bind cognate antigens where the disulfides are more formed, and in some cases, there may be no selection at all. This should be determined for each polyclonal antibody/antigen pair. These findings are relevant for experimental protocols that employ polyclonal antibodies. For example, the findings described herein may skew the results of ELISA assays or cellular immunoprecipitation experiments. The antibodies in these systems could select for antigen disulfide-bonded subsets that are not reflective of the total antigen pool and its function.

References

1. Butera D, Pijning AE, Avery NG, Coxon CH, Metcalfe C, Mimoun A, Lacroix-Desmazes S, Spiegel PC, Hogg PJ (2025) Patient anti-FVIII drug antibodies bind preferentially to a subset of FVIII covalent states. Blood Adv: accepted for publication.
2. Butera D, Wang HJ, Woon HG, Zhao YC, Ju LA, Hogg PJ (2023) Mechano-covalent protection of coagulation factor VIII by von Willebrand factor. Blood Adv 7(10):2117–2128. https://doi.org/10.1182/bloodadvances.2022008650
3. Kollman JM, Pandi L, Sawaya MR, Riley M, Doolittle RF (2009) Crystal structure of human fibrinogen. Biochemistry 48(18):3877–3886. https://doi.org/10.1021/bi802205g
4. Butera D, Hogg PJ (2020) Fibrinogen function achieved through multiple covalent states. Nat Commun 11(1):5468. https://doi.org/10.1038/s41467-020-19295-7
5. Butera D, Passam F, Ju L, Cook KM, Woon H, Aponte-Santamaria C, Gardiner E, Davis AK, Murphy DA, Bronowska A, Luken BM, Baldauf C, Jackson S, Andrews R, Grater F, Hogg PJ (2018) Autoregulation of von Willebrand factor function by a disulfide bond switch. Sci Adv 4(2):eaaq1477. https://doi.org/10.1126/sciadv.aaq1477
6. Sadler JE (1998) Biochemistry and genetics of von Willebrand factor. Annu Rev Biochem 67:395–424. https://doi.org/10.1146/annurev.biochem.67.1.395
7. Shiltagh N, Kirkpatrick J, Cabrita LD, McKinnon TA, Thalassinos K, Tuddenham EG, Hansen DF (2014) Solution structure of the major factor VIII binding region on von Willebrand factor. Blood 123(26):4143–4151. https://doi.org/10.1182/blood-2013-07-517086
8. Dong X, Leksa NC, Chhabra ES, Arndt JW, Lu Q, Knockenhauer KE, Peters RT, Springer TA (2019) The von Willebrand factor D'D3 assembly and structural principles for factor VIII binding and concatemer biogenesis. Blood 133(14):1523–1533. https://doi.org/10.1182/blood-2018-10-876300

Chapter 17

Dynamic Force Spectroscopy for Analysis of Multiple Disulfide-Bonded Protein States

Laura Moldovan and Lining Arnold Ju

Abstract

An emerging concept in chemical biology is that protein function can be regulated by the redox state of disulfide bonds. This chapter describes the dynamic force spectroscopy method for analyzing the redox regulation of distinct disulfide-bonded protein conformations. The main method described in this chapter is the biomembrane force probe (BFP), in which an ultrasoft human red blood cell is used as an ultrasensitive mechanical force probe. The BFP uses a high-speed camera and real-time imaging tracking techniques to characterize a single molecular bond with ~1 pN (10^{-12} N), ~3 nm (10^{-9} m), and ~0.5 ms (10^{-3} s) in force, spatial, and temporal resolution. As an exemplar model, we use the BFP to analyze the binding kinetics of distinct antithrombin conformations with varying degrees of reduced disulfide bonds under a range of applied forces. Through examination of measured bond lifetimes, we can identify a change in antithrombin binding affinity to standard heparin and correlate the binding kinetics with the overall proportion of reduced disulfide bonds in the antithrombin protein as validated by mass spectrometry. The methodologies described here offer a robust framework for examining other disulfide-bonded protein states.

Keywords Dynamic force spectroscopy, Mass spectrometry, Antithrombin, Heparin, Disulfide bond, Micropipette

1 Introduction

The international progress in biology is now shifting from characterizing cellular activities to subcellular and molecular behaviors. This demands nanotools with sufficiently high spatial and temporal resolutions. Over the past decade, dynamic force spectroscopies such as atomic force microscopy and optical tweezers have been the main approaches used to examine protein dynamics in purified systems [1]. As an in situ force spectroscopy, micropipette-based pico-force technique—BFP—enables the biomechanical analysis at both the single-cell and single-molecule levels [2]. More importantly, BFP manipulates both molecular binding force magnitude and "waveforms" such as tensile vs. compressive force, ramped vs. clamped force, and short vs. long duration force. With the recent upgrade of

Philip J. Hogg (ed.), *Functional Disulphide Bonds*, Methods in Molecular Biology, vol. 3016,
https://doi.org/10.1007/978-1-0716-5158-2_17,

the system to fluorescence BFP [3–5] and dual BFP [6, 7], it allows one to correlate the differential cell responses as seen in receptor affinity and conformational changes, calcium flux and exocytosis with the mechanical stimulation profile. Considering the compatibility on cell manipulation (minimum stimulation and nontoxic), BFP is an excellent biomechanical nanotool for studying single-molecule behaviors in a native cell environment. Notably, it has led to new understandings on inner workings of multiple protein nanomachines: T-cell receptor (Immunology) [5], Integrin and Syndecan (Cancer biology) [7–9], Notch (Developmental biology) [10], Glycoprotein Ib and von Willebrand factor (Hematology) [4, 11–13].

The BFP engineers a human red blood cell (RBC) into an ultra-sensitive force transducer [2, 3, 11, 14], which is aspirated by a micropipette. The probe consists of the RBC and a bead coated with the ligand at a low site density to favor single bond formation. The bead attaches to the apex of the RBC via biotin–streptavidin (SA) coupling. To interrogate receptor–ligand interaction, a receptor-bearing bead or cell is aspirated by an opposing micropipette and driven to repetitively touch with the force probe. Force spectroscopy data are derived from repeated touch cycles of contact-separation between the target cell/bead and the bead with precisely controlled time, area, compressive force, and loading rate. The temporal, spatial, and force precision are 0.7 ms, 3 nm, and 1 pN, respectively [3, 15]. The adhesion frequencies (P_a = the number of productive touches divided by the number of total touches) reflect the affinity of the receptor–ligand interaction [14].

As an emerging concept, the cleavage of allosteric disulfide bonds—a form of posttranslational redox modification has been found to regulate protein function in different biological systems [16]. In addition to measurement of molecular binding kinetics [13, 14, 17, 18], protein stiffness [8, 19], conformational changes [20, 21], and characterizing their regulations by physical factors, such as the tensile force, Brownian motion, separation distance, and diffusivity [18], the BFP has recently been used to investigate redox regulations of protein functions. BFP was used to demonstrate that antithrombin (AT) exists in multiple protein states and that the binding affinity of AT to heparin is dependent on the redox states of disulfide-bonds present in AT. This chapter outlines a novel BFP method that was used to compare the binding affinity of a subset of AT with higher percentages of unformed disulfide bonds and wild-type (WT) AT.

2 Materials (*see* Note 1)

2.1 Reagents for Biomembrane Force Probe Assay

1. Biotinylated antithrombin (AT): Wild-type (WT) AT, 0.16 mg/mL; heparin-selected subset (HS) AT, 0.30 mg/mL; and flow-through subset (FT) AT, 1.1 mg/mL.

2. Carbonate/bicarbonate buffer (pH 8.5–9): 8.4 g/L sodium carbonate (Na_2CO_3), 10.6 g/L sodium bicarbonate (Na_2HCO_3).
3. Phosphate buffer (pH 6.5–6.8): 27.6 g/L sodium phosphate monobasic ($NaH_2PO_4 \bullet H_2O$), 28.4 g/L anhydrous, sodium phosphate dibasic (Na_2HPO_4).
4. N_2-5% buffer (pH 7.2-7.4): 20.77 g/L potassium chloride (KCl), 2.38 g/L sodium chloride (NaCl), 0.13 g/L potassium phosphate monobasic (KH_2PO_4), 0.71 g/L anhydrous, sodium phosphate dibasic (Na_2HPO_4), 9.70 g/L sucrose.
5. Biotin-PEG3500-SGA (A5057-1; JenKem).
6. 3-Mercaptopropyltrimethoxysilane (MPTMS; llc 4420-74-0, United Chemical Technologies).
7. Borosilicate Glass beads (9002; Thermo Scientific).
8. Quantibrite PE Beads (340495; BD Biosciences).

2.2 Micropipette Fabrication

1. Borosilicate capillary tube OD 1.0/ID 0.7× 90 mm (G-1, Narishige).
2. Digital Flaming/Brown™ micropipette puller (P-1000, Sutter Instrument).
3. Pipette Microforge (MF-900, Narishige).

2.3 Chamber Assembly

1. Mineral Oil (M8410; Sigma-Aldrich).
2. Microscope Coverslip (40 mm x 22 mm; #1, Menzel-Glaser).
3. Micro-injector (MF34G-5; World Precision Instruments).
4. 1 mL tuberculin syringe (302100; BD).
5. Micropipette holder (HI-7; Narishige).

2.4 Biomembrane Force Probe Instrumentation

1. Acquire an inverted microscope (e.g., IX-70, Olympus) with a numerical aperture (NA) 0.55 condenser with a top lens and a LUCPlanFLN 60x objective (NA 0.7 UIS 2; Olympus) (Fig. 1A).
2. Implement the microscope onto an air anti-vibration table (1500 mm × 900 mm; Vision IsoStation, Newport), isolating the mechanical vibrations from the environment for the whole microscope (Fig. 1A).
3. Customize the microscope stage and mount: (a) a set of XYZ 3-way mechanical translator (462-XYZ-M; Newport) on the left side of the stage to manipulate the "Probe" micropipette; (b) a set of 3-way piezoelectric translator (M-105.3P; Physik Instrumente) on the top-right side to manipulate the "Target" micropipette; (c) a hydraulic micromanipulator (PH400; Karl Suss) on the bottom-right side with a remote fine control to manipulate the "Helper" micropipette (Fig. 1A) (*see* **Note 2**).

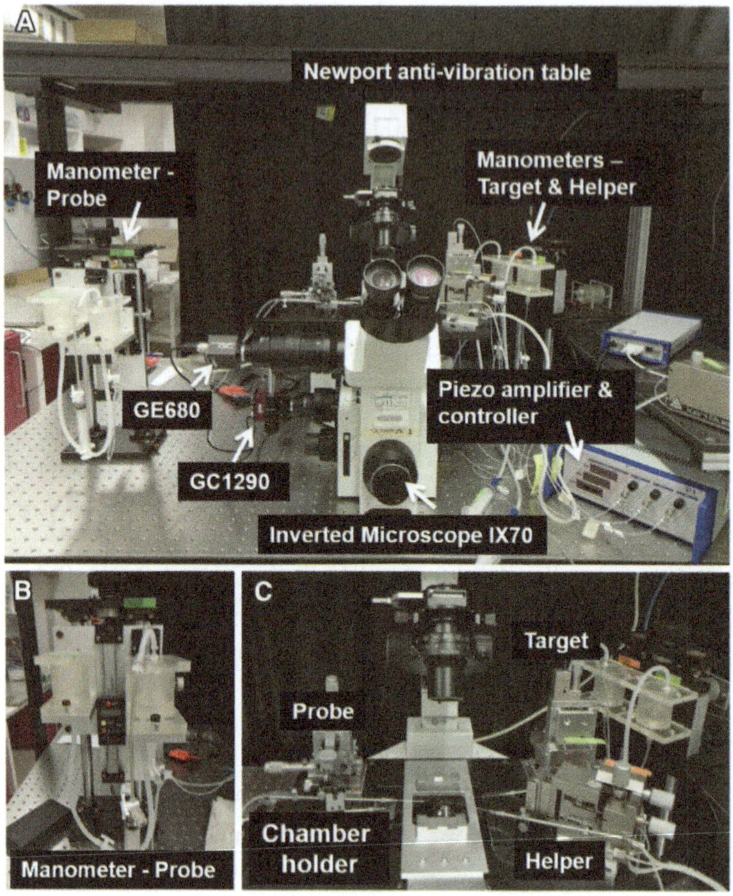

Fig. 1 Biomembrane force probe instrumentation. (**A**) The system overview. (**B**) Dual reservoir water manometer tower for adjustment of the "Probe" micropipette aspiration pressure. (**C**) The microscope stage that adapts an experimental cell chamber and three micropipette manipulators

4. Mount a 1D piezo linear actuator (P-753.1CD; Physik Instrumente) with capacitive feedback control and subnanometer precision to the M-105.3P 3-way translator.
5. Connect three micropipette holders (HI-7; Narishige) to the 462-XYZ-M, M-105.3P, and PH400, respectively.
6. Build the in-house water-pressure manometers with a height adjustment mechanism. For RBC aspiration control on the "Probe" micropipette, a fine mechanical positioner is designed to precisely control the relative height between the two reservoirs (Fig. 1B). The pressure difference is measured and used to determine the RBC's spring constant.
7. Connect the micropipette holders to their corresponding manometer reservoirs with hydraulic tubing lines.

8. Mount a high-speed CCD camera (3000 fps, 640 × 480, GigE, 1/3" CCD, mono GE680; Allied Vision Prosilica) onto the camera side port of the microscope by a video tube.
9. Mount a normal speed CCD camera (30 fps, 1280 × 960, monochrome CCD, 12 Bit ADC; GC1290, Allied Vision Prosilica) onto the eyepiece side port by a video tube.
10. Adjust halogen lamp brightness until a glass bead displays a clear diffraction pattern with a sharp edge (*see* **Note 3**).
11. Tune a customized LabVIEW program to control the 1D piezo actuator and drive the X-axial movement of the "Target" micropipette and perform image acquisition by the high-speed camera (*see* **Note 4**).

3 Methods

3.1 RBC Biotinylation for BFP

1. Collect 8–10 μL (one drop) of blood from a finger prick and add to 1 mL of the carbonate/bicarbonate buffer (pH 8.5–9). Gently vortex or pipette the mixture and centrifuge for 1 min at 900 × *g*. Discard supernatant and wash once more.
2. Weigh 2.5–3 mg of Biotin-PEG3500-SGA polymer in a small beaker, then dissolve it in the carbonate/bicarbonate buffer to make the final concentration of 8 mg/mL (*see* **Note 5**).
3. Mix 897 μL of carbonate/bicarbonate buffer, 3 μL of RBC pack, and 100 μL of Biotin-PEG3500-NHS linker solution and incubate at the room temperature (RT) for 30 min.
4. Wash the biotinylated RBCs with carbonate/bicarbonate buffer once, then with the Tyrode's buffer containing 0.5% BSA once.
5. To swell the RBCs, dilute nystatin into N_2-5% buffer to make a final concentration of 40 μg/mL.
6. Mix 5 μL of biotinylated RBCs with 71.4 μL of nystatin solution and incubate for 1 h at 0°C on ice (*see* **Note 6**).
7. Wash twice with N_2-5% buffer and store with the Tyrode's buffer in the refrigerator (4°C).

3.2 Preparation for Micropipette and Cell Chamber

1. Mount one piece of borosilicate capillary tube onto the Flaming/Brown micropipette puller.
2. Click the "Pull" button so that the capillary is heated in the middle, then pulled apart into two raw pipettes with a closed end (*see* **Note 7**).
3. Mount a raw pipette onto the Pipette Microforge and fabricate one micropipette at a time with a desired taper and orifice

size by repetitive truncations. The example micropipette orifice sizes are 1.5–2.0 μm for a bead and 2.0–2.4 μm for an RBC (*see* **Note 8**).

4. Build a homemade chamber holder, with two pieces of copper/aluminum squares (25 mm × 25 mm × 12 mm) and a handle that connects them (Fig. 1C).
5. Cut a 40 mm × 22 mm × 0.17 mm coverslip along the longitudinal axis by a diamond pen into two halves, named Coverslip 1 and 2.
6. Assemble a cell chamber by gluing two half-coverslips to the top and bottom sides of the chamber holder with grease.
7. Inject ~300 μL of Tyrode's buffer between the two coverslips. Make sure the buffer attaches to the top coverslip. Gently rotate and shake the chamber to let the buffer zone connect both ends of the chamber holder squares along the longitudinal axis.
8. Carefully inject mineral oil flanking the buffer zone to insulate the buffer from vaporization.

3.3 Bead Functionalization

1. Weigh out 50 mg of borosilicate glass bead powders and perform silanization and thiolation with 3-mercaptopropyltrimethoxysilane (MPTMS) as previously described [3].
2. Take one vial of dry MPTMS beads (~2 mg) and wash once with phosphate buffer (pH 6.5–6.8), and resuspend into 50 μL of phosphate buffer, and store at 4°C.
3. To functionalize beads with streptavidin (SA) for RBC attachment, take 10 μL of MPTMS beads with 10 μL of 4 mg/mL Streptavidin–Maleimide (SA–MAL) solution and add 180 μL of phosphate buffer.
4. Incubate the beads overnight at RT on rotar. Wash three times with phosphate buffer and finally resuspend into 100 μL of phosphate buffer and store at 4°C. These can be stored for up to 3 months in the refrigerator.
5. To coat biotinylated AT onto SA coated beads, mix 10 μL of SA coated beads with the 10 μL of biotinylated AT at 100 μg/mL and 180 μL of phosphate buffer.
6. Incubate the mixture overnight at 4°C, wash three times with phosphate buffer, and resuspend into 50 μL of phosphate buffer and store at 4°C.
7. To functionalize target beads with heparin, 5 μL of MPTMS beads were incubated with 10 μL of 10 mg/mL unfractionated heparin sodium solution (Sigma) into 85 μL of phosphate buffer at 4°C overnight.

8. Beads were washed three times with phosphate buffer and resuspended into 50 μL of phosphate buffer. These beads were made fresh each experimental day.

3.4 BFP Assembly Before an Experiment

1. Inject the three concentrated cell/bead species (biotinylated RBCs, beads coated with AT, and beads coated with unfractionated heparin (UFH)) at separate locations of the buffer zone.
2. Mount the cell chamber onto the holder stage and move it on top of the objective.
3. Use a microinjector to fill three fabricated micropipettes with Milli-Q water.
4. Assemble the "Probe," "Target," and "Helper" micropipettes to their respective micropipette holders on micromanipulators (Fig. 1C, *left*: "Probe" to grab a RBC; *top right*: "Target" to grab a bead; *bottom right*: "Helper" to deliver the bead) (*see* **Note 9**).
5. Insert the three micropipettes into the chamber buffer zone. Adjust the positions of the micropipettes and find them under the brightfield microscopy (Fig. 2A).
6. Toggle the holder stage to move the cell chamber around and scan through locations of the three injected cell/bead species.
7. Adjust the positions of the three micropipettes by toggling the knobs of the corresponding holding manipulators, then let the tip of each micropipette approach its respective cell/bead species.
8. Adjust the aspiration pressure inside each micropipette to suck a bead or a cell for all three micropipettes, then get them together in the same field of view (*see* **Note 10**).
9. Align the bead and RBC, and then carefully attach the bead to the apex center of the RBC using the "Helper" micropipette (Fig. 2B). Contact solidly and then retract slightly. Move away the helper micropipette and align the target and probe bead (Fig. 2C) (*see* **Note 11**).
10. Use the region of interest (ROI) tools in the LabVIEW program to measure the respective radii of the probe micropipette (R_p), the RBC (R_0), and the circular contact area between the RBC and probe bead (R_c).
11. Set the desired RBC spring constant by adjusting the Probe's aspiration pressure with the water manometer (*see* **Note 12**).
12. Draw a horizontal line ROI across the RBC-bead boundary to track the bead position in real time (Fig. 2D) (*see* **Note 13**).

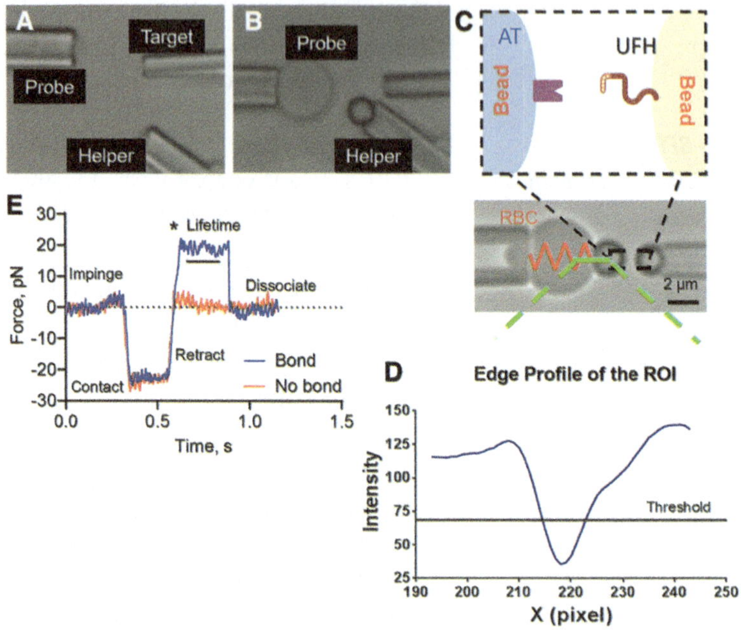

Fig. 2 Biomembrane force probe setup under microscope. (**A**) Micropipette assembly showing the "Probe" pipette (*left*), "Target" pipette (*upper right*), and "Helper" pipette (*lower right*). (**B**) Probe bead placement. A probe bead was manipulated by the "Helper" and attached to an RBC apex to form a force probe. (**C**) BFP brightfield schematic (*lower panel*) and protein functionalization (*top panel*). A micropipette-aspirated RBC with a bead attached to the apex (probe, left) was aligned against a bead held by an opposing micropipette (target, right). The probe bead was covalently linked with biotinylated human antithrombin and streptavidin for attachment of the bead to biotinylated RBC (left). The target bead was coated with unfractionated heparin (UFH) (right). Two subsets of antithrombin were tested: A flow-through subset (FT subset) and a Heparin-selected subset (HS subset). (**D**) The line ROI is placed at the bead-RBC edge for the bead position tracking, with the corresponding area highlighted in green. The intensity profile of the bead edge in (C). The ROI region in the *x*-direction is plotted as *x*-axis (in pixel number) and the light intensity (in gray scale value) averaged by binning 30 pixels along the *y*-direction. (**E**) Force vs. time traces of two representative BFP touch cycles. The target was driven to impinge, contact, and retract the probe. In a "no bond" event, the cycle ended after retraction (red). In a "bond" event (blue), the target was held (marked by $*$) at clamp force (20 pN) until the bond between probe and target dissociated, signified by the force dropping back to 0, and the lifetime of the bond was measured

13. Set the system parameters as desired, e.g., impingement force = 20 pN, loading rate = 1,000 pN/s, contact time = 0.2 s, clamped force = 5–50 pN (for force clamp assay), and spring constant κ =0.20–0.3 pN/nm.
14. Start the repeated BFP cycles for data collection.

3.5 BFP Test Cycles and Experimental Modes

1. A BFP experiment is composed of repeated test cycles that are controlled by home-written LabVIEW program. A high-speed camera continuously monitors RBC-bead deflection from the bead edge tracking as previously described [3, 12, 13, 21].

The force spectroscopy trace (Force vs. Time) for each cycle is recorded thereafter.

2. Upon the beginning of a BFP cycle, the "Target," an aspirated UFH-coated bead, is driven to approach, impinge, and contact with the "Probe," a AT-coated bead attached to the RBC, by the program-controlled piezo actuator (Fig. 2E).
3. At the end of the contact duration, the linear piezo actuator retracts the "Target" away from the "Probe."
4. (a) In the case of no bond, no tensile force is generated by the retraction (Fig. 2E, *red*), the target will return to the original position and begin the next test cycle.
   (b) In the presence of a bond, which is signified by an axial deflection of the RBC, the retracting "Target" will keep pulling the "Probe" at a constant force loading rate to a desired clamped force for the "force clamp assay" (Fig. 2E, *blue*).
5. This approach–impinge–contact–retract–clamp–dissociate test cycle will be repeated many times to acquire an ensemble of data for statistical analysis [3, 18].
6. For the adhesion frequency assay, 200 repeated cycles were performed for each bead–bead pair as "bond" and "no-bond" events are marked to calculate adhesion frequency (Fig. 3) (*see* **Note 14**).
7. For the force-clamp assay, similar BFP cycles are performed. Upon detection of a "bond," a feedback loop controls the retraction so that it would be paused at a desired clamped force (5–50 pN) until bond dissociation. After that, the "Target" is returned to the original position to complete the cycle. Lifetimes are measured from the instant when the force reaches the desired level to the instant of bond dissociation (Fig. 4).
8. After enough lifetime, events have been collected under a range of clamped forces, they can be pooled together and grouped into different force bins, which will produce an average lifetime in each force bins and altogether yield an "average lifetime vs. force" curve (Fig. 4).

3.6 Analysis of Lifetime Distributions

1. Record the parameters of all lifetime events. This includes the average force and bond lifetimes with the starting and ending times. Enough lifetime events under a range of forces should be collected. These will be grouped into different force bins, which will produce an average lifetime in each force bin.
2. The results show that bond lifetimes of HS AT–UFH have increased bond lifetimes at forces below 20 pN in comparison to both the WT AT–UFH and the FT AT–UFH (Fig. 4).

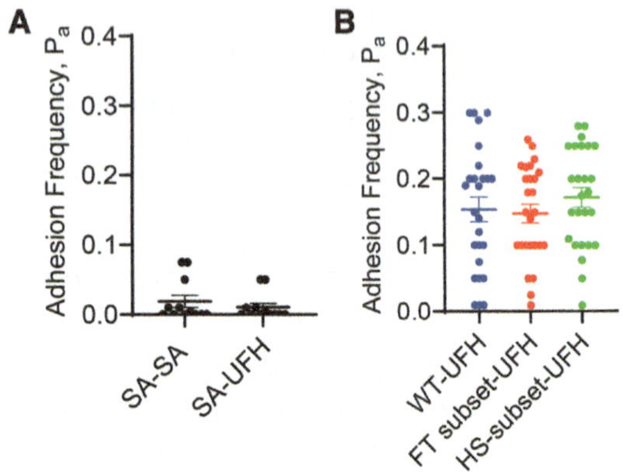

Fig. 3 Biomembrane force probe adhesion assays. (**A**) Adhesion frequencies of the negative controls were performed. Streptavidin-coated beads were brought into contact with either streptavidin or UFH and adhesion frequency was recorded. The data points are from 12 touch cycles. Each probe–target pair was tested repeatedly for 200 approach–contact–retract cycles to estimate an adhesion frequency. Errors are mean ± SEM. (**B**) Adhesion frequencies of wild-type (WT), flow-through (FT) selected subset, and heparin-selected (HS) subset were performed. The probe bead ligand coating density was adjusted to keep the adhesion infrequent (<20%) to satisfy the necessary condition for most of the adhesions (>89%) to be single bond events. The data points are from 26 pairs (200 approach–contact–retract cycles for each pair)

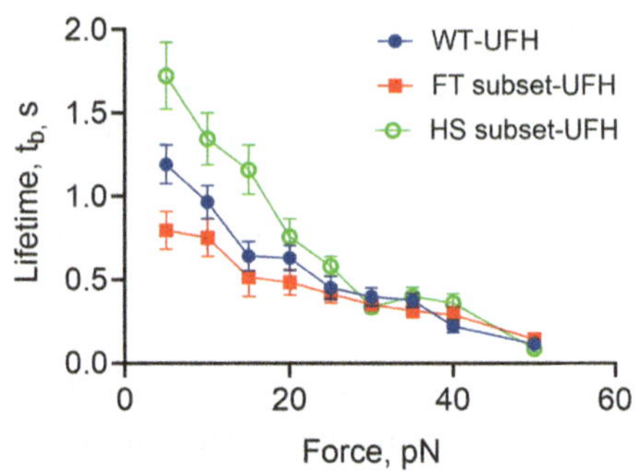

Fig. 4 Bond lifetime analysis. Lifetime of AT–UFH bonds as a function of clamp force in BFP. The AT was immobilized covalently. Results represent mean ± SEM of >20 measurements per point. SEM, standard error of the mean. WT–UFH (*n* = 216); FT subset–UFH (*n* = 217), HS subset–UFH (*n* = 209)

4 Notes

1. If not specified, the reagents are purchased from Sigma-Aldrich.
2. For stage mounting and parts connection, home-designed mechanical parts and adapters are fabricated by a machine

shop. All parts are customized using the computer-aided design software i.e., SolidWorks (Version 2012 SP5).

3. The following optical filters are recommended to add between the light source and the camera: (a) a neutral density filter to reduce the brightness for the protection of human eyes and camera CCD; (b) a diffuser to sharpen the glass bead edge.
4. All control programs are developed by the LabVIEW software (National Instruments Version 2019). The LabVIEW driver for the P-753.1CD piezo actuator can be downloaded from the manufacturer Physik Instrumente's website.
5. Biotin-PEG3500-SGA polymer should be stored in a dry ambient location at −20°C. For experimental use, take a bottle out from the freezer 30 min before the reaction and leave it at RT to warm up. The reaction needs to be accomplished as fast as possible so that the polymer has minimal exposure to the open air. After reaction, place the bottle of the polymer with loosened cap in a glass vacuum desiccator filled with the desiccant i.e., Drierite on the bottom and vacuum for 5 min, and then fill the desiccator with argon. Tighten the cap and take the bottle out. Seal the bottle with Parafilm (PM996, Bemis). Then place it into a container filled with Drierite on the bottom and store in −20°C.
6. The optimal incubation time depends on the RBC quality of the donor. If donor's RBCs are easy to lyse, reduce the incubation time; vice versa. The biotinylated RBC aliquot can be stored at 4°C for further use of up to 2 weeks.
7. According to the manual from the Flaming/Brown Micropipette Puller, the desired shape of the raw pipette has a 6–8 mm taper, 0.1–0.5 μm tip, and 60–150 MΩ resistance.
8. The Microforge has a heating unit: a glass sphere sits on a platinum wire, which partially melts when heated. The heating and cooling of the glass sphere is controlled by pressing and releasing a foot switch pedal. When the glass sphere melts, one can insert the tip of the raw pipette inside. Then let the sphere cool down and become solid, while the inserted tip part stays inside. When pulled, the pipette will break from the outside and leave its tip inside the sphere.
9. Take off the micropipette holder and put it at a position lower than the water level of the connected reservoir to allow water drip from the tip. Quickly insert the water-filled micropipette into the holder tip and make sure no air bubble gets into the micropipette during the connection, then tighten the holder screw.
10. Move around the chamber holder stage to find an open space away from the colonies of injected cell/bead species where

the experiment will be performed. Turn on the LabVIEW software that controls the GC1290 camera to have a microscopic view on a computer screen.

11. To better glue the bead onto the RBC apex, adjust the pressure of the "Helper" micropipette to gently blow the bead away.
12. The spring constant of the BFP (κ) is determined by Evans' model [14, 22]:

$$\kappa = \frac{\pi R_p \Delta p}{\left(1 - R_p/R_0\right)\ln\left[4R_0^2/\left(R_p R_c\right)\right]}, \quad (1)$$

where Δp is the pressure difference aspirated at the "Probe" micropipette tip. For accurate force determination with the Evans' model, it is required to use the optimal micropipette size R_p around 1 μm to aspirate RBC with a short tongue ($R_p \approx L_p$) as described previously [23]. When the "Probe" micropipette size is too big ($R_p \gg 1$ μm) or RBC extension by force stretching exceeds 200 nm, the BFP spring constant calculation is no longer accurate [23].

13. The bead position tracking is achieved by the peak detection algorithm in the LabVIEW. Drag the threshold line to be at around half the depth of the brightness curve. The minimum point on the brightness curve below the threshold line indicates the position of the bead boundary; thus, only one local minimum is allowed. If two or more local minima are present, they indicate the image is not optimal (likely due to the image being out of focus, or an underperformed alignment between the probe bead and the RBC).
14. Record the observed adhesion events by adding "1" for "bond" or "0" for "no bond" in a column of an Excel spreadsheet. Measure the adhesion frequency using at least three different cell pairs for each contact time to obtain a mean ± SEM.

Acknowledgments

We thank Dr. Alexander Dupuy, Catherine Chen, and Jerry Wang for helpful discussion and technical support on the BFP data acquisition and analysis.

Author Contribution L.M. and L.A.J. designed the study, prepared figures, and wrote the chapter.

Funding Information This work was supported by the Australian Research Council (ARC, DP200101970-L.A.J.;

DP240101768-L.A.J.); the Australian National Health and Medical Research Council (NHMRC, APP2003904-L.A.J); NSW Cardiovascular Capacity Building Program (Early-Mid Career Research Grant-L.A.J.); MRFF Cardiovascular Health Mission Grants (MRF2016165-L.A.J.); Tour de Cure Annual Grant Pioneering Research (RSP-391-FY2023-L.A.J); Office of Global and Research Engagement (University of Sydney-University of Glasgow Ignition Grant-L.A.J.); and Nation Heart Foundation Vanguard Grant (106979-L.A.J.). L.A.J. is a Snow Medical Research Foundation Fellow (2022SF176) and a National Heart Foundation Future Leader Fellow Level 2 (105863).

Competing Interests The authors declare no competing interests.

References

1. Chen Y, Ju L, Rushdi M, Ge C, Zhu C (2017) Receptor-mediated cell mechanosensing. Mol Biol Cell 28(23):3134–3155. https://doi.org/10.1091/mbc.e17-04-0228
2. Moldovan L, Song CH, Chen YC, Wang HJ, Ju LA (2023) Biomembrane force probe (BFP): design, advancements, and recent applications to live-cell mechanobiology. Exploration 3(4):20230004. https://doi.org/10.1002/EXP.20230004
3. Chen Y, Liu B, Ju L, Hong J, Ji Q, Chen W, Zhu C (2015) Fluorescence biomembrane force probe: concurrent quantitation of receptor-ligand kinetics and binding-induced intracellular signaling on a single cell. J Vis Exp 102:e52975. https://doi.org/10.3791/52975
4. Ju L, Chen Y, Xue L, Du X, Zhu C (2016) Cooperative unfolding of distinctive mechanoreceptor domains transduces force into signals. eLife 5:e15447. https://doi.org/10.7554/eLife.15447
5. Liu B, Chen W, Evavold Brian D, Zhu C (2014) Accumulation of dynamic catch bonds between TCR and agonist peptide-MHC triggers T cell signaling. Cell 157(2):357–368. https://doi.org/10.1016/j.cell.2014.02.053
6. Ju L, Chen Y, Li K, Yuan Z, Liu B, Jackson SP, Zhu C (2017) Dual biomembrane force probe enables single-cell mechanical analysis of signal crosstalk between multiple molecular species. Sci Rep 7(1):14185. https://doi.org/10.1038/s41598-017-13793-3
7. Ju L, McFadyen JD, Al-Daher S, Alwis I, Chen Y, Tønnesen LL, Maiocchi S, Coulter B, Calkin AC, Felner EI, Cohen N, Yuan Y, Schoenwaelder SM, Cooper ME, Zhu C, Jackson SP (2018) Compression force sensing regulates integrin αIIbβ3 adhesive function on diabetic platelets. Nat Commun 9(1):1087. https://doi.org/10.1038/s41467-018-03430-6
8. Fiore VF, Ju L, Chen Y, Zhu C, Barker TH (2014) Dynamic catch of a Thy-1–α5β1+syndecan-4 trimolecular complex. Nat Commun 5(1):4886. https://doi.org/10.1038/ncomms5886
9. Xu XR, Wang Y, Adili R, Ju L, Spring CM, Jin JW, Yang H, Neves MAD, Chen P, Yang Y, Lei X, Chen Y, Gallant RC, Xu M, Zhang H, Song J, Ke P, Zhang D, Carrim N, Yu S-Y, Zhu G, She Y-M, Cyr T, Fu W, Liu G, Connelly PW, Rand ML, Adeli K, Freedman J, Lee JE, Tso P, Marchese P, Davidson WS, Jackson SP, Zhu C, Ruggeri ZM, Ni H (2018) Apolipoprotein A-IV binds αIIbβ3 integrin and inhibits thrombosis. Nat Commun 9(1):3608. https://doi.org/10.1038/s41467-018-05806-0
10. Luca VC, Kim BC, Ge C, Kakuda S, Wu D, Roein-Peikar M, Haltiwanger RS, Zhu C, Ha T, Garcia KC (2017) Notch-Jagged complex structure implicates a catch bond in tuning ligand sensitivity. Science 355(6331):1320–1324. https://doi.org/10.1126/science.aaf9739
11. Butera D, Passam F, Ju L, Cook KM, Woon H, Aponte-Santamaría C, Gardiner E, Davis AK, Murphy DA, Bronowska A, Luken BM, Baldauf C, Jackson S, Andrews R, Gräter F, Hogg PJ (2018) Autoregulation of von Willebrand factor function by a disulfide bond switch. Sci Adv 4(2):eaaq1477. https://doi.org/10.1126/sciadv.aaq1477
12. Ju L, Chen Y, Zhou F, Lu H, Cruz MA, Zhu C (2015) Von Willebrand factor-A1 domain binds platelet glycoprotein Ibα in multiple states with distinctive force-dependent dissociation kinetics. Thromb Res 136(3):606–612. https://doi.org/10.1016/j.thromres.2015.06.019

13. Ju L, Dong J-f, Cruz MA, Zhu C (2013) The N-terminal flanking region of the A1 domain regulates the force-dependent binding of von Willebrand factor to platelet glycoprotein Ibα. J Biol Chem 288(45):32289–32301. https://doi.org/10.1074/jbc.M113.504001
14. Chen W, Zarnitsyna VI, Sarangapani KK, Huang J, Zhu C (2008) Measuring receptor–ligand binding kinetics on cell surfaces: from adhesion frequency to thermal fluctuation methods. Cell Mol Bioeng 1(4):276–288. https://doi.org/10.1007/s12195-008-0024-8
15. Ju L, Chen Y, Rushdi MN, Chen W, Zhu C (2017) Two-dimensional analysis of cross-junctional molecular interaction by force probes. In: Baldari CT, Dustin ML (eds) The immune synapse: methods and protocols. Springer New York, New York, NY, pp 231–258. https://doi.org/10.1007/978-1-4939-6881-7_15
16. Hogg PJ (2003) Disulfide bonds as switches for protein function. Trends Biochem Sci 28(4):210–214. https://doi.org/10.1016/S0968-0004(03)00057-4
17. Huang J, Zarnitsyna VI, Liu B, Edwards LJ, Jiang N, Evavold BD, Zhu C (2010) The kinetics of two-dimensional TCR and pMHC interactions determine T-cell responsiveness. Nature 464(7290):932–936. https://doi.org/10.1038/nature08944
18. Ju L, Qian J, Zhu C (2015) Transport Regulation of Two-Dimensional Receptor-Ligand Association. Biophys J 108(7):1773–1784. https://doi.org/10.1016/j.bpj.2015.02.023
19. Sarangapani KK, Marshall BT, McEver RP, Zhu C (2011) Molecular stiffness of selectins. J Biol Chem 286(11):9567–9576. https://doi.org/10.1074/jbc.M110.196485
20. Chen W, Lou J, Evans EA, Zhu C (2012) Observing force-regulated conformational changes and ligand dissociation from a single integrin on cells. J Cell Biol 199(3):497–512. https://doi.org/10.1083/jcb.201201091
21. Ju L, Lou J, Chen Y, Li Z, Zhu C (2015) Force-induced unfolding of leucine-rich repeats of glycoprotein Ibα strengthens ligand interaction. Biophys J 109(9):1781–1784. https://doi.org/10.1016/j.bpj.2015.08.050
22. Evans E, Ritchie K, Merkel R (1995) Sensitive force technique to probe molecular adhesion and structural linkages at biological interfaces. Biophys J 68(6):2580–2587. https://doi.org/10.1016/S0006-3495(95)80441-8
23. Ju L, Zhu C (2017) Benchmarks of biomembrane force probe spring constant models. Biophys J 113(12):2842–2845. https://doi.org/10.1016/j.bpj.2017.10.013

Chapter 18

Assays of Thiol Isomerase Activity

Roelof H. Bekendam, Osamede C. Owegie, Moua Yang and Robert Flaumenhaft

Abstract

Thiol isomerases are oxidoreductases that mediate disulfide bond formation in nascent proteins of the endoplasmic reticulum to ensure their structural integrity. In addition to its role in protein folding, thiol isomerases can modify allosteric disulfide bonds in both intracellular and extracellular proteins, thereby controlling protein function. The process of disulfide bond formation and cleavage is strictly regulated and responsive to redox conditions. Understanding disulfide bond regulation under different redox environments is critical to understanding physiological and pathological processes related to disulfide bond chemistry. Here, we describe protocols for the measurement of disulfide bond modulation by thiol isomerases. These assays include reductase, oxidase, denitrosylase, and sulfenylation assays. These methods can be applied to study recombinant thiol isomerases and thiol isomerases in cellular settings.

Keywords Thiol isomerase, Protein disulfide isomerase (PDI), Disulfide bond formation, Di-eosin-GSSG assay, Insulin turbidimetric assay, Oxidative refolding assay, Denitrosylation assay, Sulfenylation assay

1 Introduction

Thiol isomerases are ubiquitously expressed proteins that are highly concentrated in the endoplasmic reticulum, where they serve a role in the folding of newly formed proteins [1, 2]. Although they have several enzymatic activities, they are best known as oxidoreductases capable of forming disulfide bonds within substrates by the oxidation of free thiols and forming free thiols by reducing disulfide bonds, depending on the redox environment. In proteins and peptides, disulfide bonds can provide stability to proteins by reinforcing peptide bonds or fortifying tertiary and quaternary structures. Such disulfides are termed structural disulfide bonds and are generally introduced as proteins are being formed in the endoplasmic reticulum. In addition, allosteric disulfides, often in solvent exposed locals, can be modified post-translationally. This disulfide switching can function

Philip J. Hogg (ed.), *Functional Disulphide Bonds*, Methods in Molecular Biology, vol. 3016,
https://doi.org/10.1007/978-1-0716-5158-2_18,

as a regulatory mechanism by which protein and peptide activity is controlled and responds to the redox environment.

The PDI family of thiol isomerases currently comprises over 20 members, all of which contain at least one thioredoxin-like domain [3, 4]. Protein disulfide isomerase (PDI), the founding member and best-studied thiol isomerase, has an a–b–b'–a′ domain structure (Fig. 1). Classically, the a domains have been termed the catalytic domains as they contain the Cys–X–X–Cys motif, shown to be pivotal for disulfide bond modulation. The b domains are called the noncatalytic thioredoxin-like domains and serve as substrate binding domains (Fig. 1) [1, 5–7].

Oxidative and reductive processes of thiol isomerases are performed by the Cys–X–X–Cys motif. In the relatively oxidizing environment of the endoplasmic reticulum, thiol isomerases can transfer the disulfide bond from their Cys–X–X–Cys motif to the target substrate. The substrate becomes oxidized, whereas the thiol isomerase itself becomes reduced. The first cysteine of the active site has been implicated in reacting with the free thiol in the substrate, forming a transient mixed disulfide. The second cysteine in the active site mediates the modification and final release of the substrate (Fig. 2).

In addition to its function within the endoplasmic reticulum, thiol isomerases have been shown to traffic to subcellular and extracellular compartments [8–11]. This includes the cell surface of

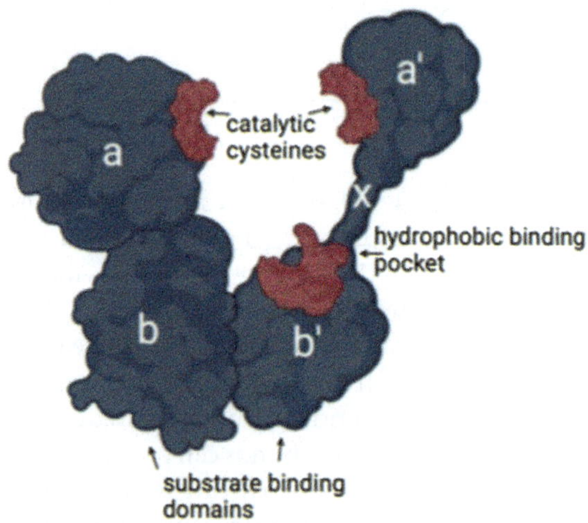

Fig. 1 Structure of protein disulfide isomerase. Protein disulfide isomerase (PDI) is a 57-kDa protein consisting of four domains. Thiol isomerases typically consist of a combination of catalytic cysteine-containing domains (**a** domain) and substrate binding domains (**b** domain). PDI has **a** and **a′** domains, both of which contain the Cys–Gly–His–Cys motif. The **b** and **b′** domains do not have the catalytic cysteines. The **b′** domain contains the hydrophobic binding site through which PDI interacts with substrates. The **b′** and **a′** domains are connected via a flexible peptide termed the x-linker (**x**)

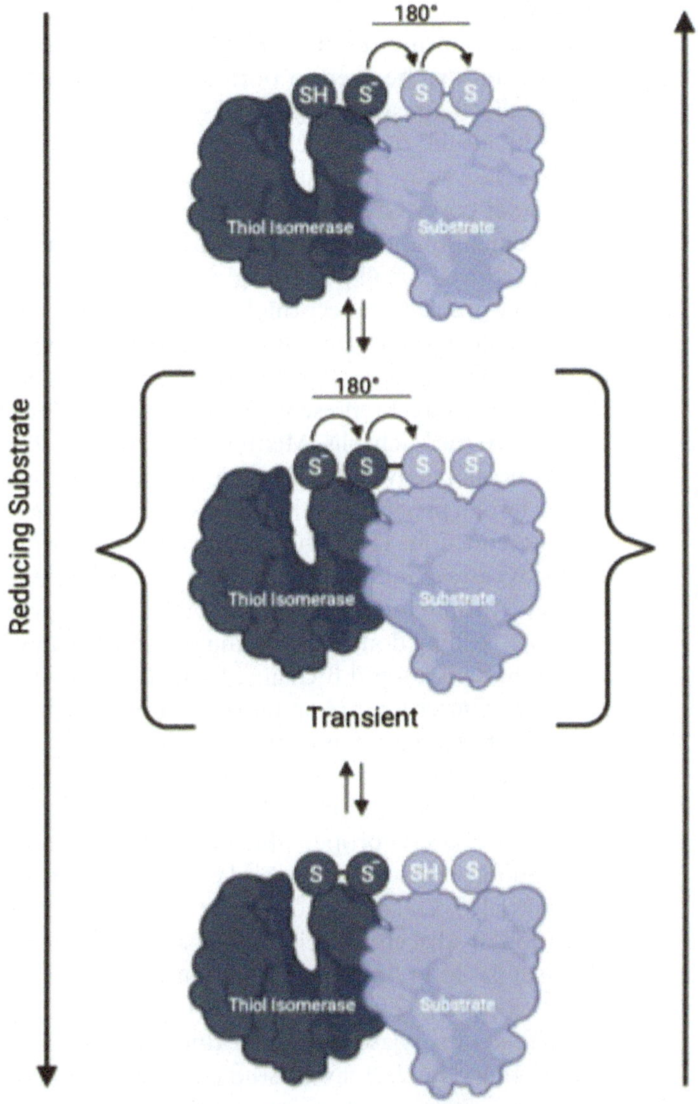

Fig. 2 Thiol isomerase-mediated disulfide exchange. Thiol isomerases reduce target substrates by attacking a disulfide bond within a substrate with the N-terminal cysteine within the Cys–X–X–Cys catalytic motif. A transient bond between the N-terminal cysteine and the substrate is formed, followed (*blue*) by C-terminal cysteine-mediated release of the substrate. This process can go in both directions, representing reduction (downward) and oxidation (upward) of the substrate

vascular cells [12, 13], hepatocytes [14], and pancreatic cells [15]. Thiol isomerases have been shown to play a pivotal role in several physiological and pathophysiological processes through disulfide bond modulation and regulate a large number of substrates [16–18].

Measurement of disulfide bond modulation through thiol isomerase activity assays has gained increasing attention. Here, we present an overview of several oxidoreductase assays to measure disulfide bond modulation of thiol isomerase substrate proteins and peptides to be used in both in vitro and cell-based systems. Specifically, we provide detailed descriptions for the di-eosin-GSSG assay, the insulin turbidimetric assay, an oxidative refolding assay, a denitrosylase assay, and a sulfenylation assay.

The di-eosin-GSSG assay was developed as a continuous direct measurement of the ability of the catalytic cysteines of thiol isomerases to reduce a small peptide, glutathione disulfide (GSSG) [19]. This assay is able to detect thiol isomerase activity in the low nanomolar range and has the highest sensitivity of any thiol isomerase assay described to date. Furthermore, the kinetics of di-eosin-GSSG cleavage are a direct representation of the thiol isomerase activity and can be used for Michaelis–Menten calculations [19, 20]. Due to the technique simplicity and a relative low cost, thiol isomerase-mediated bovine insulin precipitation has been studied extensively as a measurement of reduction. The technique has lower sensitivity than the di-eosin-GSSG assay, but does not require the analysis of the probe prior to usage. Insulin interacts with thiol isomerases (presumably via the substrate-binding domain) enabling reduction of interchain disulfide bonds that link α- and β-chains. Reduction causes precipitation of β-chains. Therefore, precipitation following reduction represents an indirect measurement of thiol isomerase activity. Substrates and competitive inhibitors that bind at the hydrophobic pockets of thiol isomerases interfere with insulin binding to thiol isomerases unlike the di-eosin-GSSG probe, which presumably binds outside of the hydrophobic binding pockets and interacts directly with the catalytic cysteines [20]. PDI also mediates the formation of disulfide bonds via oxidation of adjacent cysteines. Oxidative refolding of ribonuclease (RNase) has long been used to monitor the oxidase and isomerase activities of PDI [21, 22] and is still used today [23].

The denitrosylase activity of thiol isomerases has been clearly demonstrated but is not well-studied. The reaction of NO with thiol isomerases occurs both at the catalytic domains as well as through hydrophobic domains [24–26]. The latter is based on the observation that PDI can form N_2O_3 in its hydrophobic pockets [25]. Measurement of denitrosylation of NO-carriers, e.g., *S*-nitrosoglutathione (GSNO), can be performed using fluorescent NO reporters, such as 4-amino-5-methylamino-2′,7′-difluorofluorescein diacetate (DAF-FM) or 2,3-diaminonaphtalene (DAN). Lability of NO donors can result in a higher background, diminishing the sensitivity of the assay. The kinetics of denitrosylation represent a direct measurement of the catalytic cysteines and can be used for Michaelis–Menten calculations.

Sulfenylation of PDI can occur when a thiolate anion in the catalytic motif reacts with two-electron oxidants such as hydrogen peroxide, peroxynitrite, and lipid hydroperoxides [27, 28]. PDI can also be persulfidated [29]. Protein sulfenylation [30] and, in particular, PDI sulfenylation [28], have been shown to contribute to the prothrombotic activity of PDI. Sulfenylation can be quantified using Benzo[c] thiazine-based probe that relies on click chemistry [31].

2 Materials

All solutions should be prepared with distilled water and analytical grade reagents. Store buffers at room temperature unless otherwise specified. Follow all appropriate safety precautions and laboratory regulations, and dispose of waste according to institutional guidelines.

2.1 Recombinant Proteins

1. pT7-FLAG-SBP-1 expression vector.
2. BL21 competent *E. coli*.
3. Lysogeny Broth (10 g tryptone, 5 g yeast extract, and 10 g NaCl in 1 L of distilled H_2O) containing 50 μg/mL ampicillin (LB-Amp).
4. 1 mM isopropyl β-D-1-thiogalactopyranoside (IPTG) in PBS.
5. Lysis buffer: Bacterial-Protein Extraction Reagent (B-PER), 1X Protease Cocktail Inhibitor (PCI, stock concentration typically 100X), 5 U/mL DNAse I, and 100 μg/mL lysozyme.
6. High-capacity Streptavidin Agarose beads.
7. Washing buffer: 20 mM Tris pH 7.4, 150 mM NaCl, 2 mM ethylenediaminetetraacetic acid (EDTA), and 5 mM dithiothreitol (DTT).
8. Elution buffer: washing buffer containing 200 mM biotin.
9. Dialysis buffer: 1X phosphate-buffered saline (PBS).
10. Dialysis cassettes (12 mL volume).
11. Protein concentrators (molecular weight cutoff: 7500 Da).

2.2 Thiol Isomerase Activity Assays

1. Thiol isomerase assay buffer: 9.75 mL of 200 mM KH_2PO_4, 15.25 mL of 200 mM K_2HPO_4, and 0.2 mL of 500 mM EDTA (pH: 8.0). Adjust to 50 mL with distilled water. Adjust pH to 7.0 and filter-sterilize.
2. Di-eosin-GSSG reducing buffer (make fresh): Thiol isomerase assay buffer + 10 μM of DTT.
3. Insulin reducing buffer (make fresh): Thiol isomerase assay buffer + 1.2 mM of DTT.
4. Denitrosylation buffer (make fresh): Thiol isomerase assay buffer + 40 μM of GSH and 0.5 μM of 4-amino-5-methylamino-2′,7′-difluorofluorescein diacetate (DAF-FM).
5. 2 mM of *S*-nitrosglutathione in thiol isomerase assay buffer (make fresh).

6. Hepes Tyrodes Glucose (HTG) buffer: 134 mM NaCl, 0.34 mM sodium phosphate, 2.9 mM KCl, 12 mM sodium bicarbonate, 20 mM HEPES, 1 mM magnesium chloride, and 5 mM glucose (pH 7.3).
7. Acid Citrate Dextrose (ACD): 39 mM citric acid, 75 mM sodium citrate, 135 mM dextrose, and pH 7.4.

2.3 Di-Eosin-GSSG Probe

1. 10 mM eosin isothiocyanate (commercially available).
2. Oxidized glutathione (GSSG).
3. PD-10 column.
4. DTT.
5. Spectrophotometer.

2.4 Bovine Insulin Solution

1. Insulin solution: 4 mL of 12.5 mg/mL insulin in 50 mM Tris (pH: 8.0, *see* **Note 1**). Actively monitor the pH of the solution while adding 1 M of HCl to adjust the pH to 2.0. Once the solution clarifies, rapidly adjust the pH back to 8.0 by the addition of 1 M of NaOH (*see* **Note 2**). Adjust the volume of the solution to 5 mL to create a stock concentration of 1.5 mM (10 mg/mL). Aliquot and store the bovine insulin solution at −20°C.

2.5 Denaturing RNase A

1. Ribonuclease A from bovine pancrease, RNase A (Sigma Aldrich #R6513-50MG).
2. Dithiothreitol, DTT (BioRAD #1610611).
3. Guanidine Hydrochloride, Gu-HCl (Sigma Aldrich # G3272-25G).
4. Trizma HCl (Sigma Aldrich, T3253-1KG).
5. Ethylenediaminetetraacetic acid, EDTA, 0.5 M, pH 8.0, and 100 mL (Fisher Scientific #MT-46034Cl).
6. Glacial Acetic Acid (Sigma Aldrich #695092-500ML-GL).
7. Assay buffer-100 mM Tris–HCl (pH 8.0), 2 mM EDTA.
8. Denaturing Wash buffer-20 mM Tris–HCl (pH 8.0), 2 mM EDTA, and 1% acetic acid.
9. Desalting column.
10. Nano drop.

2.6 PDI-Mediated RNase A Refolding as a Readout of Its Oxidase and Isomerase Activity

1. N-term FLAG-tagged recombinant Human Protein Disulfide Isomerase C-term Streptavidin Binding Peptide (Stock in −80°C, different concentrations depending on batch, MW = 61 kDa, in phosphate-buffered saline, and pH 7.4).
2. Cytidine 2′:3′-cyclic monophosphate monosodium salt, cCMP (Sigma Aldrich #C9630).
3. L-Glutathione reduced, GSH (Sigma Aldrich #G6013-10G).

4. L-Glutathione oxidized, GSSG (Sigma Aldrich #G4376-5G).
5. Spectrophotometer.

2.7 In Vitro Oxidation and Labeling

1. N-term FLAG-tagged recombinant Human Protein Disulfide Isomerase C-term Streptavidin Binding Peptide (Stock in −80°C, different concentrations depending on batch, MW = 61 kDa, in phosphate-buffered saline, and pH 7.4).
2. Phosphate Buffered Saline, pH 7.4 (Fisher Scientific, #10010-023).
3. 50 mM tris(2-carboxyethyl) phosphine hydrochloride (TCEP; Sigma-Aldrich #C4706), (prepare fresh).
4. Trizma HCl (Sigma Aldrich, T3253-1KG).
5. Micro Bio-Spin P-6, Tris Buffer (Bio-Rad #732622).
6. 30% (v/v) hydrogen peroxide, H_2O_2 (Sigma-Aldrich).
7. DMSO Anhydrous (Invitrogen, #D12345, Lot#2289891).
8. 0.2 M 1-(4-pentyn-1-yl)-1*H*-2,1-benzothiazin-4(3*H*)-one 2,2-dioxide (BTD-alkyne) (MW = 263.31 g/mol), prepared in DMSO; −20°C).

2.8 Cu (I) Mediated Azide-Alkyne Cycloaddition (Click Chemistry)

1. Copper (II) Sulfate (Sigma Aldrich, #451657).
2. 3-[4-({bis[(1-Tert-butyl-1*H*-1,2,3-triazol-4-yl)methyl] amino}methyl)-1*H*-1,2,3-triazol-1-yl]propanol (BTTP) (Click Chemistry Tools, #678937); −20°C; 5 mM stocks in DMSO.
3. Azide Fluor 488 (Sigma Aldrich, #760765-1MG), made to 5 mM in DMSO, stored in −20°C.
4. Acetone (Sigma Aldrich, 650501-1L).

2.9 In-Gel Fluorescence

1. 2X Laemmli sample buffer with 10% β-mercaptoethanol (BioRad).
2. 20% w/v Sodium Dodecyl Sulfate Solution (Sigma, #05030).
3. 4X Laemmli sample buffer (BioRad, #1610741). Add 10% beta-mercaptoethanol to the 4X Laemmli Sample Buffer (BioRad, #1610710XTU).
4. Mini-Protean TGX 10% Tris–Glycine protein gels (BioRad, #4561096).
5. InstantBlue Coomassie Stain (Abcam, #ab119211).

3 Methods

3.1 Protein Purification

1. Insert the thiol isomerase gene of interest into the pT7-FLAG-SBP-1 vector (commercially available).

2. Thiol isomerase-containing vector can be expressed in the BL21 strain of *E. coli* and stored in 20% glycerol at −80°C for future use.
3. An LB-Amp plate should be streaked at 37°C overnight with the desired frozen *E. coli* stock containing the thiol isomerase of interest.
4. After overnight incubation, establish four 5 mL cultures of LB-Ampicillin by transferring bacteria from your plate to the medium and grow at 37°C.
5. Measure the OD_{600} of the samples every 30 min until it reaches maximal exponential growth (typically between 0.4 and 0.6 OD) and remove 1 mL from the culture and add 1 mM Isopropyl β-D-1-thiogalactopyranoside (IPTG) to induce protein expression, creating (+) and (−) cultures.
6. Allow cultures to incubate overnight with agitation. Evaluate samples using 10% SDS–PAGE gel.
7. Stain the gel using Coomassie Blue and assess the protein expression of the thiol isomerase.
8. Use the culture with the best expression and its corresponding (−) culture to inoculate 1 L of LB-Amp media. Incubate at 37°C with shaking.
9. Monitor the OD_{600} of the samples every 30 min until reaching maximal exponential growth (typically OD 0.4–0.6). Add IPTG to a final concentration of 1 mM, reduce the incubator temperature to 25 °C, and continue shaking overnight.
10. Weigh three GS3 bottles and note their weights.
11. Transfer the cells following overnight agitation from the incubator into the GS3 bottles and centrifuge at 3500 × *g* for 10 min.
12. Discard the supernatant and weigh the GS3 bottles containing the pellet. Calculate the weight of the pellet by subtraction of the bottle alone measurement.
13. Discard the supernatant and resuspend the pellet in 4 mL of B-PER solution containing 100 μg/mL lysozyme and 5 U/mL Dnase I per gram of pellet weight.
14. When the pellet is fully resuspended, add 1X Protease Cocktail Inhibitor.
15. Let the pellet shake at room temperature for 20 min (*see* **Note 3**).
16. While the pellet is shaking, prepare a streptavidin column by pouring 5 mL of streptavidin-containing beads onto a column. Equilibrate with 15 mL of B-PER solution.

17. Centrifuge the cell lysate at 17,000 × *g* for 30 min and load the supernatant onto the column (*see* **Note 4**).
18. Wash the column with 150 mL of washing buffer (*see* **Note 5**).
19. Elute the protein with 12 mL of elution buffer (*see* **Note 6**).
20. Collect the protein and dialyze it against 1X PBS (pH 7.4) overnight at 4°C.
21. Change the dialysis buffer and let the protein dialyze for 3 h twice.
22. Remove protein from dialysis cassette and concentrate to less than 2 mL using spin concentrators.
23. Determine protein concentration using a BCA assay and aliquot into 0.5 mg samples. Store at −80°C (*see* **Note 7**).

3.2 Di-Eosin-GSSG Assay

3.2.1 Preparation and Analysis of the Di-Eosin-GSSG Probe

The di-eosin-GSSG assay is performed according to the method described by Raturi et al with minor modifications [18].

1. To generate the di-eosin-GSSG probe, 400 μL of 1 mM GSSG should be combined with 400 μL of 10 mM eosin isothiocyanate, both prepared in thiol isomerase assay buffer.
2. Incubate the mixture overnight in the dark.
3. Equilibrate a PD-10 column with 20 mL of thiol isomerase assay buffer. During the equilibration, 2 mL tubes (15 total) should be prepared for fraction collecting. Once equilibrated, the buffer should be allowed to drain to the resin bed, and the column should be stopped (*see* **Note 8**).
4. Carefully add the mixture containing the GSSG and the eosin isothiocyanate to the top of the column.
5. Position the 2 mL tubes (15 total) for the collection of the fractions under the column.
6. Remove the stop from the column and allow the mixture containing the GSSG and eosin isothiocyanate mixture to flow into the resin.
7. Stop the column when the mixture containing the GSSG and the eosin isothiocyanate has completely entered the resin.
8. Gently add thiol isomerase assay buffer to the column, remove the stopper of the column, and continue to add thiol isomerase assay buffer until all 15 fractions are collected.
9. Add 100 μL of each of the fractions to two 2 wells in a black-walled, clear-bottomed 96-well plate.

10. Designate half the wells as −DTT and half as +DTT. Add 100 μL of thiol isomerase assay buffer to −DTT wells and 100 μL of buffer containing 20 mM DTT to +DTT wells (final concentration 10 mM DTT). Incubate for 30 min to allow complete reduction.
11. Measure fluorescence (Ex: 520 nm, Em: 550 nm) and calculate the fluorescence fold change between −DTT and +DTT samples for each fraction.
12. Discard any fractions with a fold change less than 5 (*see* **Note 9**).
13. Use the 10 mM eosin isothiocyanate stock in thiol isomerase assay buffer to prepare a twofold serial dilution ranging from 125 μM to 1 μM in a final volume of 200 μL of thiol isomerase assay buffer in a black-walled, clear-bottomed 96-well plate.
14. Add 200 μL of the usable fractions (fractions with a fold change greater than five) to the black-walled, clear-bottomed 96-well plate.
15. Measure fluorescence (Ex: 520 nm, Em: 550 nm) for the standard curve as generated by the serial dilution of 1 μM eosin isothiocyanate and a 1:40 dilution of the DTT-reduced kept fractions.
16. Estimate the concentration of probe of the usable fractions by dividing all the values by two (each GSSG molecule contains two eosin moieties), multiplying by 40 and using the standard curve as generated by the serial dilution of 10 mM eosin isothiocyanate (Fig. 3A).
17. Aliquot and store all usable fractions of di-eosin-GSSG at −20°C.

3.2.2 Assaying for Thiol Isomerase Reductase Activity Using the Di-Eosin-GSSG Probe

1. Prepare 0.5 mL tubes (2) and add 350 μL of di-eosin-GSSG reducing buffer (*see* **Note 9**).
2. Add thiol isomerase to one of the two tubes (generating thiol-isomerase-containing sample and blank sample, *see* **Note 10**), to a concentration of 100 nM (final concentration in the assay will be 50 nM) in 0.5–5 μL, depending on the concentration of thiol isomerase. A similar volume of PBS should be added to the blank sample.
3. Dilute the di-eosin-GSSG probe to a concentration of 300 nM in thiol isomerase assay buffer (*see* **Note 11**).
4. Add 100 μL of thiol isomerase sample and blank sample to individual wells (in triplicate) in a black-walled, clear-bottomed 96-well plate. Add 100 μL of diluted di-eosin-GSSG probe to each well in the plate reader (*see* **Note 12**).

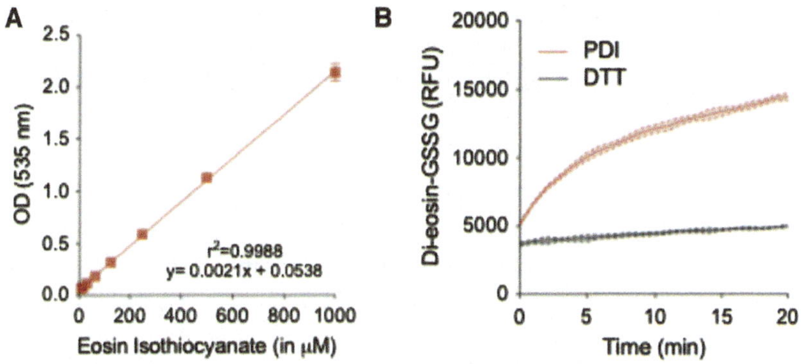

Fig. 3 Fluorescent measurement of di-eosin-GSSG cleavage by thiol isomerases. (**A**) Linear regression of the OD of different concentrations of eosin isothiocyanate solutions. Concentrations of di-eosin-GSSG probe are calculated by measuring their OD and extrapolating from the eosin isothiocyanate standard curve using the formula in the graph. The resultant value should be divided by two to correct for the fact that each di-eosin-GSSG contains two eosin moieties. (**B**) Red trace ($n = 3$, mean ± SD) shows di-eosin-GSSG cleavage (Ex: 520 nm, Em: 550 nm) by PDI (50 nM) in the presence of the reducing agent DTT (5 μM). The black trace ($n = 3$, mean ± SD) shows the effect of DTT alone on di-eosin-GSSG cleavage. The concentration of EGSH generated by thiol isomerase reductase activity can be calculated from the relative fluorescent units (RFU) and extrapolated from the eosin isothiocyanate standard curve

5. Monitor thiol isomerase-mediated cleavage of the di-eosin-GSSG probe by fluorescent measurements (Ex: 520 nm, Em: 550 nm) every minute for 20 min.

3.2.3 Assaying for Thiol Isomerase Reductase Activity on Platelets

1. Prepare washed platelets by drawing whole blood from healthy volunteers into 5% citrate.
2. Isolate platelet-rich plasma (PRP) by centrifuging blood at 200 × *g* and separating the PRP layer (top layer) from the buffy coat and the erythrocytes.
3. After 30 min of resting at 37°C, add ACD to 20% of the PRP volume (e.g., 1 mL for every 5 mL of PRP).
4. Add 1 mM prostaglandin E_1 to PRP–ACD mixture at a 1:10,000 dilution.
5. Centrifuge at 1000 × *g*, remove the supernatant, and resuspend the platelet pellet in Hepes Tyrodes Glucose (HTG) buffer to the desired amount.
6. Add freshly prepared DTT (1 mM stock solution in thiol isomerase assay buffer) to the platelets to a concentration of 5 μM.
7. Dilute the di-eosin-GSSG probe to a concentration of 1.5 μM in thiol isomerase assay buffer.
8. Add 190 μL of DTT-containing platelet mixture, untreated platelet mixture, and HTG buffer to individual wells in a black-walled, clear-bottomed 96-well plate. Add 10 μL of diluted di-eosin-GSSG probe to the plate reader (*see* **Note 12**).

9. Monitor thiol isomerase-mediated cleavage of the di-eosin-GSSG probe by recording measurements (Ex: 520 nm, Em: 550 nm) every minute for 20 min (*see* **Note 13**).

3.2.4 Data Analysis Di-Eosin-GSSG Assay

1. Generate curves of di-eosin-GSSG cleavage.
2. Use the linear part of the curve for regression analysis (Fig. 3B).
3. Calculate percentage activity using: % Activity = $(\Delta RFU_{EXPERIMENT} - \Delta RFU_{DTT\ ONLY}) / ((\Delta RFU_{PDI\ ONLY} - \Delta RFU_{DTT\ ONLY})$.
4. Calculate eosin-labeled glutathione (EGSH) generation per minute by extrapolating from the eosin isothiocyanate standard curve (Fig. 3; Ex: 520 nm, Em: 550 nm).

3.3 Insulin Turbidimetric Assay

3.3.1 Assaying for Thiol Isomerase Reductase Activity Using the Insulin Turbidimetric Assay

The insulin turbidimetric assay is performed according to the method described by Holmgren with minor modifications [32].

1. Prepare 0.5 mL tubes (two total) and add 350 μL of thiol isomerase assay buffer.
2. Add thiol isomerase to one out of the two tubes, to a concentration of 700 nM (final assay concentration 350 nM, *see* **Note 11**) in 5–10 μL, depending on the concentration of thiol isomerase. Add an equivalent volume of PBS to the other tube as a blank.
3. Dilute freshly prepared 100 mM stock solution of DTT in thiol isomerase assay buffer (*see* **Note 10**) to 1.2 mM to create the insulin reducing buffer (final assay concentration 0.3 mM).
4. Dilute the 1.5 mM insulin stock concentration to 0.4 mM (final assay concentration 0.1 mM).
5. Prepare a 96-well, clear-walled, clear-bottomed plate by adding 100 μL of the thiol isomerase sample and blank sample to individual wells in triplicate. Add 50 μL of the diluted 0.4 mM insulin to the mixture. Add 50 μL of the insulin-reducing buffer to the plate reader (*see* **Note 14**).
6. Monitor insulin precipitation at 650 nm every minute for 2 h.

3.3.2 Data Analysis Insulin Turbidimetric Assay

1. Generate curves of insulin precipitation data.
2. Identify the start of insulin precipitation data in the control (PDI only) sample.
3. Identify the OD (650 nm) after 30 min of the start of insulin precipitation data in all samples (Fig. 4; *see* **Note 15**).
4. Calculate percentage activity using: %Activity = $(OD650_{EXPERIMENT} - OD650_{DTT\ ONLY}) / ((OD650_{PDI\ ONLY} - OD650_{DTT\ ONLY})$.

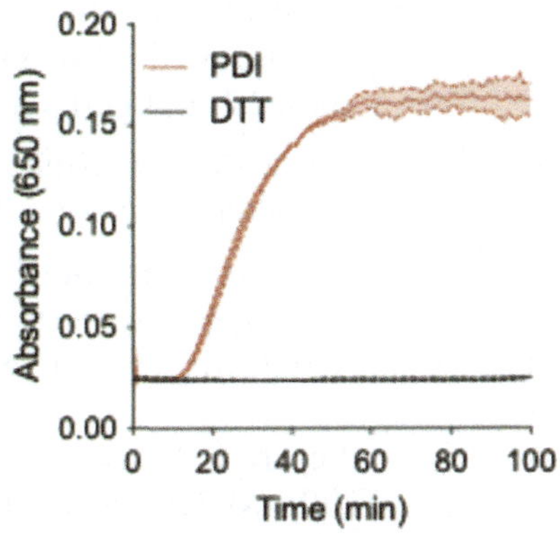

Fig. 4 Reductase activity as measured by the insulin turbidimetric assay. Red line ($n = 3$, mean ± SD) represents insulin precipitation as measured at OD_{650} by PDI (350 nM) in the presence of the reducing agent DTT (5 μM). Black line ($n = 3$, mean ± SD) represents insulin precipitation in the presence of only the reducing agent DTT (5 μM). The lag-time can be calculated from the point where absorbance increases by 0.02 from the DTT control

3.4 Thiol Isomerase Oxidase Assay

The RNase denaturation assay was modified according to the method described by Lyles and Gilbert [33].

3.4.1 Denaturing RNase A

1. Prepare a 0.5 mM stock solution of RNase A by dissolving RNase A in an appropriate buffer. Store the stock at −20°C if not using immediately.
2. To denature RNase A, prepare a 400–500 μL reaction volume containing 300 μM RNase A (from 0.5 mM stock), 6 M guanidinium hydrochloride (GuHCl), 73 mM DTT (from a 1 M stock solution), and assay buffer. Mix thoroughly by gentle pipetting (*see* **Note 16**).
3. Incubate the mixture at 37°C for 1–2 h to ensure complete denaturation. Do not rock or shake the tube during incubation.
4. Prepare a spin desalting column according to the manufacturer's protocol. Allow excess storage buffer to drain from the column by gravity.
5. Wash the column twice with 2 mL denaturing wash buffer containing 20 mM Tris–HCl (pH 8.0), 2 mM EDTA, and 1% acetic acid. Allow excess wash buffer to drain by gravity.
6. Centrifuge the column at 1000 × *g* for 2 min at room temperature. Discard the flow-through and avoid letting the resin dry out.
7. Carefully load the denatured RNase A onto the equilibrated column and centrifuge again under the same conditions.

8. Collect the flow-through containing desalted denatured RNase A (thiol-scrambled RNAse A; scrRNAse A), now free of DTT and GuHCl.
9. Measure absorbance at 280 nm using a UV–vis spectrophotometer (*see* **Note 17**).
10. Calculate the final concentration of scrRNase A using the Beer–Lambert equation and its extinction coefficient (9800 M^{-1} cm^{-1} at 280 nm) [34].

3.4.2 PDI Mediated RNase A Refolding as a Readout of Its Oxidase and Isomerase Activity

1. Prepare a 1 mL reaction mixture in a cuvette by combining 100 mM Tris–HCl (pH 8.0), 2 mM EDTA, 2 mM GSH (from a 100 mM stock in buffer), 0.2 mM GSSG (from a 100 mM stock in buffer), 1.4 μM PDI (WT or DM), and 4.5 mM cCMP (from a 100 mM stock in buffer), respectively. Mix thoroughly by gentle pipetting.
2. Set up the spectrophotometer with the following parameters: absorbance kinetic read mode, wavelength set to 296 nm, run time of 1 h, and 2-s intervals between reads.
3. Prepare a blank sample containing all reaction components except scrRNase A and use it to perform baseline correction on the spectrophotometer.
4. To initiate the reaction, quickly add 8.4 μM scrRNase A to the reaction mixture, mix gently, and immediately start data acquisition.
5. For different conditions, prepare separate reactions:
   a. For the RNase A condition, add 8.4 μM RNase A without PDI.
   b. For the scrRNase A-only condition, add 8.4 μM scrRNase A without PDI (*see* **Note 18**).

3.4.3 Data Analysis

1. Generate curves of CMP Absorbance data.
2. The linear part of the curve should be used for linear regression (Fig. 5).
3. Calculate percentage activity using: $[Vmax_{exp} - Vmax_{scrRNase\ A\ only}]/[Vmax_{WT\ PDI\ +\ scrRNase\ A} - Vmax_{scrRNase\ A}]$.

3.5 Denitrosylase Assay

The denitrosylase assay is performed according to the method described by Raturi and Mutus with minor modifications [21].

3.5.1 Assaying for Thiol Isomerase Denitrosylase Activity Using the Denitrosylase Assay

1. Prepare 0.5 mL tubes (two) and add 350 μL of thiol isomerase assay buffer.
2. Add thiol isomerase to one of the tubes to a concentration of 2 μM (final concentration in the assay will be 1 μM). Add an equivalent volume of PBS to the other tube.

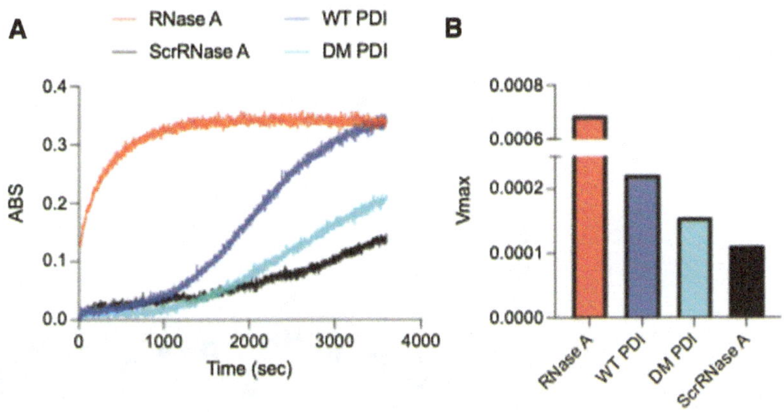

Fig. 5 PDI-mediated RNase A refolding as a readout of its oxidase and isomerase activity. (A) Red trace shows native RNase A hydrolyzation of cCMP to CMP as a positive control; no PDI is present in this reaction. Black trace shows denatured and reduced scrRNase A in the absence of PDI as a negative control showing no hydrolyzation of cCMP. Dark blue shows wildtype (WT) PDI refolding of scrRNase A and the return of RNase A activity. The dead mutant (DM) PDI is unable to refold scrRNase A in the cyan tracing. (B) Quantification of data shown in (A)

3. Dilute freshly prepared 50 mM stock solution of GSH in thiol isomerase assay buffer (*see* **Note 10**) to 40 μM. To measure NO generation, add freshly prepared 5 mM DAF-FM in DMSO to a final concentration of 2 μM to create the denitrosylation buffer.
4. Add 100 μL of thiol isomerase sample and blank sample to individual wells (in triplicate) in a black-walled, clear-bottomed 96-well plate. Add 50 μL of denitrosylation buffer, then add 50 μL of freshly prepared 2 mM GSNO. Add 50 μL of the 2 mM stock concentration of freshly prepared GSNO in thiol isomerase assay buffer (*see* **Note 19**) to each well in the plate reader.
5. Monitor thiol isomerase-mediated denitrosylation of GSNO by DAF-FM fluorescence (Ex: 485 nm, Em: 515 nm) every minute for 30 min.

3.5.2 Data Analysis Denitrosylase Assay

1. Generate DAF-FM fluorescence curves.
2. Use the linear part of the curve for regression analysis (Fig. 6).
3. Calculate percentage activity using: % Activity = ($\Delta RFU_{EXPERIMENT} - \Delta RFU_{DTT\ ONLY}$) / (($\Delta RFU_{PDI\ ONLY} - \Delta RFU_{DTT\ ONLY}$).

3.6 PDI Sulfenylation Assay

The protocol is modified from Truong and Carroll [35].

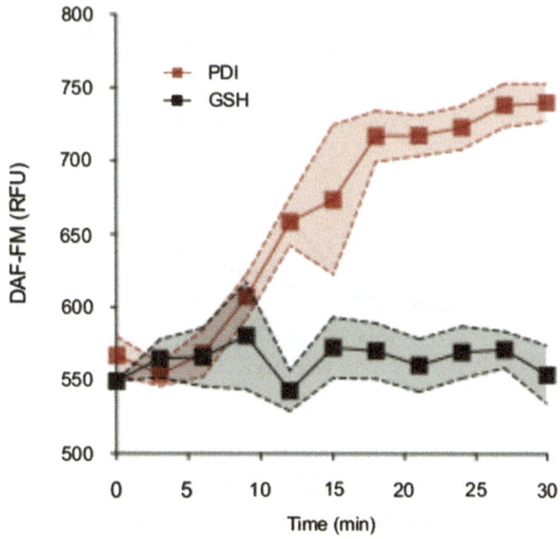

Fig. 6 Fluorescent measurement of GSNO consumption by thiol isomerases. Red trace ($n = 3$, mean ± SD) shows GSNO consumption mediated by PDI (2 μM) in the presence of the reducing agent GSH (10 μM). DAF-FM is used as the NO reporter (Ex: 485 nm, Em: 515 nm). The black trace ($n = 3$, mean ± SD) shows the effect of GSH alone on GSNO consumption

3.6.1 In Vitro Oxidation and Labeling

1. Pre-equilibrate and prepare Micro Bio-Spin P-6 columns according to the manufacturer's protocol, using a recommended volume between 10 and 75 μL. Allow excess storage buffer to drain from the column by gravity.
2. Centrifuge the column at 1000×*g* for 2 min at room temperature. Discard the storage buffer and avoid letting the column dry out.
3. Dilute PDI to 10 μM in PBS.
4. Pre-reduce the 10 μM PDI by adding 200 μM final TCEP to ensure all cysteines begin at the same redox state (*see* **Notes 20 and 21**).
5. Incubate the solution at room temperature for 15 min without rocking.
6. Desalt the PDI into Tris buffer using the pre-equilibrated Micro Bio-Spin P-6 columns.
7. Aliquot the protein according to reaction conditions, typically in 25 μL reactions, but scalable if needed.
8. Oxidize and label the protein by adding 100 μM final H_2O_2 (from a 1 mM working solution) and 100 μM final BTD-alkyne.

9. Mix the reaction by finger-flicking. Do not vortex, as vortexing can cause mixed disulfides to compete with sulfenic acid labeling.
10. Incubate the reaction at 37 °C for 60 min without rocking.

3.6.2 Cu (I) Mediated Azide-Alkyne Cycloaddition (Click Chemistry)

1. Prepare a fresh 50 mM $CuSO_4$ solution by dissolving 4.42 mg $Cu(II)SO_4$ in 560.3 μL TBS. For a 1.25 mM working stock, dilute 12.5 μL of 50 mM $CuSO_4$ into 487.5 μL TBS.
2. Prepare a fresh 50 mM TCEP solution by dissolving 14.3 mg TCEP in 1 mL TBS (MW = 286.65 g/mol).
3. Prepare a 5 mM Azide Fluor 488 stock solution in DMSO.
4. Prepare a 5 mM BTTP stock solution in DMSO.
5. Add the following reaction components to each sample, ensuring proper mixing by vortexing (final concentrations in a 25 μL total volume, scalable as needed):
   a. 0.1 mM Azide Fluor 488 (alternatively, biotin-azide may be used for other applications).
   b. 1 mM TCEP (*see* **Note 21**).
   c. 0.1 mM BTTP.
   d. 1 mM $Cu(II)SO_4$.
6. Protect samples from light due to the fluorophore's sensitivity. Incubate the reaction mixture at 37°C for 1 h with rocking.
7. Precipitate the labeled protein by adding ice-cold acetone and storing the samples at −80°C overnight.
8. Centrifuge the samples at maximum 16,000 rcf for 10–15 min in a cold room to pellet the protein.
9. Carefully discard the supernatant containing excess solvent and unreacted click chemistry reagents.
10. Allow the samples to dry completely in a chemical hood.

3.6.3 Protein Resuspension and SDS–PAGE

1. Resuspend the protein pellet in 1X PBS containing 1% v/v SDS.
2. If the protein does not dissolve completely, perform gentle sonication in a water bath Branson sonicator or incubate at 37°C to aid solubilization.
3. Add 1X Laemmli sample buffer containing 10% β-mercaptoethanol.
4. Boil the samples for 5 min at 95–100°C.
5. Pulse centrifuge to collect the sample at the bottom of the microcentrifuge tube.

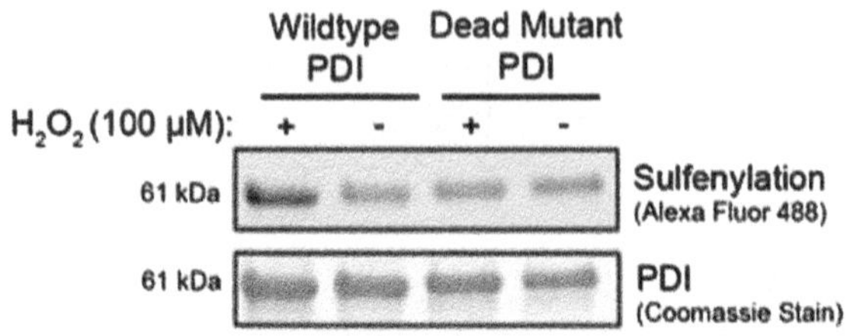

Fig. 7 In-gel fluorescence showing sulfenylation of wildtype, but not dead mutant PDI. BTD labeling of PDI active site cysteine sulfenylation using an Azide Fluor 488 probe attached via click chemistry. Only wildtype (WT) PDI showed increased labeling in the presence of H_2O_2 relative to the dead mutant (DM) PDI, where the active site cysteines are mutated to an alanine. Coomassie stain of PDI indicated similar loading of the protein

6. Run SDS–PAGE immediately or freeze the samples until electrophoresis.
7. Reduce the amount of molecular weight marker loaded if ladder fluorescence interferes with sample signals.

3.6.4 In-Gel Fluorescence and Protein Detection

1. Detect Alexa Fluor 488-labeled proteins using in-gel fluorescence imaging on the BioRad ChemiDoc MP Imager (Fig. 7).
2. Stain the gel with InstantBlue Coomassie stain according to the manufacturer's protocol to visualize total PDI levels.

4 Notes

1. By dissolving 50 mg of insulin in 4 mL of 50 mM Tris, the mixture will become cloudy. It is advisable to use a stir bar on a magnetic plate to fully dissolve the insulin and then perform the pH titration.
2. Upon lowering the pH, the insulin should dissolve completely. When the solution has turned transparent, NaOH should be added to adjust the pH back to 8.0. If not done rapidly, the insulin can precipitate, and the sample has to be discarded.
3. Monitor the turbidity of the solution. If the solution is not becoming more transparent over time, an additional dose of lysozyme and Dnase I can be added to enhance cell lysis.
4. After centrifugation, carefully transfer the supernatant to a 50 mL conical tube without disturbing the debris-rich pellet. The addition of debris to the beads can interfere with binding and the rate at which the supernatant progresses through the column.

5. The addition of detergents (such as Triton-X, NP40, or CHAPS) for the preparation of thiol isomerases should be used with caution, as they can bind to hydrophobic domains and interfere with thiol isomerase activity [36].
6. Drain the column of the washing buffer until only a small layer of washing buffer remains on top of the resin. Upon addition of 12 mL of the elution buffer, allow the column to drain for 30 s and then stop. Incubate the column for 30 min with the elution buffer and then collect the eluate in a 15 mL conical tube. Dialyze the eluate against PBS.
7. For determination of the protein concentration in the sample, use bovine serum albumin (BSA) as a control to generate a standard curve (0.25—2 mg/mL). Use a 1:4 and a 1:10 dilution of the concentrated thiol isomerase sample to ensure that the sample concentration is in the range of the standard curve. Calculate the concentration accordingly.
8. To ensure that the column never becomes dry always maintain some residual buffer on top of the resin.
9. Discard fractions demonstrating a change of less than 5-fold (DTT+/DTT−). Fractions with a less than a fivefold change contain a significant amount of residual free eosin isothiocyanate and the ratio of di-eosin-GSSG to free eosin isothiocyanate will be insufficient for proper measurement of thiol isomerase-mediated cleavage of the probe.
10. For thiol isomerase assays measuring reduction, reducing agents such as DTT and GSH need to be prepared fresh to prevent oxidation. Addition of reducing agents to premade thiol isomerase assay buffer (which can be stored for up to half a year) needs to occur immediately prior to the assay.
11. Depending on the concentration of the recombinant thiol isomerase, a volume of 1–4 μL will typically be added to the mixture, as concentrations of thiol isomerase after concentration are typically in the range of 25–100 μM. A similar volume of PBS can be added to the thiol isomerase-free control sample.
12. The di-eosin-GSSG probe should be diluted using a volume based on the calculated concentration after probe generation. It is preferable to use di-eosin-GSSG probe of the same batch for a subset of experiments to minimize differences that exist from batch to batch. Upon addition of the di-eosin-GSSG probe to the mixture containing the thiol isomerase and the reducing agent, the reaction begins immediately. It is therefore advisable to fully set up the assay and add the di-eosin-GSSG probe with the assay plate ready to be introduced into the spectrofluorometer.

13. This assay can be adapted to other cell types as desired. Di-eosin-GSSG cleavage at the cell surface represents reductase activity of all cell surface thiol isomerases and potentially other enzymes or small molecules that can interact with GSSG. Therefore, caution should be taken when interpreting these results.
14. Upon the addition of any reducing agent to the mixture containing the thiol isomerase and insulin, thiol isomerase-mediated insulin reduction will start. Therefore, it is advisable to add the reducing agent to the plate reader.
15. The insulin turbidimetric assay is a useful tool to compare different catalysts relative to each other. For quantitative analysis, however, caution should be taken. The indirect nature of the catalysis of insulin by thiol isomerases makes it a poor method for evaluation of enzyme kinetics. For different purposes, different analysis methods have been described for the insulin turbidimetric assay. For the identification of inhibitors of thiol isomerases using high-throughput screening, the described methods of a chosen time point after the start of insulin precipitation allow for IC_{50} calculation [20, 37]. Multiple inhibitors can therefore be compared relative to each other. For other purposes, such as mutational analysis around the catalytic active site, lag time (the time which it takes for insulin precipitation to start), and the rate of precipitation (measured as the maximal absorbance increase per minute) have been used [38].
16. The assay is highly dependent on the state of the denatured RNase A. It needs to be fully reduced and denatured prior to starting the assay as any further oxidation can interfere with the readout.
17. The A260/280 ratio should be around 0.6–0.9, which indicates no DTT or GuHcl contamination when calculating the concentration of RNase A. Potential contamination will require further desalting.
18. Each time, always run RNase A and scrRNase A only conditions as positive and negative controls, respectively.
19. GSNO should be freshly dissolved before every experiment and used immediately. The half-life of S-nitrosylated compounds is short (ranging from several seconds to hours).
20. Ensure all cysteines start at the same reduced state as any prior oxidation of PDI will prevent efficient H_2O_2 oxidation and BTD labeling.
21. Prepare TCEP fresh each time, as it can oxidize over time.

References

1. Hatahet F, Ruddock LW (2009) Protein disulfide isomerase: a critical evaluation of its function in disulfide bond formation. Antioxid Redox Signal 11(11):2807–2850
2. Ellgaard L, Ruddock LW (2005) The human protein disulphide isomerase family: substrate interactions and functional properties. EMBO Rep 6(1):28–32. https://doi.org/10.1038/sj.embor.7400311
3. Appenzeller-Herzog C, Ellgaard L (2008) The human PDI family: versatility packed into a single fold. Biochim Biophys Acta 1783(4):535–548. https://doi.org/10.1016/j.bbamcr.2007.11.010
4. Flaumenhaft R, Furie B (2016) Vascular thiol isomerases. Blood 128(7):893–901. https://doi.org/10.1182/blood-2016-04-636456
5. Bekendam RH, Flaumenhaft R (2016) Inhibition of protein disulfide isomerase in thrombosis. Basic Clin Pharmacol Toxicol 119(Suppl 3):42–48. https://doi.org/10.1111/bcpt.12573
6. Flaumenhaft R (2013) Protein disulfide isomerase as an antithrombotic target. Trends Cardiovasc Med 23(7):264–268. https://doi.org/10.1016/j.tcm.2013.03.001
7. Flaumenhaft R, Furie B, Zwicker JI (2015) Therapeutic implications of protein disulfide isomerase inhibition in thrombotic disease. Arterioscler Thromb Vasc Biol 35(1):16–23. https://doi.org/10.1161/ATVBAHA.114.303410
8. Araujo TLS, Fernandes CG, Laurindo FRM (2017) Golgi-independent routes support protein disulfide isomerase externalization in vascular smooth muscle cells. Redox Biol 12:1004–1010. https://doi.org/10.1016/j.redox.2017.04.034
9. Crescente M, Pluthero FG, Li L, Lo RW, Walsh TG, Schenk MP, Holbrook LM, Louriero S, Ali MS, Vaiyapuri S, Falet H, Jones IM, Poole AW, Kahr WH, Gibbins JM (2016) Intracellular trafficking, localization, and mobilization of platelet-borne thiol isomerases. Arterioscler Thromb Vasc Biol 36(6):1164–1173. https://doi.org/10.1161/atvbaha.116.307461
10. Thon JN, Peters CG, Machlus KR, Aslam R, Rowley J, Macleod H, Devine MT, Fuchs TA, Weyrich AS, Semple JW, Flaumenhaft R, Italiano Jr. JE (2012) T granules in human platelets function in TLR9 organization and signaling. J Cell Biol 198(4):561–574. https://doi.org/10.1083/jcb.201111136
11. Asquith NL, Becker IC, Scimone MT, Boccia T, Camacho V, Barrachina MN, Guo S, Freire D, Machlus K, Schulman S, Flaumenhaft R, Italiano JE (2024) Targeting cargo to an unconventional secretory system within megakaryocytes allows the release of transgenic proteins from platelets. J Thromb Haemost. 22(11):3235-3248. https://doi.org/10.1016/j.jtha.2024.07.021
12. Hahm E, Li J, Kim K, Huh S, Rogelj S, Cho J (2013) Extracellular protein disulfide isomerase regulates ligand-binding activity of αMβ 2 integrin and neutrophil recruitment during vascular inflammation. Blood 121(19):3789–3800. https://doi.org/10.1182/blood-2012-11-467985
13. Subramaniam S, Jurk K, Hobohm L, Jäckel S, Saffarzadeh M, Schwierczek K, Wenzel P, Langer F, Reinhardt C, Ruf W (2017) Distinct contributions of complement factors to platelet activation and fibrin formation in venous thrombus development. Blood 129(16):2291–2302. https://doi.org/10.1182/blood-2016-11-749879
14. Terada K, Manchikalapudi P, Noiva R, Jauregui HO, Stockert RJ, Schilsky ML (1995) Secretion, surface localization, turnover, and steady state expression of protein disulfide isomerase in rat hepatocytes. J Biol Chem 270(35):20410–20416. https://doi.org/10.1074/jbc.270.35.20410
15. Akagi S, Yamamoto A, Yoshimori T, Masaki R, Ogawa R, Tashiro Y (1988) Localization of protein disulfide isomerase on plasma membranes of rat exocrine pancreatic cells. J Histochem Cytochem 36(8):1069–1074. https://doi.org/10.1177/36.8.3292644
16. Xu S, Butkevich AN, Yamada R, Zhou Y, Debnath B, Duncan R, Zandi E, Petasis NA, Neamati N (2012) Discovery of an orally active small-molecule irreversible inhibitor of protein disulfide isomerase for ovarian cancer treatment. Proc Natl Acad Sci U S A 109(40):16348–16353. https://doi.org/10.1073/pnas.1205226109
17. Cho J, Furie BC, Coughlin SR, Furie B (2008) A critical role for extracellular protein disulfide isomerase during thrombus formation in mice. J Clin Investig 118(3):1123–1131
18. Reinhardt C, von Bruhl ML, Manukyan D, Grahl L, Lorenz M, Altmann B, Dlugai S, Hess S, Konrad I, Orschiedt L, Mackman N, Ruddock L, Massberg S, Engelmann B (2008) Protein disulfide isomerase acts as an injury response signal that enhances fibrin generation via tissue factor activation. J Clin Investig 118(3):1110–1122
19. Raturi A, Mutus B (2007) Characterization of redox state and reductase activity of protein disulfide isomerase under different redox environments using a sensitive fluorescent assay.

Free Radic Biol Med 43(1):62–70. https://doi.org/10.1016/j.freeradbiomed.2007.03.025
20. Bekendam RH, Bendapudi PK, Lin L, Nag PP, Pu J, Kennedy DR, Feldenzer A, Chiu J, Cook KM, Furie B, Huang M, Hogg PJ, Flaumenhaft R (2016) A substrate-driven allosteric switch that enhances PDI catalytic activity. Nat Commun 7: 12579. https://doi.org/10.1038/ncomms12579
21. Hillson DA, Lambert N, Freedman RB (1984) Formation and isomerization of disulfide bonds in proteins: protein disulfide-isomerase. Methods Enzymol 107:281–294. https://doi.org/10.1016/0076-6879(84)07018-x
22. Mizunaga T, Katakura Y, Miura T, Maruyama Y (1990) Purification and characterization of yeast protein disulfide isomerase. J Biochem 108(5):846–851. https://doi.org/10.1093/oxfordjournals.jbchem.a123291
23. Yang M, Zirena IH, Kennedy QP, Patel A, Merrill-Skoloff G, Sack KD, Fulcidor E, Scartelli C, Guo S, Bekendam RH, Owegie OC, Xie H, Ghiran IC, Levy O, Lin L, Flaumenhaft R (2025) Galloylated polyphenols represent a new class of antithrombotic agents with broad activity against thiol isomerases. 23(6):1850-1863. J Thromb Haemost. https://doi.org/10.1016/j.jtha.2025.01.021
24. Ramachandran N, Root P, Jiang XM, Hogg PJ, Mutus B (2001) Mechanism of transfer of NO from extracellular S-nitrosothiols into the cytosol by cell-surface protein disulfide isomerase. Proc Natl Acad Sci U S A 98(17):9539–9544
25. Sliskovic I, Raturi A, Mutus B (2005) Characterization of the S-denitrosation activity of protein disulfide isomerase. J Biol Chem 280(10):8733–8741. https://doi.org/10.1074/jbc.M408080200
26. Raturi A, Mutus B (2004) Use of 2,3-diaminonapthalene for studying denitrosation activity of protein disulfide isomerase. Anal Biochem 326(2):281–283. https://doi.org/10.1016/j.ab.2003.12.015
27. Yang M, Flaumenhaft R (2021) Oxidative cysteine modification of thiol isomerases in thrombotic disease: a hypothesis. Antioxid Redox Signal 35(13):1134–1155. https://doi.org/10.1089/ars.2021.0108
28. Yang M, Chiu J, Scartelli C, Ponzar N, Patel S, Patel A, Ferreira RB, Keyes RF, Carroll KS, Pozzi N, Hogg PJ, Smith BC, Flaumenhaft R (2023) Sulfenylation links oxidative stress to protein disulfide isomerase oxidase activity and thrombus formation. J Thromb Haemost 21(8):2137–2150. https://doi.org/10.1016/j.jtha.2023.03.034
29. Shieh M, Ni X, Xu S, Lindahl SP, Yang M, Matsunaga T, Flaumenhaft R, Akaike T, Xian M (2022) Shining a light on SSP4: a comprehensive analysis and biological applications for the detection of sulfane sulfurs. Redox Biol 56:102433. https://doi.org/10.1016/j.redox.2022.102433
30. Yang M, Li W, Harberg C, Chen W, Yue H, Ferreira RB, Wynia-Smith SL, Carroll KS, Zielonka J, Flaumenhaft R, Silverstein RL, Smith BC (2020) Cysteine sulfenylation by CD36 signaling promotes arterial thrombosis in dyslipidemia. Blood Adv 4(18):4494–4507. https://doi.org/10.1182/bloodadvances.2020001609
31. Gupta V, Yang J, Liebler DC, Carroll KS (2017) Diverse redoxome reactivity profiles of carbon nucleophiles. J Am Chem Soc 139(15):5588–5595. https://doi.org/10.1021/jacs.7b01791
32. Holmgren A (1979) Thioredoxin catalyzes the reduction of insulin disulfides by dithiothreitol and dihydrolipoamide. J Biol Chem 254(19):9627–9632
33. Lyles MM, Gilbert HF (1991) Catalysis of the oxidative folding of ribonuclease A by protein disulfide isomerase: dependence of the rate on the composition of the redox buffer. Biochemistry 30(3):613–619. https://doi.org/10.1021/bi00217a004
34. Foglia F, Carullo P, Vecchio PD (2008) The effect of trimethylamine N-oxide on RNase A stability. J Therm Anal Calorim 91:67–72
35. Truong TH, Carroll KS (2012) Bioorthogonal chemical reporters for analyzing protein sulfenylation in cells. Curr Protoc Chem Biol 4(2):101–122. https://doi.org/10.1002/9780470559277.ch110219
36. Watanabe MM, Laurindo FR, Fernandes DC (2014) Methods of measuring protein disulfide isomerase activity: a critical overview. Front Chem 2:73. https://doi.org/10.3389/fchem.2014.00073
37. Khan MM, Simizu S, Lai NS, Kawatani M, Shimizu T, Osada H (2011) Discovery of a small molecule PDI inhibitor that inhibits reduction of HIV-1 envelope glycoprotein gp120. ACS Chem Biol 6(3):245–251. https://doi.org/10.1021/cb100387r
38. Stopa JD, Neuberg D, Puligandla M, Furie B, Flaumenhaft R, Zwicker JI (2017) Protein disulfide isomerase inhibition blocks thrombin generation in humans by interfering with platelet factor V activation. JCI Insight 2(1):e89373. https://doi.org/10.1172/jci.insight.89373

Chapter 19

Probing the Mechano-Redox Control of Cell Movement Using Microfluidic Assays

Alexander Dupuy, Kaitlyn Zhang, Joyce Chiu, Lining Arnold Ju and Freda H. Passam

Abstract

The mechano-redox phenomenon—where mechanical forces and redox reactions synergistically regulate protein function—has been demonstrated in several mammalian proteins. Integrins, which mediate cellular adhesion to ligands, are particularly sensitive to redox modulation via thiol–disulfide exchange catalyzed by enzymes such as thiol isomerases. These redox-driven modifications can alter integrin conformation and ligand-binding affinity, ultimately leading to dynamic regulation of cell adhesion and de-adhesion. In this work, we present a suite of functional assays designed to investigate how manipulation of thiol exchange reactions influences key cellular processes, including thrombosis and inflammation. These assays employ either thiol-reactive enzymes or small-molecule inhibitors to modulate redox states. They include: (1) a cellular de-adhesion assay to monitor integrin–ligand interactions, (2) a neutrophil crawling assay to assess integrin-mediated motility, (3) a microfluidic thrombus formation assay to evaluate the contribution of thiol isomerases to clot development, (4) the "endo-chip"—a microfluidic platform integrating an endothelium for modelling vascular thromboinflammatory responses under flow, and (5) the microvascular-on-a-post chip, an endothelialized microfluidic device with fibrin-rich thrombosis localized on a 3D post.

Keywords Neutrophils, Platelets, Shear, Microfluidics, Thiol isomerases, Disulfides, Integrins, Thrombosis, Inflammation

1 Introduction

Protein function in the human body is tightly regulated by biochemical and biophysical mechanisms. Among these, mechanical forces and redox-based post-translational modifications are particularly critical in the vascular system, where circulating cells and plasma proteins are constantly subjected to hemodynamic shear stress and fluctuating oxidative environments. Several vascular proteins, such as von Willebrand Factor [1] and selectins [2], undergo conformational changes under mechanical force, resulting in the exposure of ligand-binding domains essential for initiating adhesive interactions.

Philip J. Hogg (ed.), *Functional Disulphide Bonds*, Methods in Molecular Biology, vol. 3016,
https://doi.org/10.1007/978-1-0716-5158-2_19,

Simultaneously, disulfide bonds within proteins can be enzymatically modified by redox reactions catalyzed by thiol isomerases, leading to significant shifts in protein conformation, binding affinity, and function [3–5].

The emerging concept of mechano-redox regulation describes the synergistic interplay between these two forces—mechanical stress and redox signaling—in modulating protein activity. This phenomenon was first described by Fernandez et al, who demonstrated that disulfide bond cleavage within the muscle protein titin could be mechanically coupled to energy transduction [6–8].

More recently, integrins—a family of cell adhesion receptors—have been identified as mechano-redox–sensitive proteins. Integrins mediate a range of adhesive processes, including firm adhesion, cell spreading, migration, and immune surveillance, by binding ligands in the extracellular matrix or on other cells. Our group has identified novel mechano-redox regulatory mechanisms in two key β-integrins: αIIbβ3, expressed on platelets [9], and Mac-1 (αMβ2), expressed on neutrophils [10]. We have shown that thiol isomerases, such as protein disulfide isomerase (PDI) and endoplasmic reticulum protein 5 (ERp5), can selectively reduce disulfide bonds within β-integrin subunits, modulating ligand-binding affinity and adhesive function in the presence of mechanical shear.

To explore the functional implications of integrin mechano-redox regulation, we developed a suite of microfluidic assays that enable precise control of biochemical and biomechanical conditions. These include:

1.1 Cellular De-Adhesion Assay

This assay was adapted from Chen et al [11] to study adhesion/de-adhesion of integrin-expressing cells to ligands under shear force. Integrin-expressing cells are introduced into a microfluidic channel and allowed to adhere to the surface, which is coated with a ligand. Following adhesion, media containing the redox reactive enzyme is perfused through the channels subjecting the cells to the combined effects of shear force and the redox enzyme. The extent of cell detachment serves as a functional readout of integrin–ligand affinity under defined redox and mechanical conditions. In this example, we have used Baby Hamster Kidney cells (BHK) cells expressing Mac-1 subjected to shear rates between 25–800 s^{-1} in the presence of oxidized or reduced PDI (Fig. 1).

1.2 Neutrophil Crawling Assay

This method, based on Buffone et al [12], investigates Mac-1–dependent neutrophil motility under shear force. Neutrophils engage ICAM-1 on inflamed endothelial substrates, enabling crawling against shear stress. Disruption of the Mac-1/ICAM-1 interaction impairs this process [13]. This assay captures neutrophil mobility using microfluidic devices, allowing quantification of neutrophil crawling under the combined effects of shear force and redox enzyme (Fig. 2).

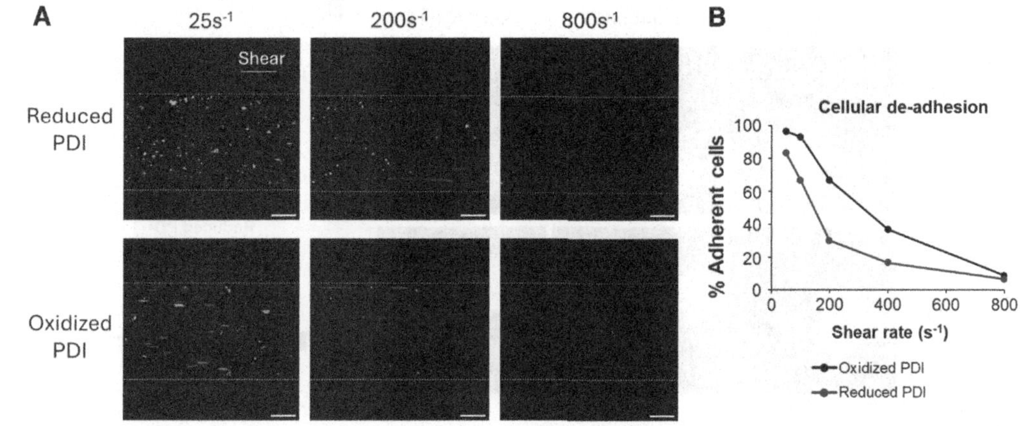

Fig. 1 Cellular de-adhesion assay. (**A**) Mac-1 expressing BHK cells, stained with Calcein AM (1 μM) for imaging, adherent on an Intercellular Adhesion Molecule-1 (ICAM-1)-coated surface. Shear rate is gradually increased until BHK cells are completely de-adhered. BHK cells were treated with either oxidized or reduced PDI (2 μM) to determine the role of disulfide modification of Mac-1, mediated by PDI, on Mac-1-dependent adhesion to ICAM-1. Scale bar represents 100 μm. (**B**) Percentage (%) of BHK cells, treated with reduced or oxidized PDI, remaining adherent after perfusion with media at increasing shear rate

1.3 Microfluidic Thrombus Formation Assay

This assay measures the effect of redox reactions on platelet adhesion under shear force. Straight microfluidic channels are coated with adhesive proteins such as fibrinogen, collagen, or von Willebrand Factor to capture platelets perfused in blood or buffer through the channel. The contribution of redox regulation can be assessed using platelets with genetic deletion of a particular thiol isomerase [14] or through pharmacological inhibition of thiol isomerases. In this example, we have used a pharmacological inhibitor of ERp5, anacardic acid [15] (Fig. 3).

1.4 Endo-Chip Model of Thromboinflammation

This model is used to study endothelial cell–cell interaction under the effect of shear and redox and is a simplified model of a vessel-on-chip. Endothelial cells are seeded into a microfluidic channel and allowed to adhere and form a monolayer. Treatment of the endothelial cells with pro-inflammatory stimuli (e.g., tumor necrosis factor-alpha (TNF-α) or phorbol 12-myristate 13-acetate (PMA) increases their expression of adhesion molecules such as ICAM-1 and selectins, as well as procoagulant factors, including tissue factor [16, 17]. This platform enables simultaneous investigation of neutrophil or platelet adhesion and thrombus formation under physiologically relevant conditions. We have applied this model to study the combined effect of redox enzymes and shear force on neutrophil crawling [10, 18]. Variations of this model have been applied to study mechanisms and to identify inhibitors of thrombosis [16], micro-occlusive, and thromboinflammatory disorders [19–21] (Fig. 4).

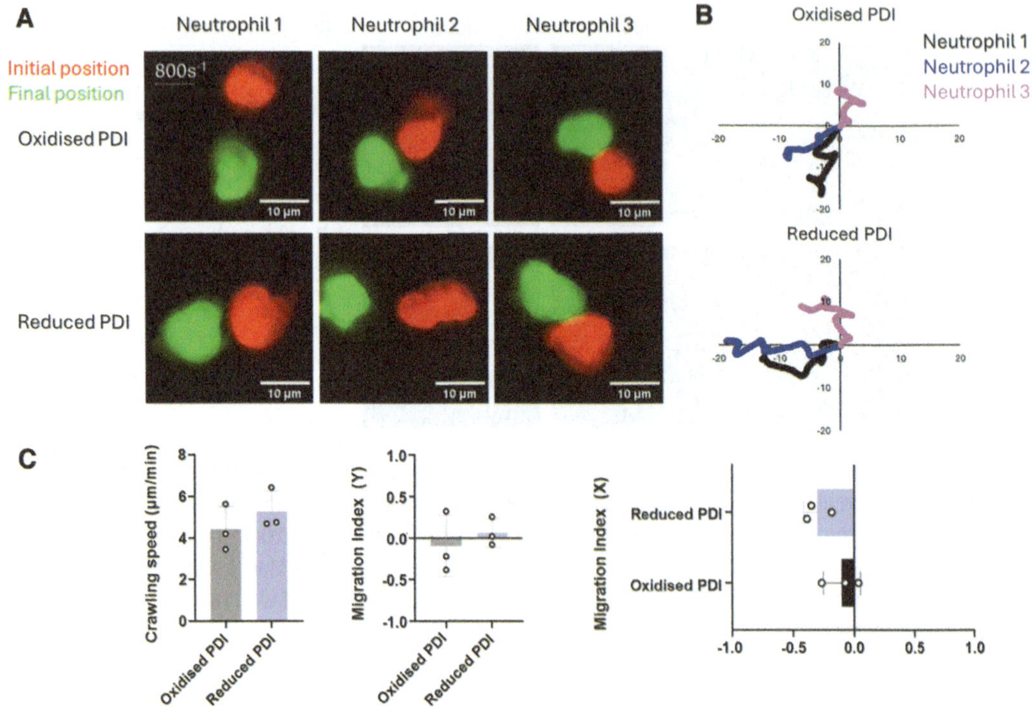

Fig. 2 Neutrophil crawling assay. (**A**) Images of neutrophils, stained with Calcein AM (1 μM) for cell tracking, at the start of perfusion (red) and end of perfusion (green). Neutrophils were treated with either oxidized or reduced PDI (2 μM) (to determine the role of disulfide modification mediated by PDI on neutrophil crawling) and then subjected to shear force by the perfusion of buffer at 800 s^{-1}. Scale bar represents 10 μm. (**B**) Cell tracks showing the movement of neutrophils, normalized to the location of neutrophils at the start of perfusion. In this example, neutrophils treated with reduced PDI migrate more in the direction of shear (X axis). (**C**) Parameters measured from the neutrophil crawling assay including crawling speed and the migration index (the proportion of movements made in a given direction) for Y and X axis. Data represent mean +/− S.D. of three neutrophils

1.5 Microvasculature-on-a-post model of thrombosis

This is another model of vessel-on-chip that introduces changes in the shear force due to flow disturbance during perfusion. Flow disturbance can be generated through constrictions of a microfluidic straight channel that promote platelet activation [22] and vWF-dependent platelet aggregation [23]. In the post model, a small post is included in the center of a straight microfluidic channel to simulate conditions of vessel stenosis, such as those created by the presence of an atherosclerotic plaque. This post is a circular prism protruding upward from the base of the channel toward the coverslip, creating a localized flow disturbance. When the post is endothelialized and perfused with recalcified whole blood, fibrin generation, and platelet aggregation rapidly localize at the post following the direction of the shear force [17] (Fig. 5).

The aforementioned five assays form a comprehensive toolkit for dissecting the role of mechano-redox mechanisms in integrin function and vascular biology.

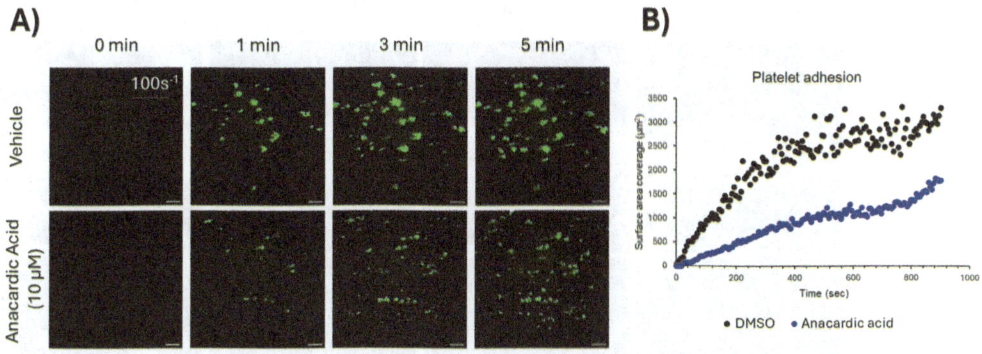

Fig. 3 Microfluidic platelet adhesion assay. (**A**) Whole blood, stained with Calcein AM (1 μM) to image platelets and treated with either ERp5 inhibitor anacardic acid (10 μM) or vehicle control was perfused across a collagen type III-coated surface at a shear rate of 100 s^{-1} for 15 min. Scale bar is 20 μm. (**B**) Platelet adhesion over time, represented as surface area coverage in μm^2 in the presence of vehicle (DMSO) or anacardic acid

2 Materials

2.1 Cellular De-Adhesion Assay

1. Polydimethylsiloxane (PDMS) microfluidic devices (*see* **Note 1**).
2. Baby Hamster Kidney (BHK) cells (*see* **Note 5**).
3. PBS (137 mM NaCl, 2.7 mM KCl, 10 mM Na_2HPO_4, and 1.8 mM KH_2PO_{34}).
4. PBS with calcium and magnesium (137 mM NaCl, 2.7 mM KCl, 10 mM Na_2HPO_4, 1.8 mM KH_2PO_4, 1 mM $CaCl_2$, and 1 mM $MgCl_2$).
5. Recombinant human ICAM-1/Fc.
6. Calcein AM (1 mg/mL in DMSO).
7. Bovine serum albumin (BSA) (0.5% w/v in PBS).

2.2 Neutrophil Crawling Assay

1. PDMS microfluidic devices (*see* **Note 1**).
2. Isolated neutrophils (*see* **Note 4**).
3. Hank's buffered saline solution (*see* **Note 4**).
4. Recombinant human ICAM-1/Fc.
5. Reduced PDI (*see* **Note 6**).
6. Oxidized PDI (*see* **Note 6**).
7. *N*-formyl-methionyl-leucyl-phenylalanine (fMLF) (5 mM in absolute ethanol).
8. Calcein AM (1 mg/mL in DMSO).
9. FIJI ImageJ.

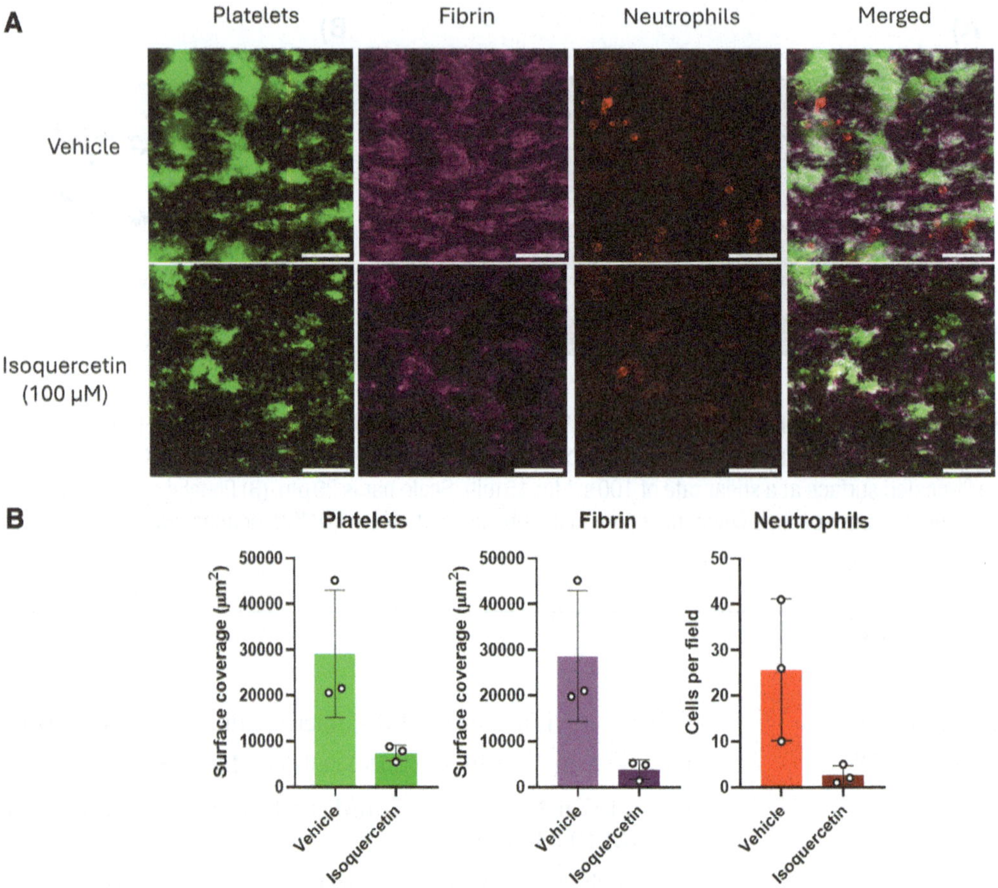

Fig. 4 Endo-chip model. (**A**) Microfluidic devices were coated with human umbilical vein endothelial cells (HUVECs), treated with TNF-α, fixed with paraformaldehyde, and perfused with whole blood for 15 min. Whole blood was labeled with an anti-platelet, anti-fibrin, and anti-neutrophil fluorescently labeled antibodies for visualization of platelets, fibrin, and neutrophils respectively. Whole blood was treated with the PDI inhibitor isoquercetin (100 μM) or vehicle control to determine the role of PDI in thrombus formation under shear force. Scale bar is 50 μm. (**B**) Representative data from three biological replicates of platelet adhesion and fibrin deposition (represented as surface coverage at the end of perfusion in μm^2) and neutrophil adhesion (represented as total cells per field of view). (Reproduced from [18] with permission)

2.3 Microfluidic Thrombus Formation Assay

1. PDMS microfluidic devices (*see* **Note 1**).
2. Coating proteins.
   a. Collagen type I (1 mg/mL, diluted using H_2O or 0.02 N acetic acid).
   b. Collagen type III (0.1 mg/mL, diluted using H_2O or 0.02 N acetic acid).
   c. Fibrinogen (4.3 mg/mL, diluted using PBS).
   d. Fibronectin (1 mg/mL, diluted using PBS).
   e. Von Willebrand Factor (0.23 mg/mL, diluted using H_2O).

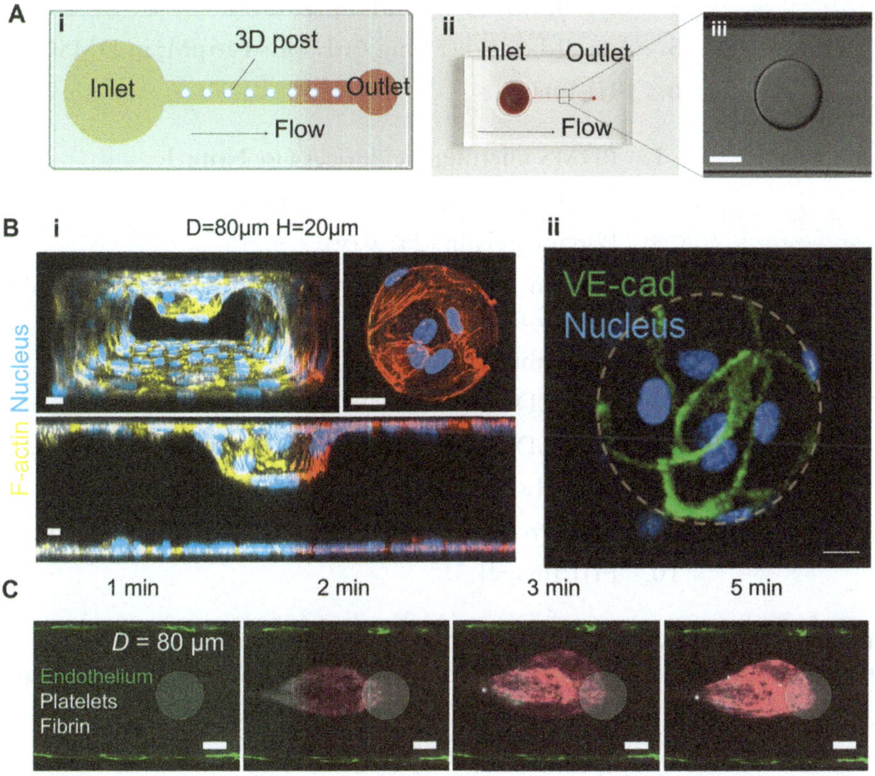

Fig. 5 Microvasculature-on-a-post model. (**A**) microfluidic post design (i) and final device (ii) showing a microfluidic straight channel with a post consisting of a centralized cylinder with a diameter of 80 μm and height of 20 μm (iii), scale bar represents 40 μm. (**B**) Endothelialization of the microfluidic post channel to create the microvascular-on-a-post. The device is completely covered with a monolayer of HUVECs, showing actin and nuclear staining (i), scale bar represents 20 μm. The post is also endothelialized, as visualized with endothelial gap junction marker VE-Cadherin (ii), scale bar represents 20 μm. (**C**) Platelet and fibrin accumulation over time developing at the endothelialized post under shear 100 s^{-1}. Scale bar represents 40 μm. (Reproduced from [17] with permission)

3. Human whole blood,
   a. Collect blood via venipuncture from healthy donors with approval from your Institute's Human Research Ethics Committee.
   b. For experiments aimed at examining the coagulation system, collect whole blood anticoagulated with sodium citrate to a final concentration of 0.32%–0.38% w/v. The blood will need to be recalcified with $CaCl_2$ to a final concentration of 4–10 mM prior to perfusion to restore coagulability.
   c. For experiments aimed at examining platelet function in whole blood in the presence of calcium, without examining the coagulation system, collect whole blood anticoagulated with clexane (enoxaparin) or heparin to a final concentration of 20 IU/mL.

4. Calcein AM (1 mg/mL in DMSO).
5. Anacardic acid (50 mg/mL, or ~146 μM, in DMSO).
6. FIJI ImageJ.

2.4 Endo-Chip

1. PDMS microfluidic devices (*see* **Note 1**).
2. Human fibronectin or rat tail collagen type I.
3. Porcine gelatin (2% w/v).
4. Human Umbilical Vein Endothelial Cells (HUVECs) (*see* **Note 7**).
5. Recombinant TNF-alpha.
6. Anti-CD41 antibody (FITC conjugated).
7. Anti-CD66b antibody (Alexa Fluor 647 conjugated).
8. Anti-fibrin antibody (Alexa Fluor 594 conjugated).
9. Isoquercetin (30 mg/mL or ~64.6 μM in DMSO).
10. FIJI ImageJ.

2.5 Microvasculature-on-a-Post

1. PDMS microfluidic devices (channel: width 200 μm, height 70 μm; post geometry: diameter 80 μm, height 20 μm; perfusion rate 36 μL/min = 100 s^{-1}).
2. Human fibronectin.
3. Porcine gelatin (2% w/v).
4. HUVECs (*see* **Note 7**).
5. Anti-CD41 antibody (FITC conjugated).
6. Anti-fibrin antibody (Alexa Fluor 594 conjugated).
7. Phorbol 12-myristate, 13-acetate (10 μM or ~6.17 mg/mL in DMSO).

2.6 Microfluidics Setup

1. 2 mL glass Luer lock syringe.
2. 24 mm (W) × 50 mm (L) × 0.17 mm (H) #1 borosilicate rectangular coverslips.
3. Medical grade Tygon tubing, 0.8 mm internal diameter, and 1.6 mm external diameter.
4. 18-gauge blunt-end syringe needle.
5. Programmable infuse/withdraw syringe pump.
6. Three-way stopcock, two female, and one rotating male Luer lock connector.
7. Fluorescent microscope or confocal microscope.

3 Methods

3.1 Microfluidics Setup

1. Prepare PDMS microfluidic device (*see* **Note 1**) and place it on the microscope stage.
2. Secure a glass syringe to the syringe pump. Connect the glass syringe to a piece of tubing with a Luer connector (female) inserted at the end, attached to the male end of a three-way Luer valve (two female, one male) (*see* **Note 2**).
3. Attach a metal pin connector (made by bending a blunt-end syringe needle by 90°) to the tubing (*see* **Note 2**).
4. Insert the metal pin connector into the outlet of the PDMS microfluidic device.
5. Place a strip of tape on both ends of the microfluidic device to secure the device to the microscope stage and minimize drift of the channel from the field of view during image acquisition (Fig. 6).

3.2 Cellular De-Adhesion Assay

1. Perfuse the solution of ICAM-1/Fc at 10 μg/mL through the microfluidic channel (*see* **Note 3**).
2. Allow the ICAM-1/Fc to incubate on the microfluidic device for 2 h at room temperature.
3. Detach the Mac-1-expressing BHK cells from cell culture flasks by first washing the cells with PBS, then incubating the cells with 5 mM EDTA in PBS for 5–10 min. Once cells have become rounded, tap the flask to detach the cells from the surface.
4. Remove the cell suspension from the flask and centrifuge cells at 200 × *g* at room temperature for 5 min. Remove the supernatant and resuspend the cell pellet in PBS with calcium and magnesium, adjusting the cell concentration to 10^7/mL. Stain the cells with Calcein AM at 1 μg/mL.
5. Treat the BHK cell suspensions with reduced or oxidized PDI (*see* **Note 6**) at 10 μM for 15 min.
6. Wash the channel to remove unbound ICAM-1/Fc by perfusing PBS containing BSA at 0.5% w/v. Allow the PBS to incubate on the channel for 15 min to block.
7. Attach the microfluidic device to a syringe pump setup on a fluorescent microscope (*see* **Note 2**).
8. Perfuse the BHK cell suspension through the microfluidic channel at a shear rate of 100 s^{-1} by starting the syringe pump with the required flow rate. Once sufficient number of BHK cells have entered the channel, stop the perfusion by stopping the syringe pump and allowing the cells to incubate on the channel for 15 min.

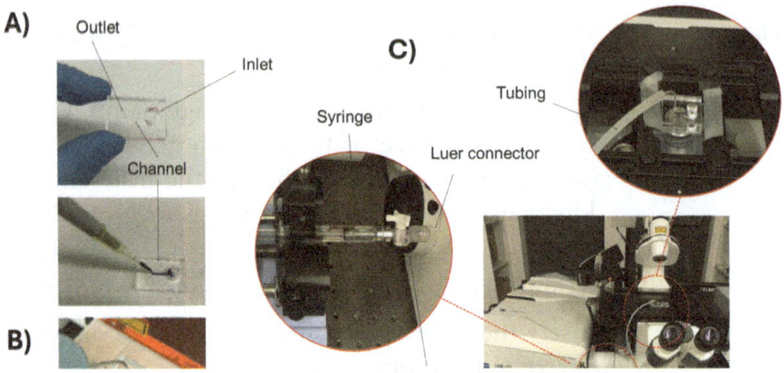

Fig. 6 Microfluidics setup. (**A**) The design of a microfluidic device consisting of a channel, in this case a straight channel, imprinted on PDMS, with holes punched out at opposite ends of the channel to create an inlet (for sample introduction) and outlet (to connect a syringe pump to the device and introduce shear force). (**B**) Demonstration of a metal pin connector (produced from a blunt-end syringe needle) attached to the outlet of the microfluidic device and tubing. (**C**) Setup of the microfluidic device and syringe pump on a microscope. A glass syringe is placed on a syringe pump, connected to a piece of tubing with a Luer connector (female) inserted at the end, attached to the male end of a three-way Luer valve (two female, one male)

9. Start the syringe pump at a shear rate of 25 s^{-1} until nonadherent cells have detached, then take a fluorescent image at excitation/emission wavelengths of 490/520 nm for an initial cell count.
10. Double the flow rate on the pump and image the cells after 1 min. Repeat this step until the desired endpoint shear rate has been achieved. Top up the inlet of the channel with fresh PBS, calcium and magnesium, as the channel empties.

3.3 Neutrophil Crawling Assay

1. Perfuse the solution of ICAM-1/Fc at 10 μg/mL through the microfluidic channel (*see* **Note 3**).
2. Allow the ICAM-1/Fc to incubate on the microfluidic device for 2 h at room temperature.
3. Adjust the concentration of the isolated neutrophils to 10^6/mL.
4. Stain the neutrophils with Calcein AM at 1 μg/mL.
5. Wash the channel by perfusing HBSS containing BSA. Allow the HBSS to incubate on the channel for 15 min to block.
6. Attach the outlet of the microfluidic device to a syringe pump setup on a fluorescent microscope (*see* **Note 2**).
7. Stimulate an aliquot of the neutrophil suspension with fMLF at 5 μM. Remove the HBSS from the inlet of the microfluidic device and pipette the stimulated neutrophil suspension into the inlet.

8. Gently perfuse the neutrophil suspension through the microfluidic channel at a shear rate of 100 s^{-1} by starting the syringe pump with the required flow rate. Once sufficient number of neutrophils have entered the channel, stop the perfusion by stopping the syringe pump and allowing the neutrophils to incubate on the channel for 15 min.
9. Remove the neutrophils from the inlet and replace with HBSS containing 5 μM fMLF. Start the syringe pump with a flow rate corresponding to the desired shear rate. Top up the inlet with fresh HBSS containing 5 μM fMLF as the inlet begins to empty.
10. Image the neutrophils over 10 min, taking fluorescent images at excitation/emission wavelengths of 490/520 nm every 2 s.

3.4 Microfluidic Thrombus Formation

1. Perfuse a solution of the desired protein through the microfluidic device (*see* **Note 3**). Allow the microfluidic device to incubate for 2 h at room temperature. *See* **Note 3** for more details on coating proteins.
2. Remove the coating protein from the inlet. Wash the channels by perfusing the solvent used to dilute the coating protein through the channel.
3. Connect the microfluidic device to a syringe pump setup on a fluorescent microscope (*see* **Note 2**).
4. Stain whole blood with Calcein AM at 1 μg/mL.
5. Aliquot the blood and treat one aliquot with an inhibitor and the other with an equal volume of vehicle as a control. For Fig. 3, whole blood was treated with the ERp5 inhibitor anacardic acid at a final concentration of 10 μM.
6. Remove the solvent from the inlet and replace with whole blood.
7. Set the syringe pump to a flow rate corresponding to the desired shear rate (*see* **Note 3**). Start the pump and find the focal plane to which platelets are adherent.
8. Image the adherent platelets at regular intervals, taking fluorescent images at excitation/emission wavelengths of 490/520 nm.

3.5 Endo-Chip

1. Perfuse a solution of fibronectin at 50 μg/mL through plasma-bonded microfluidic devices (*see* **Note 3**). Allow the microfluidic device to incubate with the fibronectin for 2 h at room temperature.
2. Remove the fibronectin from the inlet and replace with PBS. Perfuse the PBS through the microfluidic device.
3. From a 80-90% confluent T25 flask of HUVECs (*see* **Note 7**), detach HUVECs by removing HUVEC culture medium,

washing the HUVEC surface with 5 mL of PBS, and adding 1–2 mL of Trypsin/EDTA. Incubate the flask at 37°C for 30–60 s.

4. Once cells are rounded, tap the flask to complete detachment of the HUVECs. Neutralize trypsin with equal volume of HUVEC culture media, and transfer the HUVEC suspension to a 15 mL centrifuge tube. Centrifuge the HUVEC suspension at 200 × *g* for 5 min.
5. Discard the supernatant and resuspend the cell pellet with ~30–50 μL of HUVEC culture medium.
6. Perform a cell count and adjust the cell concentration to 15–20 × 10^6/mL.
7. Pipette the cell suspension through the outlet of the microfluidic device until the channel is filled with the HUVEC suspension. Check under a light microscope to determine sufficient coverage of the coverslip surface with HUVECs.
8. Place the microfluidic device in a cell culture incubator at 37°C, 5% v/v CO_2 for 30 min.
9. Fill the inlet of the channel with HUVEC culture medium and perfuse through the channel to wash off nonadherent HUVECs (*see* **Note 3**). Place the microfluidic device back in the cell culture incubator. Repeat this step every 30–60 min until a confluent HUVEC monolayer forms (*see* **Note 9**). For treatments, perfuse HUVEC culture medium containing biochemical stimulus such as TNF-α (0.5–10 ng/mL).
10. Perfuse PBS through the microfluidic device to wash HUVECs.
11. Stain citrate-anticoagulated whole blood with anti-CD41 antibody-FITC (for platelets), anti-CD66b antibody-AF647 (for neutrophils), and anti-fibrin-AF594 antibodies at a final concentration of 0.5 μg/mL. Allow antibodies to incubate for at least 10 min prior to perfusion.
12. Treat the stained whole blood with isoquercetin at a final concentration of 100 μM or vehicle control (equal volume of DMSO).
13. Connect the microfluidic device to a syringe pump setup on a fluorescent microscope (*see* **Note 2**).
14. Recalcify the whole blood with $CaCl_2$ to a final concentration of 10 mM and perfuse through the microfluidic device at 100 s^{-1}.
15. Image the subsequent thrombus formation at the excitation/emission wavelengths of the fluorophores corresponding to each directly-labeled antibody.

3.6 Microvasculature-on-a-Post

1. Perfuse a solution of fibronectin at 100 μg/mL through plasma-bonded microfluidic devices containing the post geometry (*see* **Note 3**). Allow the microfluidic device to incubate with the fibronectin for 2 h at room temperature.
2. Remove the fibronectin from the inlet and replace with PBS. Perfuse the PBS through the microfluidic device.
3. From a 80-90%% confluent T25 flask of HUVECs (*see* **Note 7**), detach HUVECs by removing HUVEC culture medium, washing the HUVEC surface with 5 mL of PBS, and adding 1–2 mL of Trypsin/EDTA. Incubate the flask at 37° C for 30–60 s.
4. Once cells are rounded, tap the flask to complete detachment of HUVECs. Neutralize trypsin with equal volume of HUVEC culture media, and transfer the HUVEC suspension to a 15 mL centrifuge tube. Centrifuge the HUVEC suspension at 200 × *g* for 5 min.
5. Discard the supernatant and resuspend the cell pellet with ~30–50 μL of HUVEC culture medium.
6. Perform a cell count and adjust the cell concentration to 15–20 × 10^6/mL.
7. Pipette the cell suspension through the outlet of the microfluidic device until the channel is filled with the HUVEC suspension, then flip the microfluidic channel with the coverslip facing upwards.
8. Place the microfluidic device in a cell culture incubator at 37°C, 5% v/v CO_2 for 15 min.
9. Place the microfluidic device coverslip down, then pipette the cell suspension through the outlet once more and place the device back in the cell culture incubator for 15 min.
10. Fill the inlet of the channel with HUVEC culture medium and perfuse through the channel to wash off nonadherent HUVECs (*see* **Note 3**). Top up the inlet with fresh HUVEC culture media and place the device in an enclosed petri dish (55× 15 mm). Place a tissue soaked in H_2O into the petri dish and place the petri dish in the cell culture incubator. Allow the microfluidic device to incubate for 16 h.
11. Replace the media in the inlet with fresh HUVEC culture media treated with either PMA at a concentration of 100 nM or vehicle control (equal volume DMSO). Perfuse the media through the channel and place the microfluidic device back in the cell culture incubator and allow the device to incubate for 2 h.
12. Stain citrate-anticoagulated whole blood with the anti-CD41-FITC antibody to stain platelets, and directly labeled anti-fibrin-AF594 antibody, at a final concentration of 0.5 μg/mL. Allow antibodies to incubate for at least 10 min prior to perfusion.

13. Connect the microfluidic device to the syringe pump setup on a fluorescent microscope (*see* **Note 2**).
14. Perfuse PBS through the microfluidic device to wash the HUVECs and locate the post geometry on the brightfield setting of the microscope.
15. Recalcify the stained whole blood with $CaCl_2$ to a final concentration of 4 mM and perfuse through the microfluidic device at a shear rate of 600 s^{-1}.
16. Image the microfluidic device at the excitation/emission wavelengths of each fluorophore corresponding to the respective directly-labeled antibody.

3.7 Data Analysis

3.7.1 Cellular De-Adhesion Assay

1. Count the number of cells in the field of view after 1 min of perfusion at 25 s^{-1}. This will be the initial count of cells used for normalization.
2. Count the number of cells in the field of view at each time point.
3. Divide the number of cells counted at each time point by the initial count of cells and multiply by 100. This is the percentage of cells that remain adherent at each shear rate.
4. Plot the percentage of adherent cells against the corresponding shear rate.

3.7.2 Neutrophil Crawling Assay

1. If the scale of the image has not been calibrated yet, adjust the scale of the image on ImageJ. Create a duplicate of the image on ImageJ for analysis.
2. Crop the field of view to encapsulate one crawling neutrophil.
3. On ImageJ, adjust the threshold of the image by going to Image > Adjust > Threshold. On the threshold window, move the sliders such that a mask is generated around the neutrophil.
4. Convert the image stack to a binary: Click Apply, tick the "Calculate threshold for each image" and "Black background (of binary masks)" options, then click OK.
5. Go to Analyze > Set Measurements, then select the "Centroid" and "Limit to threshold" options.
6. Go to Image > Stacks > Measure Stack.
7. Copy the data from the Results window and paste into Microsoft Excel. The X and Y columns represent the X and Y coordinates of the centroid of the neutrophil.

8. To generate a neutrophil track, subtract the initial X and Y coordinates from the rest of the data to normalize the centroid to zero, then plot the Y coordinates of the centroid against the X coordinates.
9. To determine the distance traveled between each frame, use the formula below:

$$\text{Distance} = \sqrt[2]{\left((X_{n+1}^2 - X_n^2) + \left(Y_{n+1}^2 - Y_{n+1}^2\right)\right)},$$

where:
n is the current frame,
X is the X displacement of the neutrophil for a given frame,
Y is the Y displacement of the neutrophil for a given frame.

10. To determine neutrophil crawling speed, use the formula below:

$$\text{Crawling speed } (\mu\text{m/min}) = \frac{\text{Total Distance } (\mu\text{m})}{\text{Total time (min)}},$$

where:
Total Distance is the sum of distances traveled between each frame for a given image stack,
Total Time is the sum of the times between each frame for a given image stack.

11. To determine the migration index for the X direction, use the formula below:

$$\text{Total Distance}_X = \sum_{n=1}^{k} | X_{n+1} - X_n |,$$

$$X \text{ Migration Index} = \frac{X_{\text{Final}} - X_{\text{Initial}}}{\text{Total Distance}_X},$$

where:

Total Distance_X is the total distance traveled in the X-direction,
k is the total number of frames in the image stack,
X_n is the X-coordinate of the centroid at a given frame n,
X_{Final} is the final X-coordinate of the centroid of the image stack,
X_{Initial} is the initial X-coordinate of the centroid of the image stack,

For the Y migration index, use the corresponding Y-coordinates in these equations.

3.7.3 Microfluidic Thrombus Formation, Endo-Chip, and Microvasculature-on-a-Post

1. If the scale of the image has not been calibrated yet, adjust the scale of the image on ImageJ. Create a duplicate of the image on ImageJ for analysis.
2. On ImageJ, adjust the threshold of the image by going to Image > Adjust > Threshold. On the Threshold window, move the sliders such that a mask is generated around the thrombi.
3. Convert the image stack to a binary: Click Apply, tick the "Calculate threshold for each image" and "Black background (of binary masks)" options, then click OK.
4. Go to Analyze > Set Measurements, then select the "Area" option.
5. Go to Image > Stacks > Measure Stack.
6. Copy the data from the Results window and paste onto Microsoft Excel.
7. Plot the area (thrombus size) against the frame number.
8. For neutrophil counts, where samples were stained for neutrophils, manually count the number of neutrophils per frame, using the neutrophil stain as a guide to identify neutrophils based on morphology.

4 Notes

1. Microfluidic devices are fabricated from PDMS made from a Sylgard 184 kit. PDMS is cast into silicon wafer molds fabricated by soft lithography and, unless otherwise stated, produces microfluidic channels with dimensions of 100 μm height, 1000 μm width, and 10 000 μm length. For these devices, a flow rate of 10 μL/min will result in a shear rate of 100 s^{-1}. The molds used in these protocols were fabricated by the Sydney Nano Institute. The method used for casting PDMS is described below:

1. Materials

- Sylgard 184 PDMS kit.
- #1 borosilicate coverslips (22 × 50 mm).
- Silicon wafer molds.
- Plasma cleaner.
- Oven.
- Vacuum desiccator.
- 1 mm biopsy punch.
- 6 mm biopsy punch.
- Razor blades.

2. Method

- Mix the base material with the curing agent of the Sylgard 184 kit at a 10:1 ratio.
- Degas the mixture in a vacuum desiccator until all the bubbles are visibly removed.
- Carefully pour the mixture onto the silicon wafer mold without introducing any new bubbles.
- Use a pipette to remove any excess bubbles by carefully drawing them up into the pipette tip.
- Transfer the molds to an oven and allow the PDMS to cure at 60°C for at least 4 h.
- Cut out the PDMS channels into ~2 × 3 cm rectangles. Punch out an inlet and an outlet at opposite ends of the channel using a 6 mm and 1 mm biopsy punch respectively.
- Place the PDMS channels and borosilicate coverslips in a plasma cleaner.
- Expose the PDMS channels to oxygen plasma on the highest setting at 1.72 Torr for 1 min.
- Remove the PDMS channels from the plasma cleaner and place on plasma-treated coverslips, channel side down, to form the final PDMS device. Press gently to remove air bubbles.
- Incubate the final PDMS device in an oven at 60°C for 15 min.

2. A basic microfluidic setup involves connecting a microfluidic device to a metal pin bent at a 90° angle. This pin can either be made from an 18-gauge blunt-end syringe needle or purchased from a supplier directly. The pin is then connected to tygon tubing, the opposite end of which is connected to a luer-to-tubing adapter. The luer lock adapter is then locked onto a luer lock syringe, which is placed on a syringe pump. Use the Luer valve to flush the tubing clean with water after each microfluidics run without disconnecting the tubing from the setup. The final setup is assembled on a microscope to image the microfluidic device. See Fig. 6 for an example.
3. Coating a microfluidic device involves filling the inlet of the device with a solution of coating protein, placing the tip of a pipette on the outlet of the device with the pipette plunger depressed, then drawing the solution through the channel by slowly releasing the plunger. Optimal concentrations of coating proteins and shear rate need to be determined on a per-experiment basis. Recommended starting concentrations and shear rates for proteins used in the microfluidic thrombus formation assay are provided below:

- Collagen type I: 250 μg/mL, 100–1000 s^{-1}
- Collagen type III: 10 μg/mL, 100–1000 s^{-1}
- Fibrinogen: 60 μg/mL, 100–1000 s^{-1}
- Fibronectin: 5–100 μg/mL, 100 s^{-1}
- Von Willebrand Factor: 50 μg/mL, 1000–10 000 s^{-1}

4. Neutrophils can be isolated by any method, such as magnetic bead separation or density gradient centrifugation. The method used for neutrophil isolation in this protocol was histopaque density gradient centrifugation:

1. Materials
   - Histopaque 1077.
   - Histopaque 1119.
   - Hank's buffered saline solution (HBSS, prepared fresh and 0.22 μm filtered, 140 mM NaCl, 5 mM KCl, 0.3 mM Na_2HPO_4, 0.4 mM KH_2PO_4, 6 mM D-glucose, 4 mM $NaHCO_3$, 1 mM $CaCl_2$, 1 mM $MgCl_2$, 0.5% w/v BSA, and pH 7.4).
   - 0.2% w/v NaCl.
   - 1.6% w/v NaCl.

2. Method
   - Collect human whole blood by venipuncture and anticoagulated with clexane or heparin at 20 IU/mL.
   - Pipette 12 mL of Histopaque 1119 into a 50 mL centrifuge tube.
   - Tilt the centrifuge tube slightly and carefully layer 12 mL of Histopaque 1077 on top of the Histopaque 1119 by pipetting on the wall of the tube closest to the floor.
   - Gently layer 24 mL of human whole blood on top of the Histopaque layers.
   - Centrifuge the layered blood-histopaque at 600 × *g* for 30 min with maximum acceleration and no brake.
   - Discard the plasma and histopaque. Collect the pink band (neutrophils) directly above the red blood cell pellet and transfer to a fresh 50 mL centrifuge tube.
   - Dilute the collected neutrophils with HBSS until a final volume of 50 mL is achieved.
   - Centrifuge the neutrophil suspension at 600 × *g* for 10 min.
   - Discard the supernatant and resuspend the cell pellet with 1 mL of HBSS. Transfer to a 15 mL centrifuge tube.
   - Add 7 mL of 0.2% w/v NaCl to the neutrophil suspension to lyse the contaminating red blood cells. Invert to mix and

allow to incubate for 30 s. Neutralize the lysis reaction by adding 7 mL of 1.6% w/v NaCl and centrifuge at 600 × *g* for 5 min. Repeat this step no more than 2 additional times.

- Resuspend the cell pellet with HBSS and perform a cell count on a hemocytometer. Keep the neutrophil suspension on ice until ready to use for experiments.

5. The de-adhesion assay can be used with a variety of adherent cells transfected to bind ligands that do not normally interact with the cells being used. The protocols in this chapter use BHK cells, which do not express β2 integrins. These cells were transfected to express the Mac-1 integrin, allowing adhesion to ICAM-1. The de-adhesion has also been used with Chinese Hamster Ovary cells. An example of the transfection method is provided below:

1. Materials
   - Lipofectamine 2000.
   - Opti-MEM.
   - ITGAM cDNA cloned into pcDNA3.1/HygroB(+) vector.
   - ITGB2 cDNA cloned into pcDNA3.1/Neo(+).
   - FSPI-HF restriction enzyme.
   - SSPI-HF restriction enzyme.
   - 10× Cutsmart buffer.
   - TE buffer (10 mM Tris–HCl, 0.1 mM EDTA, pH 8.0).
   - Sodium acetate (3 M).
   - Absolute Isopropanol.
   - Ethanol (70% w/v).
   - Distilled H_2O.
   - Hygromycin B.
   - G418.
   - DMEM high glucose (10% FCS, 2 mM L-glutamine, 100 μg/mL streptomycin, 100 U/mL penicillin).
   - 0.5 M EDTA, pH 8.0.
   - Antihuman CD11b-APC.
   - Antihuman CD18-AF488.

2. Method
   - Plate 0.3×10^6 BHK cells into the wells of a 12-well plate. Culture the cells in DMEM high glucose until 60%–80% confluent.
   - Linearize the *ITGAM* and *ITGB2* plasmids by restriction enzyme digest: In a 1.5 mL centrifuge tube, incubate 20 μg of plasmid in 60 μL of CutSmart buffer. For the *ITGAM*

plasmid, incubate with SSPI-HF restriction enzyme; for the *ITGB2* plasmid, incubate with FSPI-HF restriction enzyme, to a final concentration of 150 U/mL. Incubate at 37°C for 2 h.

- Add 6 μL of sodium acetate and mix gently by pipetting.
- Add 66 μL of 100% isopropanol, mix by vortexing, and leave the tube on ice for 15 min.
- Centrifuge the tube at 17 000 × *g* for 10 min at 4°C.
- Gently discard the supernatant without disturbing the DNA pellet, then add 200 μL of 70% w/v ethanol.
- Centrifuge the tube at 17 000 × *g* for 5 min at 4°C and discard the supernatant. Allow the DNA pellets to air-dry under a class II biosafety cabinet until the DNA pellets become translucent.
- Resuspend the pellets in DNAse-free, sterile H_2O and determine the DNA concentration using a NanoDrop spectrometer.
- Remove the supernatant of the plated BHK cells then wash the cells with OptiMEM. Cover the surface of the BHK cells with 2 mL of OptiMEM.
- Add 5 μg of the linearized *ITGAM* and *ITGB2* plasmids to 50 μL of Opti-MEM.
- Dilute 5 μg of Lipofectamine 2000 in 50 μL of Opti-MEM and add to the DNA-OptiMEM mixture. Allow the mixture to incubate at room temperature for 15 min.
- For each well of BHK cells, add 10 μL of the DNA-Lipofectamine mixture dropwise. Swirl the plate with each drop to mix. Allow the cells to incubate overnight at 37°C, 5% v/v CO_2.
- Discard the Opti-MEM media and replace with DMEM high glucose. Allow the cells to incubate overnight at 37°C, 5% v/v CO_2 overnight.
- Discard the DMEM high glucose and replace with fresh DMEM high glucose containing G418 and Hygromycin B at 500 μg/mL and allow the cells to incubate at 37°C, 5% v/v CO_2. Monitor the cells for cell death and replace the media as the cells detach. Once the cells start growing, continue with general cell culture.
- Once confluent, check the cells for Mac-1 expression by detaching cells from the culture vessel by trypsinisation, then incubate the cells with anti-CD18-AF488 and anti-CD11b-APC antibodies at 1 μg/mL. Determine % expression by flow cytometry, comparing to nontransfected BHK cells as a negative control.

6. PDI can be reduced or oxidized with DTT or oxidized glutathione, respectively. Below is an example protocol:

1. Materials
    - Recombinant PDI.
    - Oxidized glutathione.
    - DTT.
    - 7K zebaspin columns.
2. Method
    - Incubate PDI with either DTT at 10 mM for 30 min (for reduction) or oxidized glutathione at 200 μM for 16 h (for oxidation) at 25°C.
    - Prepare a zebaspin column as per manufacturer's instructions.
    - Add the treated PDI directly to the column, then centrifuge at 1500 × *g* for 2 min.
    - Collect the filtrate and use the PDI within 4 h.

7. The method for endothelialization can be used with a variety of endothelial cells. This protocol uses HUVECs cultured in M199 media; an example protocol for standard HUVEC culture is provided below:

1. Materials
    - Human Umbilical Vein Endothelial Cells.
    - HUVEC culture medium:
        1. For the Endo-chip, use M199 medium containing 20% v/v FCS, 2 mM L-glutamine, 100 U/mL Penicillin, 100 μg/mL Streptomycin, 1 mM Sodium Pyruvate, 1× MEM non-essential amino acids, 1.34 mM $NaHCO_3$, and 0.3% v/v Endothelial cell growth supplement.
        2. For the microvasculature-on-a-post model, use Endothelial Growth Media-2, which is Endothelial Basal Media containing 2% v/v FBS, 0.2 μg/mL hydrocortisone, 4 ng/mL hFGF-B, 2 ng/mL VEGF, 5 ng/mL R3-IGF-1, 75 μg/mL ascorbic acid, 10 ng/mL hEGF, 1 ng/mL heparin, 30 μg/mL gentamicin, and 15 ng/mL amphotericin.
    - Porcine gelatin (2% w/v)
2. Method:
    - Coat the surface of a T25 flask with porcine gelatin (2% w/v) or rat tail collagen type I (150 μg/mL) for 1 h at room temperature.

- Flash-thaw a 1 mL vial of HUVECs by swirling the vial in a 37°C water bath.
- Transfer the thawed HUVECs to 9 mL of HUVEC culture medium in a 15 mL centrifuge tube.
- Centrifuge the HUVEC suspension at 200 × *g* for 5 min.
- Discard the supernatant and resuspend the HUVEC pellet with 5 mL of HUVEC culture medium.
- Remove the gelatin/rat tail collagen type I from the T25 flask and transfer the 5 mL HUVEC suspension to the flask. Place the flask in a cell culture incubator at 37°C, 5% v/v CO_2.
- Replace culture medium every 2 days until HUVECs are 80-90%% confluent.

8. Make sure to verify that the centroid values accurately portray the movement of the neutrophils. This can be done by visually comparing the cell tracks with the movement of the cell. Migration indices should be a value between −1 and 1.
9. Prior to commencing experiments with endothelialized microfluidic devices, check that the endothelial cells have formed a confluent monolayer under a brightfield microscope.

References

1. Ju L, Chen Y, Zhou F et al (2015) Von Willebrand factor-A1 domain binds platelet glycoprotein Ibα in multiple states with distinctive force-dependent dissociation kinetics. Thromb Res 136:606–612
2. McEver RP (2015) Selectins: initiations of leucocyte adhesion and signalling at the vascular wall. Cardiovasc Res 107:331–339
3. Passam FH, Lin L, Gopal S et al (2015) Both platelet- and endothelial cell-derived ERp5 support thrombus formation in a laser-induced mouse model of thrombosis. Blood 125:2276–2285
4. Bowley SR, Fang C, Merrill-Skoloff G et al (2017) Protein disulfide isomerase secretion following vascular injury initiates a regulatory pathway for thrombus formation. Nat Commun 8:14151
5. Sun K, Zhang Y, Yang A et al (2025) Extracellular thiol isomerase ERp5 regulates integrin αIIbβ3 activation by inhibition of fibrinogen binding. Platelets 36:2455743
6. Wiita AP, Ainavarapu SRK, Huang HH et al (2006) Force-dependent chemical kinetics of disulfide bond reduction observed with single-molecule techniques. Proc Nat Acad Sci U S A 103:7222–7227
7. Wiita AP, Perez-Jimenez R, Walther KA et al (2007) Probing the chemistry of thioredoxin catalysis with force. Nature 450(7166):124–127
8. Baldus IB, Gräter F (2012) Mechanical force can fine-tune redox potentials of disulfide bonds. Biophys J 102:622–629
9. Passam F, Chiu J, Ju L et al (2018) Mechano-redox control of integrin de-adhesion. eLife 7:e34843
10. Dupuy A, Aponte-Santamaria C, Yeheskel A et al (2023) Mechano-redox control of Mac-1 de-adhesion by PDI promotes directional movement under flow. Circ Res 132:e321926
11. Chen J, Salas A, Springer TA (2003) Bistable regulation of integrin adhesiveness by a bipolar metal ion cluster. Nat Struct Mol Biol 10:995–1001
12. Buffone Jr A, Anderson NR, Hammer DA (2019) Human neutrophils will crawl upstream on ICAM-1 if Mac-1 is blocked. Biophys J 117:1393–1404
13. Sumagin R, Sarelius IH (2010) Intercellular adhesion molecule-1 enrichment near tricellular endothelial junctions is preferentially associated with leukocyte transmigration and signals for reorganization of these junctions to accommodate leukocyte passage. J Immunol 184:5242–5252

14. Lay AJ, Dupuy A, Hagimola L et al (2023) Endoplasmic reticulum protein 5 attenuates platelet endoplasmic reticulum stress and secretion in a mouse model. Blood Adv 7:1650–1665
15. Horibe T, Torisawa A, Okuno Y et al (2014) Discovery of protein disulfide isomerase P5 inhibitors that reduce the secretion of MICA from cancer cells. ChemBioChem 15:1599–1606
16. Jain A, Van Der Meer A, Papa A-L et al (2016) Assessment of whole blood thrombosis in a microfluidic device lined by fixed human endothelium. Biomed Microdevices 18:s10544
17. Zhang Y, Aye S-S-S, Cheng V et al (2023) Microvasculature-on-a-post chip that recapitulates prothrombotic vascular geometries and 3D flow disturbances. Adv Mater Interfaces 10:23003234
18. Dupuy A, Qi M, Fenwick JCL, Yates DM, Coleman PR, Gamble JR, Ju LA, Passam FH (2025) Endo-chip laser-induced thrombus formation: a vessel-on-chip model for in vitro testing of antithrombotic agents. Blood Vessel Thromb Hemost. 18;2(4):100096. https://doi.org/10.1016/j.bvth.2025.100096. PMID:41030610
19. Dupuy A, Liu X, Kong Y, Qi M, Perdomo J, Fenwick J, Tieng J, Johnston B, Shi QS, Larance M, Zhang Y, Ju LA, Coleman P, Gamble J, Gardiner EE, Poncz M, Tran H, Chen V, Passam FH (2025) Endothelial cell activation enhances thromboinflammation in vaccineinduced immune thrombotic thrombocytopenia. Blood Adv. 24;9(12):2891–2906
20. Qiu Y, Lin J, Wang A et al (2025) Clinically relevant clot resolution via a thromboinflammation-on-a-chip. Nature 641:1298–1308. https://doi.org/10.1038/s41586-025-08804-7
21. Satta S, Lai A, Cavallero S et al (2021) Rapid detection and inhibition of SARS-CoV-2-Spike mutation-mediated microthrombosis. Adv Sci 8:2103266
22. Nesbitt WS, Westein E, Tovar-Lopez FJ et al (2009) A shear gradient-dependent platelet aggregation mechanism drives thrombosis formation. Nat Med 15:665–673
23. Westein E, van der Meer AD, Kuijpers MJE et al (2013) Atherosclerotic geometries exacerbate pathological thrombus formation poststenosis in a von Willebrand factor-dependent manner. Proc Nat Acad Sci U S A 110:1357–1362

Check for updates

Chapter 20

Flow Cytometry Assessment of Procoagulant Platelets Using A Dithiol Reactive Probe

Chuen Wen Tan, Morgane Bourcy, Leonardo Pasalic and Vivien M. Chen

Abstract

Flow cytometry assessment of platelets using the combination of GSAO [(4-(*N*-(*S*-glutathionylacetyl) amino)phenylarsonous acid], a dithiol reactive probe, and P-selectin, a platelet activation marker, is a novel and powerful assay in the identification and quantification of the procoagulant subpopulation of platelets that has the capacity to support thrombin generation. In the following chapter, we provide the flow cytometry protocols aimed at the study of procoagulant platelets under resting and agonist-stimulated conditions in whole blood and washed platelet of both human and murine samples.

Keywords Procoagulant platelets, Flow cytometry, GSAO, Whole blood, Washed platelets

1 Introduction

It is increasingly recognized that procoagulant platelets are distinct subpopulation of platelets essential for the assembly of coagulation factors and generation of thrombin that overlap with, but are not identical to the coated platelets originally described by Dale and colleagues [1, 2]. Procoagulant platelets have distinct biological characteristics and appear to be functionally important in health and disease, thus ability to accurately measure this subpopulation is of interest. Procoagulant platelets are characterized by exposure of phosphatidylserine (PS) on the outer membrane, ability to bind activated coagulation factors (e.g., FVa/Xa) to the membrane surface, mitochondrial membrane depolarization, sustained elevation of intracellular calcium, and increased membrane permeability [2]. Although PS exposure is a prerequisite for formation of procoagulant platelets, detection on PS exposure alone is not sufficiently specific to identify procoagulant platelets and development of an investigative modality to identify procoagulant platelets with high sensitivity and specificity, particularly in clinical samples, would promote further understanding of this platelet subset in both clinical and research settings.

Philip J. Hogg (ed.), *Functional Disulphide Bonds*, Methods in Molecular Biology, vol. 3016,
https://doi.org/10.1007/978-1-0716-5158-2_20,

GSAO [(4-(*N*-(*S*-glutathionylacetyl)amino)phenylarsonous acid] is a dithiol reactive probe that has been characterized as a cell death marker (*see* Fig. 1a) [3]. When tagged with a receptor compound, such as the fluorophore AlexaFluor 647 at the γ-glutamyl residue, GSAO is able to cross the membrane of procoagulant platelets and is then retained in the cytoplasm of these platelets by covalently binding to proteins containing closely spaced dithiols resisting washout [4]. Trivalent arsenicals cross-link dithiols forming stable cyclic dithioarsinite complexes that are effectively irreversible in biological milieu. This metalloid has only very weak affinity for single thiols. Intracellular dithiol protein targets significantly outnumber extracellular targets, thus the intensity of labeling is at least 1000-fold in platelets in which labeled-GSAO enters the cytosol. GSCA [4-(*N*-[(*S*-glutathionyl) acetyl]amino)benzoic acid] contains an inert carboxylic acid group replacing the chemically reactive arsenical in GSAO. GSCA has the same biodistribution as GSAO but does not react with proteins, so washes out of cells and acts as a control compound (*see* Fig. 1b). GSAO and GSCA labeling is compatible with a number of different reporter groups, including hydrophilic fluorophores and biotin. Conjugates of GSAO with fluorescein, oregon green, and the Alexa Fluor (AF488, AF546, AF647, AF750) and Cy (Cy5, Cy7) dyes have been tested in our hands. For the purposes of this methods paper, we have used GSAO-AF647 as the example reporter.

We have established a method using combination markers of GSAO and P-selectin on a flow cytometry platform to characterize procoagulant platelets in both human and murine (mouse) blood. We have previously demonstrated that these markers correlate to the other markers of procoagulant platelets, including phosphatidylserine externalization, sustained intracellular calcium flux, loss of mitochondrial membrane potential, binding of coagulation factors to the platelet surface, and thrombin generation [4, 5]. Platelets that uptake GSAO and express surface P-selectin represent the procoagulant subpopulation. Platelets with uptake of GSAO without P-selectin represent apoptotic platelets and those with surface P-selectin alone are activated, but not procoagulant [4, 5]. In this chapter, we

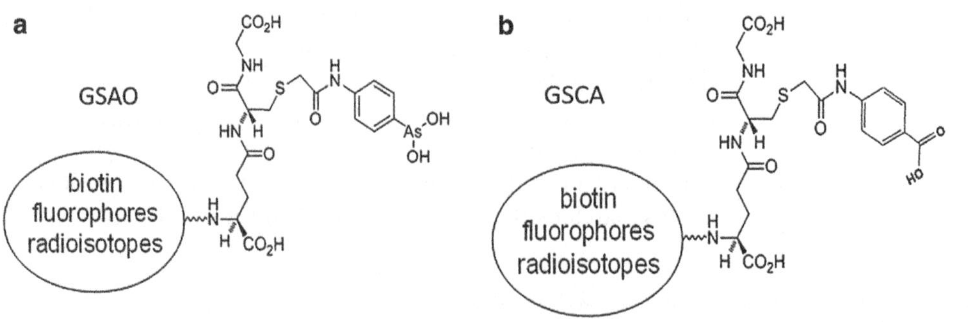

Fig. 1 (**a**) Structure of GSAO. (**b**) Structure of GSCA

provide the protocols for the assessment of procoagulant platelets under resting and agonist-stimulated conditions in whole blood and washed platelet samples.

2 Preparation of GSAO and GSCA

1. GSAO and GSCA can be conjugated using standard succinimidyl ester reaction with primary amines with a tag of choice [6].
2. For conjugation with AlexaFluor 647, a solution of AlexaFluor succinimidyl ester is added to GSAO or control GSCA in 0.1 M bicarbonate buffer, pH 8.3, and incubated for 180 min in the dark in the presence of argon (or nitrogen). The molar ratio of fluorophore to pendant is 1:1. Unreacted fluorophore is quenched by adding glycine to 10 mM (*see* **Note 1**).
3. Store the conjugates in the dark at 4°C to maintain the fluorophore.
4. GSAO ability to label procoagulant platelets is dependent on the trivalent arsenical reaction to dithiol targets in a redox reaction. Thus, GSAO needs to be kept in a reduced state. Each aliquot needs to be stored in oxygen-free conditions under argon (or nitrogen) to prevent oxidation of the trivalent arsenical. Under these conditions, labeled compound should be active for several months. GSCA does not need to be maintained under oxygen-free conditions (*see* **Note 2**).
5. Stocks of GSAO and GSCA can be made and kept as higher concentrations, e.g., GSAO at 500 μM in the labeling solution, and be diluted to the desired concentration as required using Hank's balanced salt solution (HBSS) (1×), pH adjusted to 7.35.

3 Preparation of Human Samples

3.1 Materials

3.1.1 Blood Collection and Processing

1. Evacuated sterile blood collection tubes containing 0.109 M (3.2% w/v) sodium citrate (for whole blood assay).
2. Evacuated sterile blood collection tubes containing acid citrate dextrose (ACD) (85 mM sodium citrate, 135.9 mM dextrose, 41.6 mM citric acid) as anticoagulant (for washed platelet assay).
3. 21-gauge butterfly needles.

3.1.2 Equipment

1. 96-well clear round bottom polypropylene microplate, not treated, individually wrapped, with lid, sterile.
2. Adjustable (single and multichannel) pipettes.
3. 5 mL round bottom polystyrene flow tubes (FACS tubes).
4. 1.5 mL microcentrifuge tubes.
5. 14 mL round bottom falcon tubes (for washed platelet assay).
6. 3 mL transfer pipettes (for washed platelet assay).
7. Flow cytometer (*see* **Note 3**).

3.1.3 Reagents for Procoagulant Platelet Assay

1. Hank's balanced salt solution (HBSS) (10×), no calcium, no magnesium, and no phenol red (Life Technologies, 14185-052). Diluted with water to HBSS (1×) and pH adjusted to 7.35 and stored at 4°C.
2. Wash buffer (HBSS with 0.35% human serum albumin, pH 7.35).
3. Gly-Pro-Arg-Pro amide (GPRP) (Sigma Aldrich, G5779) made up to 40 mM in HBSS pH 7.35, stored at −20°C.
4. $CaCl_2$ solution made up to 100 mM in HBSS pH 7.35, stored at 4°C.
5. Agonists as per experimental protocol, e.g.:
   - Thrombin from bovine plasma, 1000 U dissolved in water for injection at 1000 U/mL, stored at −80°C.
   - Equine tendon collagen (type I), 1 mg/mL.
6. Antibodies and isotype controls:
   - CD41a BV510 (BD Biosciences, Clone HIP8, 563250).
   - CD45 BUV395 (BD Biosciences, Clone HI30, 563792) (CD45 conjugated to FITC can be used as substitute, if the flow cytometer used has no UV laser).
   - CD62P PE (eBioscience, Clone Psel.KO2.3, 12-4714-42).
   - IgG1 K isotype control PE (eBioscience, 12-4714-42).
   - GSAO and GSCA conjugated to AlexaFluor 647 (Life Technologies). GSAO needs to be stored under inert gas (*see* **Note 4**).
7. Fixative PamFix (platelet solutions, PSR-001) stored at 4°C.

3.1.4 Reagents for Washed Platelet Isolation

1. Tyrode's buffer (137 mM NaCl, 2.5 mM KCl, 0.5 mM $MgCl_2$, 12 mM $NaHCO_3$, 0.36 mM Na_2HPO_4, 10 mM HEPES, 5.5 mM glucose, and 0.35% human serum albumin) (*see* **Note 5**).
2. Prostaglandin E1 (PGE_1, Sigma-Aldrich, P5515).

3.2 Methods

3.2.1 Blood Collection

1. Obtain whole blood by standard venepuncture using a 21 G butterfly needle with no or minimal stasis (*see* **Note 6**) into a citrate tube (for whole blood assay) or ACD tube (for washed platelet assay).
2. Discard the first tube collected.
3. Ensure the tube is completely filled and mix by gentle inversion four to five times.
4. The tube(s) should be transported in a vertical position to the laboratory at room temperature within 15 min of collection to minimize artificial platelet activation.

3.2.2 Washed Platelet Preparation

1. Collect blood into 4 ×8.5 mL ACD tubes.
2. Centrifuge the whole blood at 200 × *g* for 15 min without braking.
3. Collect plasma-rich plasma (PRP) and dilute (1:1) with Tyrode's buffer containing PGE_1 at a final concentration of 1 µM (no more than 10 mL of sample in each round bottom 14 mL falcon tubes) (*see* **Note 7**).
4. Mix by gentle inversion.
5. Pellet by centrifugation at 650 × *g* for 15 min without braking and pour off supernatant.
6. Resuspend the pellet gently with Tyrode's buffer containing 1 µM of PGE_1, pool samples together, and pellet again by centrifugation at 650 × *g* for 15 min, without braking.
7. Resuspend to the required count in Tyrode's buffer to have 166.66×10^3 platelets/µL and let rest for 20 min before performing procoagulant platelet assay.

3.2.3 Procoagulant Platelet Assay

1. Thaw (37°C) sufficient HBSS (1x), pH 7.35 and prepare wash buffer [HBSS (1x) with 0.35% human serum albumin, pH 7.35].
2. Allow PamFix to reach room temperature (in dark).
3. Label FACS tubes and dispense 3 mL of wash buffer in to each FACS tube.
4. Prepare the reaction buffer [HBSS (1x), pH 7.35 containing 2.5 mM GPRP, 2.5 mM $CaCl_2$] (*see* Table 1). Omit GPRP for washed platelets.
5. Prepare working dilutions of agonist [*see* **Note 9**) in HBSS (1x), pH 7.35] and keep on ice until use, e.g.:
   - Thrombin (final concentration: variable).
   - Collagen (final concentration: variable).

Table 1
Reaction buffer composition for procoagulant platelet assay in whole blood and washed platelets

Reagent	*n* (wells)	Volume (μL)	
		For whole blood	For washed platelets
HBSS (1×)	1	24.7	27.8
GPRP (40 mM)	1	3.1	-
$CaCl_2$ (100 mM)	1	2.2 (*see* **Note 8**)	
TOTAL	1	30	
(Final reaction volume after addition of 5 μL of agonist and 15 μL whole blood or washed platelets is 50 μL)			

HBSS Hank's balanced salt solution, *GPRP* Gly-Pro-Arg-Pro

Table 2
Stain cocktail for procoagulant platelet assay in human whole blood and washed human platelets

Reagent	PE isotype control		Procoagulant platelet stain mix	
	For whole blood	For washed platelets	For whole blood	For washed platelets
CD41a BV510 (μL)	1.5	1.5	1.5	1.5
CD45 BUV395 (μL)	2.5	-	2.5	-
GSCA AF647 (50 μM) (μL)	1	1	-	-
GSAO AF647 (50 μM) (μL)	-	-	1	1
PE isotype control (μL)	0.625	0.625	-	-
CD62P-PE (μL)	-	-	0.625	0.625
HBSS (1×) (μL)	74.375	76.875	74.375	76.875
TOTAL	80	80	80	80

Reagent volumes for human assay: BV510 brilliant violet 510, *BUV395* brilliant ultraviolet 395, *FITC* fluorescein isothiocyanate, *PE* phycoerythrin, *GSCA* 4-(*N*-[(*S*-glutathionyl) acetyl]amino)benzoic acid, *GSAO* 4-[*N*-(*S*-glutathionylacetyl)amino] phenylarsonous acid, *AF647* AlexaFluor 647, *HBSS* Hank's balanced salt solution

6. Prepare the stain mix according to Table 2 (*see* **Note 10**).
7. Prepare 96-well reaction plate (*see* **Note 11**):
    1. Dispense 30 μL of reaction buffer to each reaction well (column 2) just before blood collection or during platelet isolation. Allow to reach room temperature (*see* **Note 12**).
    2. Add agonist or vehicle control (HBSS) in 5 μL to each reaction well (column 2).

3. Add 170 μL of HBSS to wells adjacent to reaction wells (column 3) (*see* **Note 13**).
4. Dispense 30 to 50 μL of citrated whole blood or washed platelets gently without introducing bubbles into column 1 wells (*see* **Note 14**).

8. Transfer 15 μL of whole blood or washed platelets (column 1) to each reaction well (column 2) using multichannel pipette and gently mix four times (if using electronic pipettor set, mix volume to 30 μL, moderate speed).
9. Incubate for 10 min at room temperature.
10. Dispense stain mix to stain wells (column 4) (80 μL/well). Place in dark.
11. At the end of the 10 min incubation period, transfer HBSS (150 μL) (from column 3) to each reaction well (column 2) and mix gently (*see* **Note 15**). Transfer 20 μL of diluted stimulated whole blood to the wells containing the stain mix (column 4). Transfer another 20 μL from a dual agonist stimulated well to the well containing control stain. Mix gently.
12. Incubate for 15 min in dark at room temperature.
13. During staining, dispense 205 μL of PamFix to unoccupied wells (column 5).
14. Add 200 μL of PamFix to all stain and control wells and mix gently.
15. Incubate for 5 min in dark at room temperature.
16. After gentle mix, transfer the fixed and stained cell suspension into the FACS tubes containing 3 mL wash buffer.
17. Centrifuge, 1500 × *g* for 8 min, medium acceleration and deceleration.
18. Decant the supernatant (up-end tube in single action to remove supernatant, pressing rim of tube against kimiwipe to remove bulk of residual supernatant before returning to upright position).
19. Resuspend the pellet in the residual fluid by giving the FACS tube a few gentle flicks.
20. Add 0.5 to 1.2 mL of wash buffer to the FACS tubes (*see* **Note 16**).
21. Leave in dark at room temperature for at least 1 h before analysis by flow cytometry.

4 Preparation of Murine (Mouse) Samples (*see* Note 17)

4.1 Materials

4.1.1 Blood Collection and Processing

1. Sodium citrate (3.2%).
2. Acid citrate dextrose (ACD) (85 mM sodium citrate, 135.9 mM dextrose, 41.6 mM citric acid).
3. 23-gauge needles.
4. Ketamine Ceva injection (100 mg/mL).
5. Ilium Xylazil-100 (100 mg/mL).

4.1.2 Equipment

1. 96-well clear round bottom polypropylene microplate, not treated, individually wrapped, with lid, sterile.
2. Adjustable (single and multichannel) pipettes.
3. 5 mL round bottom polystyrene flow tubes (FACS tubes).
4. 1.5 mL microcentrifuge tubes.
5. 14 mL round bottom falcon tubes (for washed platelet assay).
6. 3 mL transfer pipettes (for washed platelet assay).
7. 1.2 mL microtiter tube, bulk, nonsterile (ThermoFisher scientific, 3492).
8. Flow cytometer (*see* **Note 3**).

4.1.3 Reagents for Procoagulant Platelet Assay

1. Hank's balanced salt solution (HBSS) (10x), no calcium, no magnesium, and no phenol red (Life Technologies, 14185-052). Diluted with water to HBSS (1x) and pH adjusted to 7.35 and stored at 4°C.
2. Wash buffer (HBSS with 0.35% human serum albumin, pH 7.35).
3. Gly-Pro-Arg-Pro amide (GPRP) (Sigma Aldrich, G5779) made up to 40 mM in HBSS pH 7.35, stored at −20°C.
4. $CaCl_2$ solution made up to 100 mM in HBSS pH 7.35, stored at 4°C.
5. Agonists as per experimental protocol, e.g.:
   - Thrombin (final concentration: variable)
   - Collagen related peptide (final concentration: variable)
6. Antibodies and isotype controls:
   - CD45 BV421 (BD Biosciences, Clone 30-F11, 563890) (*see* **Note 18**).
   - CD42b Dylight 488 (Emfret Analytics, X488).
   - CD62P PE (eBioscience, Clone Psel.KO2.3, 12-4714-42).
   - IgGl K isotype control PE (eBioscience, 12-4714-42).

- GSAO and GSCA conjugated to AlexaFluor 647 (Life Technologies). GSAO needs to be stored under inert gas (*see* **Note 4**).

7. Fixative PamFix (platelet solutions, PSR-001) stored at 4°C.

4.1.4 Reagents for Washed Platelet Isolation

1. Tyrode's buffer (137 mM NaCl, 2.5 mM KCl, 0.5 mM $MgCl_2$, 12 mM $NaHCO_3$, 0.36 mM Na_2HPO_4, 10 mM HEPES, 5.5 mM glucose, and 0.35% human serum albumin) (*see* **Note 5**).
2. Prostaglandin E1 (PGE_1, Sigma-Aldrich, P5515).

4.2 Methods

4.2.1 Blood Collection

1. Anesthetize mice by injection of a mixture of xylazine and ketamine. In our laboratory, C57BL/6J mice are anesthetized with 100 mg of ketamine and 20 mg of xylazine per kilo of body weight.
2. Collect blood by cardiac puncture after a thoracotomy using a 23 G needle into a syringe preloaded with citrate anticoagulant with a ratio of 1 part citrate to 9 parts blood (e.g., 0.1 mL citrate plus 0.9 mL blood = 1.0 mL total sample) (*see* **Note 19**).
3. Remove the needle from the syringe, gently transfer blood in Eppendorf, and mix by gentle inversion four to five times. Use blood within 15 min of collection to minimize artificial platelet activation.

4.2.2 Washed Platelet Preparation

1. Collect blood as above, but transfer blood in a 1.2 mL microtiter tube (*see* **Note 20**). Blood can be collected into a syringe preloaded with ACD with a ratio of 1 part ACD to 5.6 parts blood (*see* **Note 21**).
2. Dilute blood with Tyrode's buffer containing PGE1 at a final concentration of 1 μM before a centrifugation at 240 ×*g* for 2 min without braking.
3. Carefully collect PRP and dilute the leftover blood with Tyrode's buffer containing PGE1 at a final concentration of 1 μM before a centrifugation at 240 ×*g* for 2 min without braking. Repeat this step three to four times to collect the maximum of PRP.
4. Pool PRP and centrifuge it at 1960 ×*g* for 1 min.
5. Wash platelet pellet with Tyrode's buffer containing 1 μM of PGE1, perform platelet count and centrifuge at 1960 ×*g* for 1 min.
6. Resuspend to the required count in Tyrode's buffer to have 166.66×10^3 platelets/μL and let rest for 20 min before performing procoagulant platelet assay.

4.2.3 Procoagulant Platelet Assay (see Note 22)

1. Thaw (37°C) sufficient HBSS (1x), pH 7.35, and prepare wash buffer [HBSS (1x) with 0.35% human serum albumin, pH 7.35].
2. Allow PamFix to reach room temperature (in dark).
3. Label FACS tubes and dispense 3 mL of wash buffer in to each FACS tube.
4. Prepare the reaction buffer [HBSS (1x), pH 7.35 containing 2.5 mM GPRP, 2.5 mM $CaCl_2$] (*see* Table 1). Omit GPRP for washed platelets.
5. Prepare working dilutions of agonist (*see* **Note 11**) in HBSS (1x), pH 7.35, and keep on ice until use, e.g.:
   - Thrombin (final concentration: variable).
   - Collagen related peptide (final concentration: variable).
6. Prepare the stain mix according Table 3 for murine (mouse) samples (*see* **Note 10**).
7. Prepare 96-well reaction plate (*see* **Note 11**):
   a) Dispense 30 μL of reaction buffer to each reaction well (column 2) just before blood collection or during platelet isolation. Allow to reach room temperature (*see* **Note 12**).
   b) Add agonist or vehicle control (HBSS) in 5 μL to each reaction well (column 2).

Table 3
Stain cocktail for procoagulant platelet assay in murine (mouse) whole blood and washed murine (mouse) platelets

Reagent	PE isotype control		Procoagulant platelet stain mix	
	For whole blood	For washed platelets	For whole blood	For washed platelets
CD42b DyLight 488 (μL)	0.5	0.5	0.5	0.5
CD45 BV421 (μL)	2	-	2	-
GSCA AF647 (50 μM) (μL)	1	1	-	-
GSAO AF647 (50 μM) (μL)	-	-	1	1
PE isotype control (μL)	0.625	0.625	-	-
CD62P-PE (μL)	-	-	0.625	0.625
HBSS (1×) (μL)	75.875	77.875	75.875	77.875
TOTAL	80	80	80	80

Reagent volumes for murine (mouse) assay. BV421 brilliant violet 421, *FITC* fluorescein isothiocyanate, *PE* phycoerythrin, *GSCA* 4-(*N*-[(*S*-glutathionyl) acetyl]amino)benzoic acid, *GSAO* 4-[*N*-(*S*-glutathionylacetyl)amino] phenylarsonous acid, *AF647* AlexaFluor 647, *HBSS* Hank's balanced salt solution

c) Add 170 µL of HBSS to wells adjacent to reaction wells (column 3) (*see* **Note 13**).

d) Dispense 30 to 50 µL of citrated whole blood or washed platelets gently without introducing bubbles into column 1 wells (*see* **Note 14**).

8. Transfer 15 µL of whole blood or washed platelets (column 1) to each reaction well (column 2) using multichannel pipette and gently mix four times (if using electronic pipettor set, mix volume to 30 µL, moderate speed).
9. Incubate for 10 min at room temperature.
10. Dispense stain mix to stain wells (column 4) (80 µL/well). Place in dark.
11. At the end of the 10 min incubation period, transfer HBSS (150 µL) (from column 3) to each reaction well (column 2) and mix gently (*see* **Note 15**). Transfer 20 µL of diluted stimulated whole blood to the wells containing the stain mix (column 4). Transfer another 20 µL from a dual agonist stimulated well to the well containing control stain. Mix gently.
12. Incubate for 15 min in dark at room temperature.
13. During staining, dispense 205 µL of PamFix to unoccupied wells (column 5).
14. Add 200 µL of PamFix to all stain and control wells and mix gently.
15. Incubate for 5 min in dark at room temperature.
16. After gentle mix, transfer the fixed and stained cell suspension into the FACS tubes containing 3 mL wash buffer.
17. Centrifuge, 1500 ×*g* for 8 min, medium acceleration and deceleration.
18. Decant the supernatant (up-end tube in single action to remove supernatant, pressing rim of tube against kimiwipe to remove bulk of residual supernatant before returning to upright position).
19. Resuspend the pellet in the residual fluid by giving the FACS tube a few gentle flicks.
20. Add 0.5 to 1.2 mL of wash buffer to the FACS tubes (*see* **Note 16**).
21. Leave in dark at room temperature for at least 1 h before analysis by flow cytometry

5 Flow Cytometric Acquisition and Analysis

1. Start acquisition at least 1 h after fixation and complete not longer than 4 h post-fixation.
2. Set threshold to the lowest available value on forward scatter (FSC).
3. Define initial platelet gate based on side scatter (SSC) vs CD41a (or CD42b for murine samples) plot (*see* **Note 23**). Further refine the platelet gate, defined as CD41a+ (or CD42b+ for murine samples)/CD45− (see Fig. 4 (*see* **Note 24**).
   - Biexponential scaling is helpful for visualizing all events in a heterogeneous complex cellular matrix like whole blood (*see* **Note 25**).
4. Collect at least 10,000 platelets events for analysis.
5. Aim for event rate of around 2000 total events per second to minimize coincident events. This rate will depend on a particular flow cytometer.
6. Analyze the events from the platelet gate on a CD62P PE vs GSAO AF647 fluorescence (*see* **Note 26**).
   - Use the isotype PE and GSCA AF647 control sample to set the threshold for CD62P positive and GSAO positive events, respectively (*see* **Note 27**). The thresholds used to define positive population are derived from the GSCA/isotype PE control. Occasionally, there can be some "smearing" with GSCA. However, even in these instances, predominant homogeneous population can be easily seen and should be used to set the threshold rather than the "tail" (*see* Fig. 4).
7. The flow cytometric acquisition can also be performed on a 96-well plate (*see* **Note 28**). The ability to perform acquisition on a 96-well plate is dependent on the flow cytometer. To set up the plate on LSRFortessa™ (BD Biosciences), transfer 250 μL

Table 4
Flow cytometric settings for acquisition on 96-well plate

Settings	
Sample flow rate (μL/s)	0.5
Sample volume (μL)	200
Mixing volume (μL)	100
Mixing speed (μL/s)	200
Number of mixes	3
Wash volume (μL)	300

of sample from each FACS tubes onto the 96-well plate. Run the plate using standard mode with settings as shown in Table 4. Collect 10,000 or as many possible platelets events from each well for analysis. If platelet events are not sufficient from a single well, data from two or more wells to can be combined.

6 Notes

1. We confirm the efficiency of labeling of dying cells for each fresh GSAO conjugate in Fas ligand treated Jurkat A3 cells correlating against an independent cell death marker, such as Sytox Green or Propidium Iodide.
2. If reduction of signal over time is seen, often this will be due to oxidation of the GSAO compound. [Note: redox status of trivalent arsenical can be tested using a DTNB (5,5'-dithio-bis(2-nitrobenzoic acid)-based assay. DTNB reacts with free thiols at neutral pH, producing a strong yellow 5-thio-2-nitrobenzoate dianion.]
3. The minimum and optimum flow cytometer configuration (number of lasers and detection photomultiplier tubes) will depend on the complexity of each individual panel and experiment. Older instruments with a single excitation laser (488 nm) are usually adequate for a two to three colors experiment, but the setup can be more complicated due to compensation issues. Collecting uncompensated data and performing compensation using a software package can simplify this process. Newer instruments equipped with at least three lasers (violet, 405 nm; blue, 488 nm; and red, 631 nm) provide increased flexibility with respect to the number of parameters measured often with no or minimal spill over.
4. GSCA is a GSAO control compound that has an inert carboxylic acid group instead of the chemically reactive As(III) in GSAO. GSCA has similar biodistribution as GSAO but does not react with intracellular proteins, and thus it washes out of cells. GSAO needs to be stored under inert gas to prevent oxidization.
5. Correct pH of the buffer to 7.4. Glucose and albumin to be added to the buffer just before use.
6. Collect blood preferably from a forearm vein and remove tourniquet after blood starts to flow will usually fulfil the no or minimal stasis requirement.
7. Platelet pellet resuspension is easier in a round bottom tubes.
8. Free calcium in citrated blood can be restored to the physiological level by supplementing blood with exogenous $CaCl_2$

(*see* Fig. 2). A quick calculation will give you a final concentration of 4.4 mM in the reaction mix with whole blood. However, due to chelation of calcium in citrated whole blood the final concentration will be 2.5 mM.

9. Depending on the aim of the experiment, often a dose response curve with increasing concentrations of agonists can be used.
10. CD45 is omitted in the stain mix preparation for the washed platelet assay.
11. An experiment template for the 96-well plate (*see* Fig. 3) indicating location of reaction wells, wells containing HBSS, stain cocktail, and PamFix is very useful. (e.g., column 1 for whole blood or washed platelets, column 2 for reaction buffer, column 3 for HBSS, column 4 for stain mix, and column 5 for fixative). With this, setting an extra step, such as inhibition step before agonist stimulation, can be easily performed by adding a column containing the inhibitor.
12. Allow all reagents to come to room temperature prior to reaction.
13. This will be used to transfer HBSS to the reaction wells to dilute the agonist and Ca^{++} in the reaction wells after 10 min incubation period with whole blood or washed platelets.
14. Do not attempt to expel the contents of the pipette tip beyond the first stop to avoid introducing bubbles. Dispense blood within 15 min of collection or washed platelets directly after the resting step.

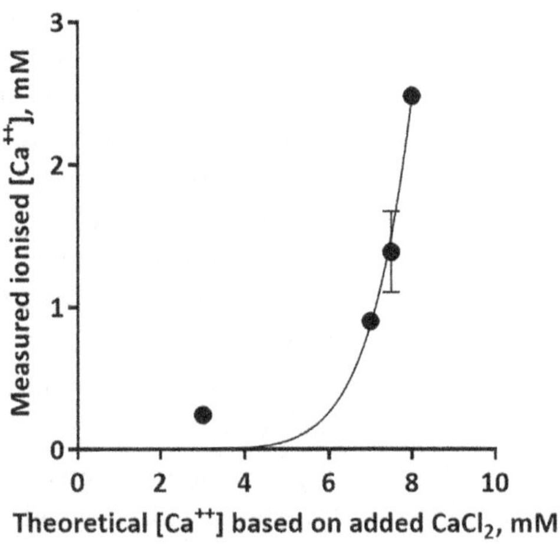

Fig. 2 Free $[Ca^{++}]$ in citrated blood can be restored to the physiological level by supplementing blood with exogenous $CaCl_2$

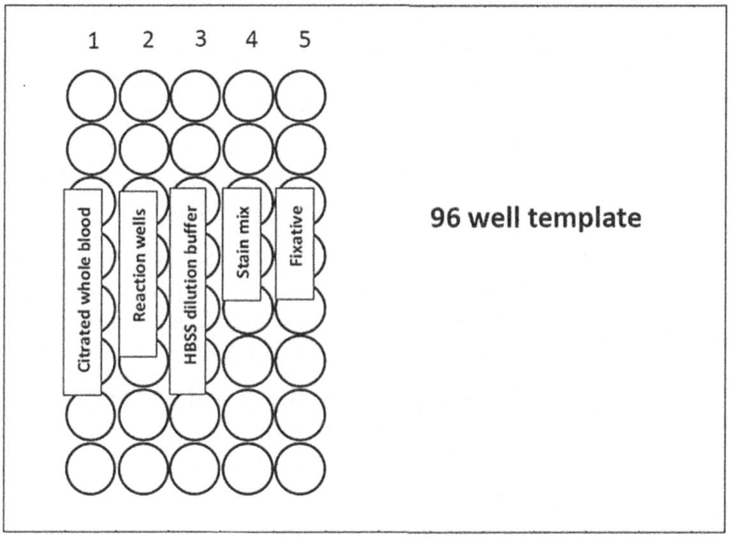

Fig. 3 Layout of the 96-well plate for procoagulant platelet assay preparation. The plate is set up as demonstrated. The reaction wells should be prepared just prior to blood collection. All reactions, staining, and fixing are performed at room temperature

15. Cells tend to settle a bit by the end of 10 min, so direct the pipette tip to several different directions to produce even suspension.
16. The resuspension volume depends of the sample. To minimize coincident events during the flow cytometric acquisition, the event rate should be around 2000 total events per second. This rate can be adjusted with the acquisition speed but depends of the sample concentration as well. Some pretesting might be performed to set up the resuspension volume. However, the sample is more stable in a higher resuspension volume. For the flow cytometric acquisition on a 96-well plate, a minimum of 1 mL is recommended to aim the right acquisition rate.
17. The method discussed is for mouse samples, but rat blood can also be used as a substrate with species specific antibodies.
18. Only the leucocyte and platelet markers are different since this CD62P clone antibody reacts with human and murine species and GSAO is a species independent intracellular marker.
19. The ratio of citrate to blood is crucial, therefore, it is preferable to a priori choose to draw a set amount of blood from the mouse rather than taking the maximum possible from the individual mouse.
20. This type of tube is narrower than an Eppendorf which facilitates PRP isolation.

21. Washed platelets are prepared from blood collected on ACD instead of citrate to minimize platelet activation. However, if procoagulant assay were to be performed on whole blood and isolated platelets from the same mouse, consider to perform only one blood collection by cardiac puncture on citrate and to split the blood to have a whole blood preparation and a blood fraction to isolate platelets.
22. The procoagulant platelet assay for mouse is essential similar to that for human apart from the use of collagen related peptide instead of collagen in mouse and the differences in stain mix preparation (*see* Table 3).
23. The initial platelet gating based on SSC vs CD41a (or CD42b for murine samples) plot is not mandatory. Platelet gate can be defined directly from CD45 vs CD41a (or CD42b for murine samples) plot.
24. The platelet population shifts to the right with regards CD41a (or CD42b for murine samples) fluorescence with stimulation (brighter). With stimulation, sometimes a "tail" (CD41a+ or CD42b+/CD45+) is seen in the platelet gate that demonstrates nonspecific binding (*see* Fig. 4), and this "tail" should not be included under the platelet gate.
25. SSC and FSC voltages are set so that both platelet and leukocyte events can be visualized at the same time.
26. With a 561 nm (50 mW) laser, there is minimal spill over between PE and AF647 (640 nm, 50 mW). However, there is slight spill over from BV510 into PE. Compensation controls can be set up with compensation beads, excepting the GSAO conjugate compensation control which must be set up with dual agonist treated blood gated on platelets.
27. Some flow cytometry analysis softwares have both straight and curvilinear quadrant functions. When applicable, curvilinear quadrants often provide a clearer separation of the population of interest. Commonly with stimulated samples, there can be some "smearing" with the highly positive CD62P population although the predominant homogenous CD62P+/GSAO− population can still be easily identified. Curvilinear quadrants hence provide a clearer separation for the procoagulant platelet population (defined as CD62P+/GSAO+) from CD62P+/GSAO− population compared to straight quadrants (*see* Fig. 4).
28. Validation studies have been performed to compared the flow cytometric acquisition from tube and plate. The results are equivalent.

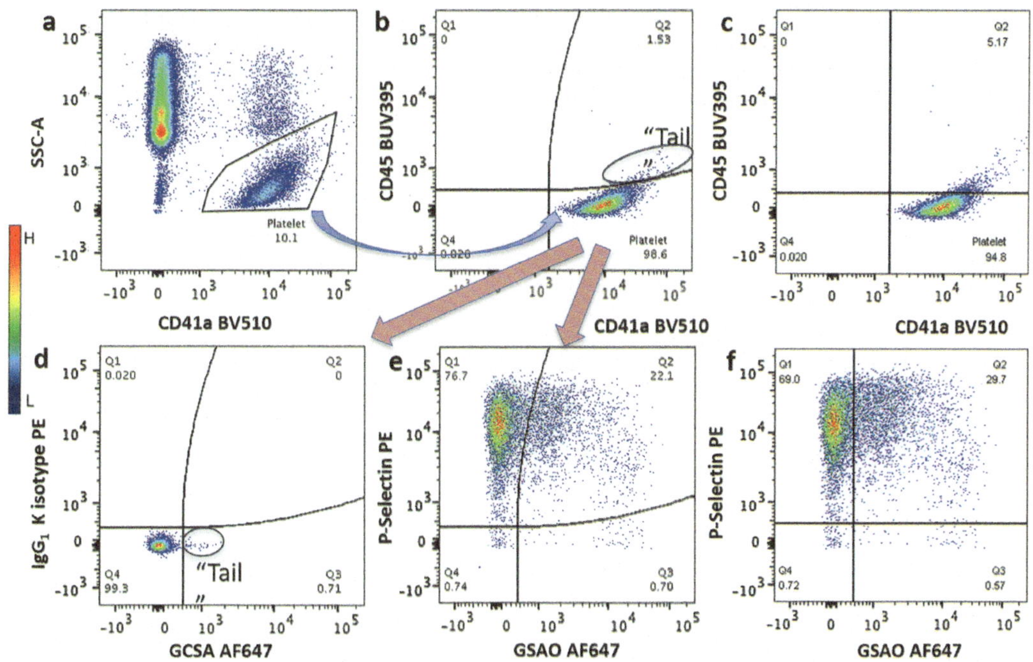

Fig. 4 Example of flow cytometry of procoagulant platelets. Whole blood samples were stained with CD45 BUV395, CD41a BV510, CD62P PE, and GSAO AF647 or controls (isotype PE and GCSA AF647). (**a**) Platelets are first identified using polygonal gate on SSC-A vs CD41a plot as low SSC and CD41a+ and subsequent analyses are based on this gate. (**b**) The "platelet gate" is further refined in the lower right quadrant as CD41a+/CD45− using curved quadrants and subsequent analyses are based on this gate. The "tail" population (CD41a+/CD45+) is excluded. (**c**) Applying straight quadrants to the same platelet population and thresholds for both the axes shown in Fig. 4b yields slightly different results and the use of curvilinear quadrants provides a better separation of the platelet population. (**d**) Thresholds for CD62P and GSAO are set by isotype PE and GSCA AF 647 controls. Occasionally, there can be some "smearing" with GSCA ("tail") but the predominantly homogenous population in left lower quadrant can be easily identified and should be used to determine the threshold rather than the "tail." (**e**) Procoagulant platelets are identified in the right upper quadrants as CD62P+/GSAO+ in a stimulated sample (e.g., dual thrombin and collagen). Curved quadrants are used as there can be some "smearing" with the highly positive CD62P population from the left upper quadrant into the right upper quadrant and curved quadrants could distinguish these populations better. (**f**) Applying straight quadrants to the same platelet population and thresholds for both the axes shown in Fig. 4e yields slightly different procoagulant platelet percentages and the use of curvilinear quadrants provides a clearer separation between CD62P+/GSAO+ and CD62P+/GSAO− populations

References

1. Alberio L, Safa O, Clemetson KJ et al (2000) Surface expression and functional characterization of alpha-granule factor V in human platelets: effects of ionophore A23187, thrombin, collagen, and convulxin. Blood 95:1694–1702
2. Agbani EO, Poole AW (2017) Procoagulant platelets: generation, function, and therapeutic targeting in thrombosis. Blood 130:2171–2179
3. Park D, Shon IH, Hua M et al (2014) Visualization of proteins and cells using dithiol-reactive metal complexes. In: Gasser G (Ed.) Inorganic chemical biology. Wiley, Chichester, UK, 215–232
4. Hua VM, Abeynaike L, Glaros E et al (2015) Necrotic platelets provide a procoagulant surface during thrombosis. Blood 126:2852–2862

5. Pasalic L, Wing-Lun E, Lau JK et al (2018) Novel assay demonstrates that coronary artery disease patients have heightened procoagulant platelet response. J Thromb Haemost 16:1198–1210
6. Park D, Xie B-W, Van Beek ER et al (2013) Optical imaging of treatment-related tumor cell death using a heat shock protein-90 alkylator. Mol Pharm 10:3882–3891

Check for updates

Chapter 21

Preparation of a Dithiol-Reactive Probe for PET Imaging of Cell Death

Ivan Ho Shon, Michael P. Gotsbacher, Jennifer Guille, Divesh Kumar, Rachel Codd and Philip J. Hogg

Abstract

Conjugates of 4-(*N*-(*S*-glutathionylacetyl)amino)phenylarsonous acid (GSAO) with optical or radionuclide probes are able to image cell death in vivo. GSAO conjugates are retained in the cytosol of dying and dead cells via the formation of covalent bonds between the As(III) ion and the thiol groups of proximal cysteine residues. Here, we describe the method for preparing a NODAGA–GSAO conjugate (hereafter referred to as cell death indicator, CDI-DX001) and its radiolabeling with gallium-68 (^{68}Ga-CDI-DX001) for positron emission tomography (PET) imaging of cell death.

Keywords Gallium-68, GSAO, Bifunctional chelator, Positron emission tomography

1 Introduction

4-(*N*-(*S*-Glutathionylacetyl)amino)phenylarsonous acid (GSAO) is a trivalent As(III) peptide, which has been found to activate the mitochondrial permeability transition pore. GSAO is toxic to proliferating cells and inhibits angiogenesis in vivo [1], but is nontoxic to quiescent endothelial cells in vitro. Conjugation of the γ-glutamyl residue of GSAO with fluorophores or radionuclides results in loss of its anti-angiogenic effect and gain in the ability to identify dying cells. When the plasma membrane integrity has been compromised the GSAO conjugate enters and binds to intracellular proteins, predominantly 90 kDa heat shock proteins (Hsp90) [2]. The As(III) motif of GSAO cross-links the unpaired thiols of Cys597 and Cys598 of Hsp90 forming a stable cyclic dithioarsinite. GSAO conjugates with fluorophores have been shown to detect cyclophosphamide-induced tumor cell death in murine orthotopic human mammary tumors, with a significant increase in the GSAO fluorescence signal in vivo or ex vivo [3]. In a murine traumatic brain injury model, fluorophore-labeled GSAO was able to image cell death in vivo and

Philip J. Hogg (ed.), *Functional Disulphide Bonds*, Methods in Molecular Biology, vol. 3016,
https://doi.org/10.1007/978-1-0716-5158-2_21,

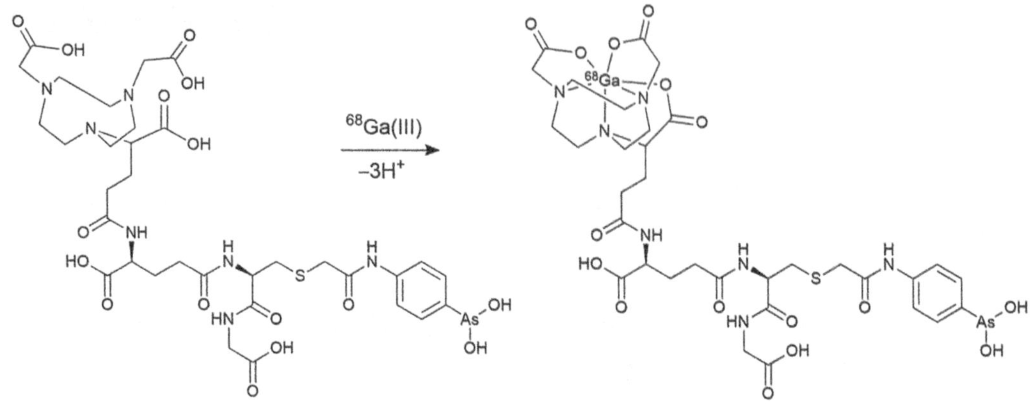

CDI-DX001 ^{68}Ga-CDI-DX001

Fig. 1 Structure of CDI-DX001 and ^{68}Ga-CDI-DX001

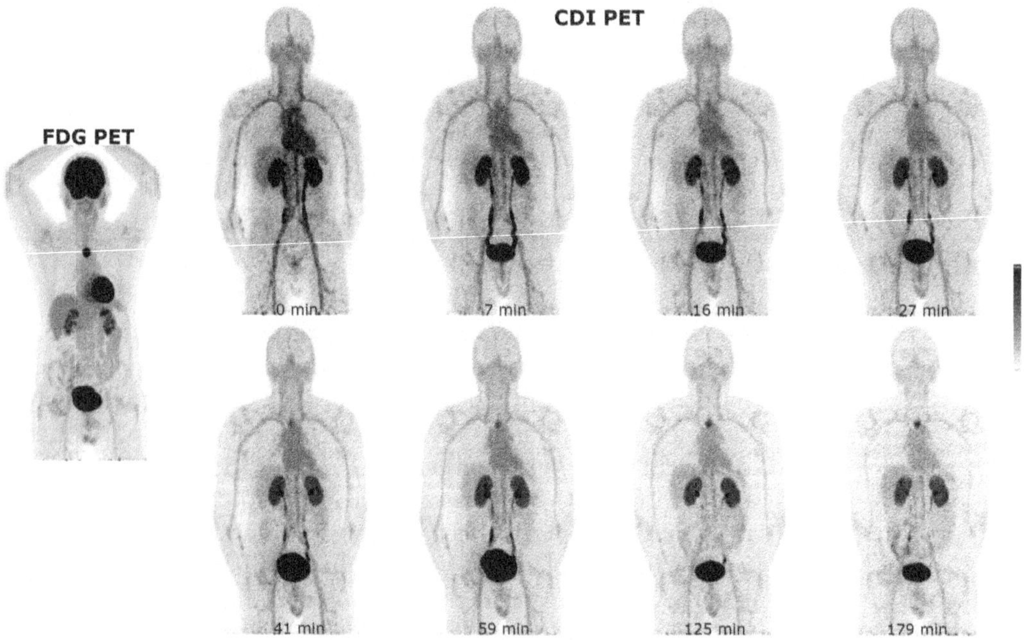

Fig. 2 Maximum intensity projection images of FDG PET and eight sequential ^{68}Ga-CDI-DX001 PET scans (performed from 0 to 179 min following ^{68}Ga-CDI-DX001 administration) in a patient with esophageal carcinoma (the tumor is arrowed on the FDG PET CT and the ^{68}Ga-CDI-DX001 PET CT performed at 59 min). All images are scaled from SUV 0 to 7. Reproduced under creative commons license CC BY [6]

ex vivo [4]. GSAO was initially radiolabeled with indium-111 (^{111}In) using diethylenetriaminepentaacetic acid (DTPA) as a bifunctional chelator and shown to detect dead and dying cells in subcutaneous lung and colorectal carcinoma xenografts in a murine model

[2]. Subsequently, GSAO was radiolabeled with gallium-67 using 1,4,7,10-tetraazacyclododecane-1,4,7,10-tetraacetic acid (DOTA) and, when compared with ^{111}In-DTPA-GSAO, was found to have lower uptake in normal organs and higher tumor uptake [5]. For positron emission tomography (PET) imaging applications, GSAO was conjugated to 2,2'-(7-(1-carboxy-4-((2,5-dioxopyrrolidin-1-yl)oxy)-4-oxobutyl)-1,4,7-triazonane-1,4-diyl)diacetic acid (NODAGA), hereafter referred to as cell death indicator (CDI-DX001) to enable radiolabeling with ^{68}Ga (^{68}Ga-CDI-DX001) (Fig. 1). A first in human study of ^{68}Ga-CDI-DX001 demonstrated that it is safe, well tolerated has low radiation dosimetry, excellent biodistribution and imaging characteristics (Fig. 2) and uptake correlates with histological cell death [6]. Proof of concept studies to assess baseline tumor cell death and changes in cell death in response to treatment are underway.

2 Materials

Prepare all solutions using ultrapure water (prepared by purifying deionized water, to attain a sensitivity of 18.2 MΩ-cm at 25°C) unless otherwise specified. For development and preclinical work, analytical grade reagents may be used. For human studies, use pharmacopeia or equivalent grade reagents. Reagents should be selected with the lowest possible metal content. Prepare and store all reagents at room temperature (unless indicated otherwise). This method should only be performed by skilled personnel with appropriate training and licensed to handle radioisotopes. There is the potential for significant radiation exposure to the operator(s) and appropriate shielding and radiation safety advice and procedures are essential. All relevant radiation protection guidelines, regulations and legislative requirements, and waste disposal regulations should be followed throughout this protocol with due diligence.

2.1 CDI-DX001

1. GSAO: GSAO was prepared as previously described [2].
2. NODAGA–NHS was purchased from CheMatech (Dijon, France) as supplied containing approximately one equivalent each of trifluoroacetic acid (TFA) and hexafluorophosphoric acid.

2.2 CDI-DX001 Purification

1. High performance liquid chromatography (HPLC) was carried out on a Shimadzu LC-20 series LC system with two LC-20AP pumps, a SIL-10AP autosampler, an SPD-20A UV/VIS detector, and a Shimadzu ShimPack GIS-C18 column (150 × 10.0 mm i.d., 5 μm, 4 mL min^{-1}) (system A).
2. Shimadzu LabSolutions software (Ver. 5.73) was used for data acquisition and processing.

2.3 CDI-DX001 Purity

1. Liquid chromatography-mass spectrometry (LC–MS) was conducted using an Agilent system (Santa Clara, CA, USA) consisting of a 1260 series quaternary pump with an inbuilt degasser, 1200 series autosampler, thermostated column compartment, diode array detector, and fraction collector, a 6120 series single quadrupole mass spectrometer and an Agilent Zorbax Eclipse XDB-C18 column (150 × 4.6 mm i.d., 5 μm) at 30°C (system B).
2. The drying gas flow, temperature, and nebulizer were set to $12\,L\,min^{-1}$, 350°C, and 35 psi, respectively.
3. Agilent OpenLAB chromatography data system (CDS) ChemStation Edition (C.01.05) was used for data acquisition and processing.
4. Electrospray ionization (ESI) was used to analyze aliquots (5 μL) in positive ion mode with a 3500 V capillary voltage.
5. Nuclear magnetic resonance (NMR) spectroscopy (^{1}H and ^{13}C) spectra were recorded in 5-mm Pyrex tubes (Wilmad, NJ, USA) using a Varian 400-MR NMR spectrometer (Lexington, MA, USA) at a frequency of 399.73 MHz (^{1}H) or 100.51 MHz (^{13}C) at 24°C operated with VnmrJ 3.1 software (Agilent Technologies, Santa Clara, CA, USA). The spectral data are reported in ppm (δ) and referenced to residual solvent (deuterated dimethyl sulfoxide [DMSO-d_6] 2.50/39.52 ppm).

2.4 ^{68}Ga Radiolabeling of CDI-DX001

1. Sodium acetate buffer: 1.5 M $CH_3COONa \bullet 3H_2O$, pH 4.5. 10.21 g $CH_3COONa \bullet 3H_2O$ was dissolved in water (Water Ultrapur, Merck). The pH was adjusted to pH 4.5 with glacial acetic acid and water was added to a total volume of 50 mL.
2. Ascorbic acid solution: 0.25 M ascorbic acid. 44 mg ascorbic acid was dissolved in 1 mL water (Water Ultrapur, Merck) (*see* **Note 1**).
3. Phosphate buffer: 384 mM sodium phosphate, pH 7.9. Anhydrous NaH_2PO_4 (235 mg) and anhydrous Na_2HPO_4 (2450 mg) were each weighed into two separate beakers (20 mL) and each solid was dissolved in ~20 mL water. The solutions were quantitatively transferred into one volumetric flask (50 mL) and the solution was made to 50 mL with water.
4. HCl: 5.5 M HCl. 15.66 mL of 5.5 M HCl (32% w/w) was added to 9.33 mL water.
5. NaCl/HCl elution solution: 5 M NaCl/5.5 M HCl. 200 μL of 5.5 M HCl was added to 8 mL 5 M NaCl.
6. Barbed female luer thread attachment (Value Plastics).
7. Cation exchange cartridge (Agilent Bond Elut SCX, 100 mg, 1 mL) (*see* **Note 2**).
8. $^{68}Ge/^{68}Ga$ generator (*see* **Note 3**).

9. 0.22 μm filter (Millipore GV).
10. 10 mL evacuated glass reaction vials.

2.5 ^{68}Ga-CDI-DX001 Quality Assurance

1. HPLC was performed on Shimadzu LC-20 series LC system with two LC-20AP pumps, a SPD-20A UV/VIS detector, LabLogic Dual Scan-RAM radiometric detector, and LabLogic Laura software (version 4.2.11.129 SP2) with a Raptor ARC-18 column (150 × 4.6 mm i.d., 5 μm, 0.6 mL min^{-1}) (system C).
2. DMP/DMSO. 2,3-Dimercapto-1-propanol (DMP) in dimethyl sulfoxide (DMSO)—dissolve 5 μL DMP in 495 μL DMSO.
3. 0.9% w/v NaCl in water.
4. ITLC paper (PALL Life Sciences).
5. Dose calibrator (Capintec CRC-55tR).

3 Methods

3.1 Synthesis of CDI-DX001

1. The synthesis of CDI-DX001 was adapted from a previously published method for the synthesis of GSAO-Cy5.5 [2].
2. Sodium bicarbonate and ultrapure water were purged with nitrogen for 30 min prior to use.
3. The reaction setup and purification were performed under an inert atmosphere of nitrogen.
4. GSAO (20.0 mg, 36.5 μmol) was dissolved in 0.1 N sodium bicarbonate (7.4 mL) at 4°C and stirred for 10 min.
5. NODAGA–NHS (34.5 mg, 47.0 μmol) was dissolved in anhydrous dimethylformamide (DMF) (1 mL) and added to the reaction mixture dropwise over 1 h.
6. The reaction mixture was stirred for 4 h, acidified by the addition of 1 M hydrochloric acid (1 mL), shock-frozen in liquid nitrogen and freeze-dried.

3.2 CDI-DX001 Purification

1. The resulting residue was redissolved in de-aerated water (4 mL), filtered (0.45 μm), and purified by RP-HPLC system A. A gradient of 2%–20% mobile phase B (0.2% TFA in acetonitrile) in mobile phase A (0.2% TFA in ultrapure water) was applied from 0 to 25 min.
2. CDI-DX001 eluted at 20.6 min. The sample was collected by hand and each fraction was instantly purged with nitrogen.
3. The pooled fractions were frozen at −20°C and freeze-dried to give 7.3 mg of white powder (21.6% yield).
4. CDI-DX001 was dispensed in aliquots of 54 μg per 100 μL water and stored at −20°C.

3.3 CDI-DX001 Purity

1. The purity of the compound (>95%) was verified by injecting a solution of CDI-DX001 (5 μL; approx. 17 mM, in water) onto LC–MS (system B) at 2–2–50% mobile phase B [0.1% formic acid (FA) in acetonitrile] in mobile phase A (0.1% FA in mass spectrometry grade water) over 0–5–45 min.
2. CDI-DX001 eluted at 19.4 min.
3. Absorbance was measured at 210 nm and 254 nm and the respective area under the curve (AUC) was used to determine compound purity as percent of total AUC compared with background.

3.4 Radiolabeling of CDI-DX001 with ^{68}Ga

1. The barrel of the BondElute SCX column was cut, so that when inserted, the barbed female luer thread rested just above the column media to create a cartridge (hereafter referred to as the SCX cartridge) (Fig. 3) (*see* **Note 4**).
2. The SCX cartridge was primed with 1 mL 5.5 M HCl and then flushed with 10 mL of water.
3. The SCX cartridge was purged with air.
4. One vial of 54 μg CDI-DX001 was thawed and mixed with 100 μL of ascorbic acid solution (*see* **Note 5**), 250 μL of sodium acetate buffer and 3.5 mL water and the mixture transferred to a 10 mL evacuated glass reaction vial.

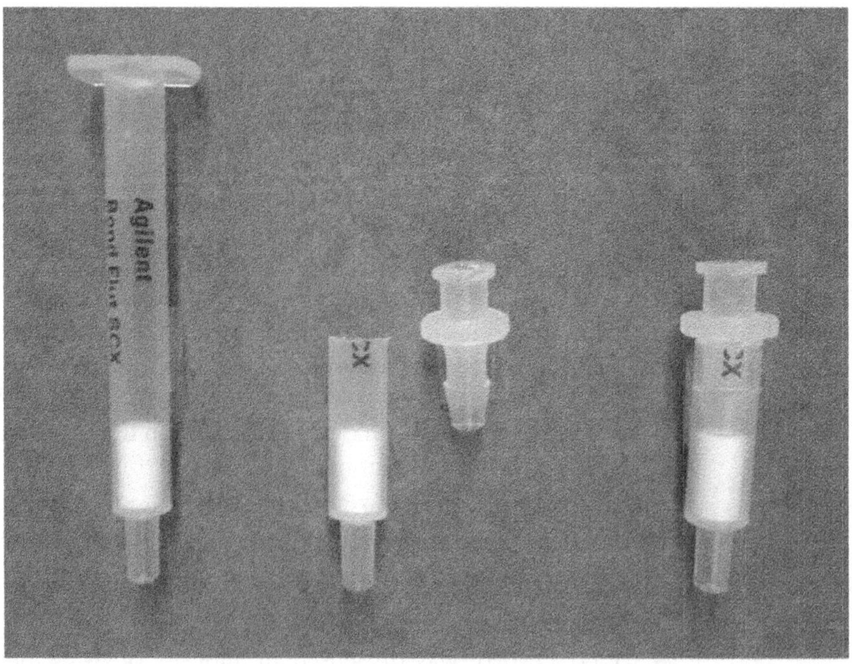

Fig. 3 Assembly of SCX cartridge

5. The ^{68}Ge/^{68}Ga generator was eluted according the supplier's instructions onto the primed SCX cartridge.
6. The SCX cartridge was purged with air.
7. The contents of the SCX cartridge was eluted into the reaction vial with 500 μL of the NaCl/HCl elution solution followed by 0.5 mL air (*see* **Note 6**).
8. The contents of the reaction vial was briefly mixed and allowed to react at room temperature for 10 min. The pH of the reaction mixture should be approximately 4.5.
9. 3 mL of phosphate buffer was added to the reaction vial.
10. The contents of the reaction vial was withdrawn with a 10 mL syringe and passed through a 0.22 μm filter into a new sterile vial yielding the final product for injection (*see* **Note 7**).
11. A closed system has also been used for this radiolabeling procedure (Fig. 4) (*see* **Note 8**).

3.5 ^{68}Ga-CDI-DX001 Purity

1. Radiochemical purity of ^{68}Ga-CDI-DX001 (approximately 100 μL sample of the final product) was assessed by HPLC system C at 9–9–60% mobile phase B (acetonitrile) in mobile phase A (0.1% TFA in utlrapure water) over 0–6–10 min using radiometric detection. The AUC of ^{68}Ga-CDI-DX001 peak over the sum of all radiometric peaks greater than three times background was used to determine radiochemical purity. Absorbance was also measured at 210 nm and 280 nm, however, the molar quantities were below the limits of reliable

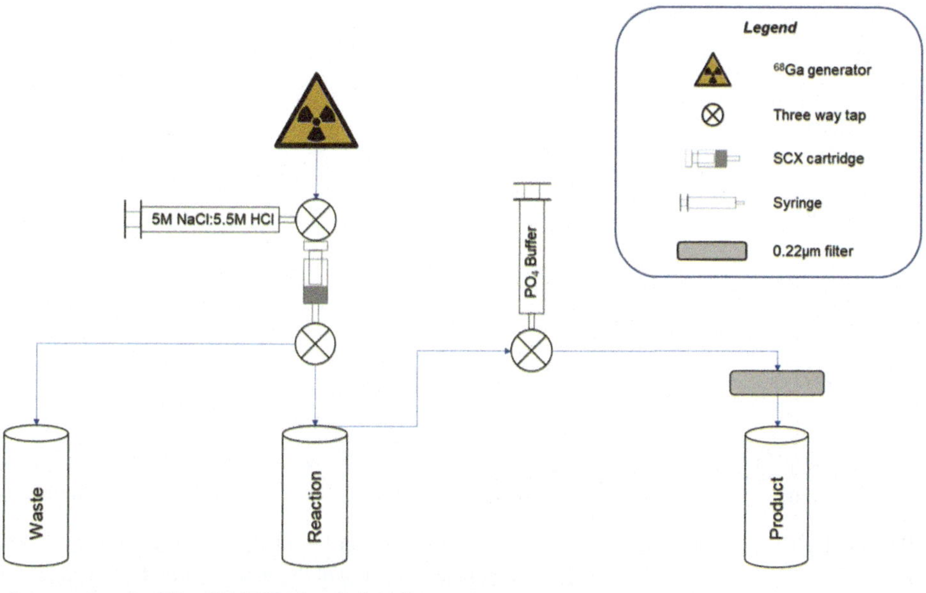

Fig. 4 Schematic of a ^{68}Ga CDI-DX001 radiolabeling system

absorbance detection and was, therefore, not used for assessment of purity.

2. ^{68}Ga-CDI-DX001 was eluted with a retention time of approximately 3 min 55 s (Fig. 5). The release criteria used for radiochemical purity of ^{68}Ga-CDI-DX001 in the final product was ≥ 91% [7].
3. Further assessment of the radiochemical purity of ^{68}Ga-CDI-DX001 was performed by reacting 200 μL of the final product with 5 μL of DMP/DMSO solution at room temperature with occasional agitation for 10 min. Approximately 100 μL of this mixture was assessed by HPLC system C at 9–9–60% mobile phase B (acetonitrile) in mobile phase A (0.1% TFA in utlrapure water) over 0–6–10 min using radiometric detection. The DMP-^{68}Ga-CDI-DX001 peak (with a retention time of approximately 9 min and 30 s) over the sum of all radiometric peaks greater than three times background should be ≥91% (*see* **Note 9**) (Fig. 6). However, this is not included in the required release criteria to minimize loss of product due to decay.
4. Assessment of colloidal contaminants was performed by instant thin layer chromatography developed in 0.9% NaCl. Colloidal contaminants remained at the origin while ^{68}Ga-CDI-DX001 had Rf>0.5. The release criterion used for colloid contaminants was total radioactivity with Rf≥0.5 was ≥90%.
5. Half-life was determined by a minimum of four measurements over 10 min performed on a dose calibrator. The release

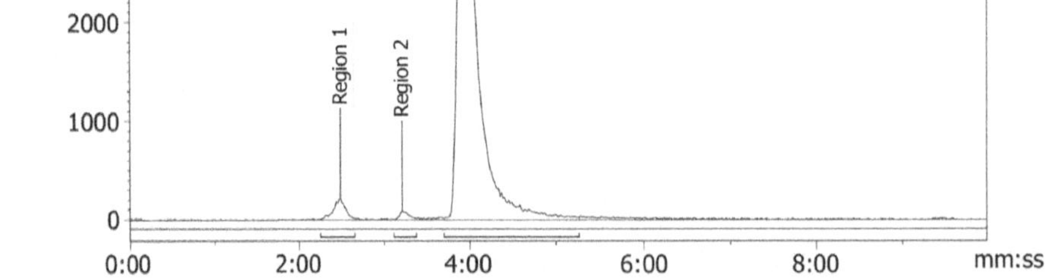

Fig. 5 Radiometric HPLC chromatogram of the final product. Region 1 = unchelated ^{68}Ga, region 2 = oxidation products, and region 3 = ^{68}Ga-CDI-DX001. The radiochemical purity of the final product based on this analysis was 96%

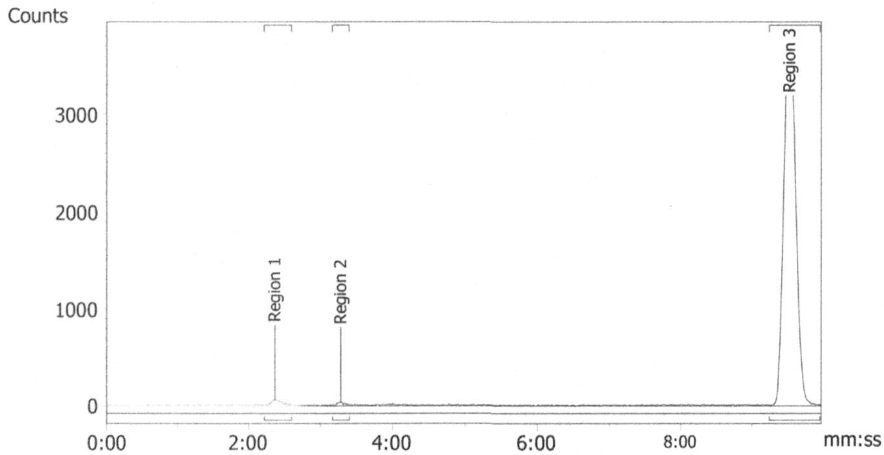

Fig. 6 Radiometric HPLC chromatogram of the final product (same product as in Fig. 5) following reaction with DMP. Region 1 = unchelated ^{68}Ga, region 2 = oxidation products, and region 3 = DMP-^{68}Ga-CDI-DX001. The radiochemical purity of the final product based on DMP reactivity analysis was 98%

criterion used was a calculated half-life between 64 and 72 min (*see* **Note 10**).

3.6 Sterility and Pyrogenicity Testing

Sterility and pyrogenicity were initially tested in an appropriately accredited laboratory on three serial syntheses to confirm that for the process sterility and pyrogenicity are within pharmacopeia guidelines [7]. Subsequent random testing of preparations is performed at regular intervals.

4 Notes

1. Made fresh.
2. This method using the cation exchange cartridge was first described by Mueller and colleagues [8].
3. The method has been performed successfully with IRE Galli Eo and Eckert and Ziegler $^{68}Ge/^{68}Ga$ generators. All $^{68}Ge/^{68}Ga$ generators must be tested prior to use and undergo routine quality assurance as for routine clinical use [9].
4. The barbed female luer thread should be fitted firmly and securely into the cut barrel of the BondElute SCX column to create a sealed cartridge that is air and liquid tight (Fig. 3).
5. Ascorbic acid was used as a free radical scavenger as GSAO is sensitive to radiolysis and oxidation
6. To minimize leaching of metal ions from the needles, use of B. Braun Stericam needles is recommended.

7. No postpurification of the product was performed as ^{68}Ga-CDI-DX001 was not significantly retained on C-18 cartridges and a suitable biocompatible postpurification cartridge/solvent system has not been identified. Despite this, the method described produced ^{68}Ga-CDI-DX001 of high radiochemical purity and specific activity, exceeding current release requirements for ^{68}Ga radiopharmaceuticals.
8. A sterile, closed radiolabeling system is preferred for preparation for human use and also for minimization of the risk of radioactive contamination to the operator and environment. This may also be automated using a radiochemistry synthesis module.
9. As DMP binds with very high affinity to the phenylarsonous moiety of ^{68}Ga-CDI-DX001, this will abolish the usual ^{68}Ga-CDI-DX001 peak with a retention time of approximately 3 min and 55 s and result in a new peak with a retention time of approximately 9 min and 30 s. This provides specific information about the radiochemical purity of the active GSAO and is able to distinguish between ^{68}Ga-CDI-DX001 and other products, such as oxidized degradation products of GSAO.
10. The half-life determination is required to confirm the absence of significant ^{68}Ge breakthrough.

References

1. Don AS, Kisker O, Dilda P, Donoghue N, Zhao X, Decollogne S, Creighton B, Flynn E, Folkman J, Hogg PJ (2003) A peptide trivalent arsenical inhibits tumor angiogenesis by perturbing mitochondrial function in angiogenic endothelial cells. Cancer Cell 3(5):497–509
2. Park D, Don AS, Massamiri T, Karwa A, Warner B, Macdonald J, Hemenway C, Naik A, Kuan KT, Dilda PJ, Wong JW, Camphausen K, Chinen L, Dyszlewski M, Hogg PJ (2011) Noninvasive imaging of cell death using an hsp90 ligand. J Am Chem Soc 133(9):2832–2835
3. Park D, Xie BW, Van Beek ER, Blankevoort V, Que I, Lowik CW, Hogg PJ (2013) Optical imaging of treatment-related tumor cell death using a heat shock protein-90 alkylator. Mol Pharm 10(10):3882–3891. https://doi.org/10.1021/mp4003464
4. Xie BW, Park D, Van Beek ER, Blankevoort V, Orabi Y, Que I, Kaijzel EL, Chan A, Hogg PJ, Lowik CW (2013) Optical imaging of cell death in traumatic brain injury using a heat shock protein-90 alkylator. Cell Death Dis 4:e473. https://doi.org/10.1038/cddis.2012.207
5. Ho Shon I, Kumar D, Sathiakumar C, Berghofer P, Van K, Chicco A, Hogg PJ (2020) Biodistribution and imaging of an hsp90 ligand labelled with ^{111}In and ^{67}Ga for imaging of cell death. EJNMMI Res 10(1):4. https://doi.org/10.1186/s13550-020-0590-x
6. Ho Shon I, Hennessy T, Guille J, Gotsbacher MP, Lay AJ, McBride B, Codd R, Hogg PJ (2022) A first-in-human study of [68Ga] Ga-CDI: a positron emitting radiopharmaceutical for imaging tumour cell death. Eur J Nucl Med Mol Imaging 49(12):4037–4047. https://doi.org/10.1007/s00259-022-05880-z
7. Council of Europe (2017) Gallium (68Ga) Edotreotide injection. European Pharmacopeia. 9th edn. Strasbourg, France, p 1150-2
8. Mueller D, Klette I, Baum RP, Gottschaldt M, Schultz MK, Breeman WA (2012) Simplified NaCl based ^{68}Ga concentration and labeling procedure for rapid synthesis of ^{68}Ga radiopharmaceuticals in high radiochemical purity. Bioconjug Chem 23(8):1712–1717. https://doi.org/10.1021/bc300103t
9. Council of Europe (2017) Gallium (68Ga) Chloride solution for radiolabelling. European Pharmacopoeia. 9th edn. Strasbourg, France, p. 1148-9

Subject Index

Philip J. Hogg (ed.), *Functional Disulphide Bonds*, Methods in Molecular Biology, vol. 3016, https://doi.org/10.1007/978-1-0716-5158-2,

The manufacturer's authorised representative in the EU is Springer Nature Customer Service Centre GmbH, Europaplatz 3, 69115 Heidelberg, Germany. If you have any concerns regarding our products, please contact ProductSafety@springernature.com

Printed and bound by CPI Group (UK) Ltd, Croydon, CR0 4YY
15/09/2026
02217000-0006